HYDROLOGY

HYDROLOGY

Editor

Franca Masin

scitus academics

Hydrology

Edited by **Franca Masin**

This Book is Printed on Acid-Free Paper. ♾

ISBN: 9781681179704
Library of Congress Control Number: 2017949106

www.scitusacademics.com

Notice

Printed in United States of America.

Preface

Hydrology is the science that deals with the processes governing the depletion and replenishment of water resources of the earth's land areas. It also refers to the cycling of liquids such as hydrocarbons on other planets. It focuses on the distribution of water in the subsurface, surface and atmosphere, the chemistry of that water, and the effects of climate on the water cycle. Measurement is fundamental for assessing water resources. Research in this field provides a better understanding of the processes involved in the hydrologic cycle, and insight for environmental engineering, policy and planning. Particularly noteworthy for the department are our research achievements in numerical and physical models for water and sediment flow in rivers and reservoirs. An important research subject is the studying of global and regional changes in the hydrological cycle. This can have various causes such as climate changes and other human interference, e.g. extensive forest clearance. With the help of process and field studies together with statistical and physical modeling, one can study how the hydrological cycle and water balance are affected.

This volume 'Hydrology' provides some of the scientific, technologic, and policy aspects of the field of Hydrology. It covers the study of the movement, distribution, and quality of water throughout earth, and thus addresses both the hydrologic cycle and water resources. It provides a better understanding of the processes involved in the hydrologic cycle, and insight for environmental engineering, policy and planning. The studies also encompass water quality modeling, in particular pollution of field- and groundwater, and the transport and degradation of pollutants. This book will be of valuable for practicing professionals and students, mathematical modelers, hydrogeologists and water resources specialists.

Table of Contents

CHAPTER 1

Evaluation of a Simple Hydraulic Resistance Model Using Flow Measurements Collected in Vegetated Waterways

Fredrik Huthoff[1,2], Menno W. Straatsma[3], Denie C. M. Augustijn[1], Suzanne J. M. H. Hulscher[1]

[1]Department of Water Engineering & Management, Faculty of Engineering Technology, University of Twente, Enschede, The Netherlands; 2HKV Consultants, Lelystad, The Netherlands; 3Department of Earth System Analysis, Faculty of Geosciences, Utrecht University, Utrecht, The Netherlands.

ABSTRACT

A simple idealized model to describe the hydraulic resistance caused by vegetation is compared to results from flow experiments conducted in natural waterways. Two field case studies are considered: fixed-point flow measurements in a Green River (case 1) and vessel-borne flow measurements along a cross-section with floodplains in the river Rhine (case 2). Analysis of the two cases shows that the simple flow model is consistent with measured flow velocities and the present vegetation characteristics, and may be used to predict a realistic Manning resistance coefficient. From flow measurements in the river floodplain (case 2) an estimate was made of the equivalent height of the drag dominated vegetation layer, as based on measured flow characteristics. The resulting height corresponds well with the observed height of vegetation in the floodplain. The expected depth-dependency of the associated Manning resistance coefficient for could not be detected due to lack of data for relatively shallow flows. Furthermore, it was shown that topographical variations in the floodplain may have an important impact on the flow field, which should not be mistaken as roughness effects.

1. INTRODUCTION

Various studies have shown that if vegetation penetrates a significant part of the water column then a constant Manning resistance value is no longer adequate to describe the hydraulic resistance for varying flow conditions [1-5]. These studies have shown that the hydraulic resistance due to vegetation (in terms of Manning's n) tends to decrease with increasing water level. No methodology to describe such behavior is generally accepted despite intense research efforts in recent years. This is partly due to the empirical nature of proposed relationships and the difficulties in deriving generally applicable methods due to the complexity of flow around vegetation. In particular, hydraulic resistance parameters determined in the field are unavoidably contaminated with additional external influences, such as geometrical variations of the channel, density currents, sediment interactions or surface waves [6]. Furthermore, collecting flow data in natural vegetated waterways is in itself a tricky task: the presence of vegetation obstructs detailed flow velocity sampling [7]. In particular when vegetation is abundant and has a large impact on the flow field, accurate data sampling is difficult. As a result, studies where flow characteristics are measured in vegetated waterways are relatively scarce [8], they are case-specific and, due to the complexity of the environment, are difficult to interpret (see [9], where flume studies are used as a reference to field measurements).

Most experimental studies on hydraulic resistance of vegetation are conducted in laboratory flumes, where it is possible to minimize hydraulic impacts due to other external influences [10-13]. That way, investigations of vegetative hydraulic resistance allow isolation of the impact of specific vegetation characteristics (such as stem width, height, flexibility). These studies have revealed that flow through vegetation is difficult to describe based on geometrical conditions of the vegetation alone, even if vegetation is described in a simplified way as cylindrical stems. A recurring complication of these hydraulic resistance models is the need for a general representation of energy losses associated with turbulent mixing patterns. Such energy-loss representations may enter the flow models in the form of a turbulent mixing length [14-17]. Flow descriptions that include the mixing length concept only have practical value if the mixing length is directly related to measurable quantities. Such relations have been proposed [16,18,19], but due to lack of theoretical justification their general applicability remains questionable. Other, more general, approaches include k-ε turbulence models, which explicitly describe the transport of turbulent energy [20-22]. However, these methods require considerably larger computational effort, in particular if applied in models on river-reach scales.

Alternatively, a simple vegetation resistance model has been proposed by Huthoff et al. [23] that includes only measurable quantities. The model requires knowledge of a vegetation drag coefficient C_D, the average vegetation height k, stem diameter D and a representative spacing between neighboring plants s (see also [24]). Among these parameters, only the drag coefficient cannot be directly obtained from vegetation dimensions, but is to be determined from flow experiments. It appears that the C_D-value is species-specific [25] and is a function of the Reynolds number [26], stem aspect ratio [27], vegetation distribution density [28,29] and foliage and streamlining effects [30,31].

In the current work we investigate whether the simplified model proposed in [23] is consistent with flow measurements in the field, and whether it provides a potential candidate for integration with river-reach flow models. Two case-studies with flow in vegetated waterways are considered (Figure 1). First, measured flow velocities and vegetation characteristics in a Green River are evaluated against predictions of the hydraulic resistance model (case 1). It is shown that the model provides an acceptably accurate estimate of the average flow velocity, as based on general vegetation characteristics. Next, flow velocities are measured in a natural floodplain at different discharge magnitudes (case 2). In this case study, no detailed information about the vegetation characteristics was available. Therefore, we evaluate whether the measured dynamic behavior of the hydraulic resistance is consistent with the vegetation resistance model. It is shown that the proposed model describes the vegetation resistance well, but that because of the weak dynamic behavior of the hydraulic resistance, and the lack of data at shallow flows, also a simple wall-roughness model may be used for the considered flow conditions.

2. CASE 1: FIXED-POINT MEASUREMENTS

2.1. Study Location

In February 2005, during a high-discharge event in the Rhine river a Green River[1] was deployed to convey surplus river discharge. Due to its relative homogeneous geometrical boundaries, the Green River is suitable for measurements of the hydraulic effect of the present vegetation. The aerial picture in Figure 2 shows the study area, including the bridge from where measurements were performed. The flow direction in this picture is from east (right) to west (left). Figure 3 shows the presence of vegetation on the channel bed of the Green River, as observed during dry conditions. Note that the Green River includes a central sub-channel (indicated in Figure 2 and also just visible in right panel of Figure 3). The sub-channel is always wet but is closed at both ends, and therefore does not convey water.

2.2. Methodology and Results

The bridge that spans the Green River was used to tether a float from, to record the flow velocities below. Figure 4 shows a picture of the bridge on the 17th of February 2005, the day that data were collected. Three locations

Figure 1. A map of the Netherlands indicating locations of the two case studies (Driel and Bimmen).

Figure 2. Aerial picture of the Green river at Driel. Also, the central (sub-) channel is indicated.

Figure 3. The measurement location at Driel during dry conditions (pictures: M.W. Straatsma).

along the bridge were used for measurements: one in the central sub-channel of the Green River (Location 2) and two locations well-separated to either side of the central sub-channel (Locations 1 and 3). These latter two locations have similar bed coverage characteristics (as seen in Figure 3) and are thus expected to be hydraulically equivalent.

An RD Instruments Acoustic Doppler Current Profiler (ADCP) was used to record flow velocity profiles. The ADCP resolves the three-dimensional flow vector at various user-defined depths. Due to the size of the ADCP itself, and the inability of the ADCP to measure flow velocities immediately below the device, flow velocities were measured downwards from 46 cm below the water surface, at depths 10 cm apart. An average ensemble interval of 6 s was used to determine the mean streamwise velocity. Near the bed, large fluctuations in flow velocities were measured (Figure 5). The fluctuations are likely due to the presence of vegetation. Vegetation parameters were independently measured after the flood, giving an average height of k = 37.5 cm, an average stem width of the vegetation of D = 3.7 mm, an average surface density of m = 51 stems/m^2, leading to an average spacing between plants of s = 13.6 cm.

Figure 5 shows the measured flow velocity profiles for the three locations along the bridge. The graphs also show the profiles of the average streamwise velocities, together with 16- and 84-percentile error boundaries. These boundaries correspond to a 1σ standard deviation if errors were distributed normally. The flow depth h is measured independently by the ADCP, as stated above the graphs in Figure 5. Also stated are the depth-averaged flow velocities in the surface layer U_s. representing flow above the vegetation.

The graphs in Figure 5 show that, between the three measurement locations, the depth-averaged flow velocity in the deepest location (Location 2) is slightly smaller than for the other two locations. This may seem an unexpected result, because for equal bed resistance one would expect larger flow velocities at larger depths.

However, it appears that at Location 2 not the entire water column contributes to discharge capacity. As mentioned before, the sub-channel at Location 2 is closed at both ends, thus partially obstructing flow. Another effect that may cause the flow in the central channel to slow down is the presence of a small bridge (see Figure 3). This small bridge was entirely flooded during the highdischarge event in February 2005 (see Figure 4), thus obstructing flow near the central sub-channel. Consequently, we consider the flow measurements at Location 1 and Location 3 in Figure 5 more representative of the hydraulic response due to bed vegetation. In the remainder, the measurements from Location 2 in the central sub-channel are therefore discarded for the analysis of hydraulic resistance due to vegetation.

Finally, the energy slope i was determined by tracking a float from a fixed location on the river bank (for float tracking methodology, see [32]). This resulted in an average value for the energy slope of $i = 9.2 \times 10^{-5}$, having an error of about 20%.

2.3. Comparison to Model Predictions

Here we consider the simple model for the hydraulic resistance of vegetation as proposed by Huthoff et al. [23] (also see [24]). In this model, the flow velocity in the vegetation layer (or resistance layer) U_r is described

Figure 4. The measurement location at Driel during high discharge on February 17, 2005 (picture: M.W. Straatsma).

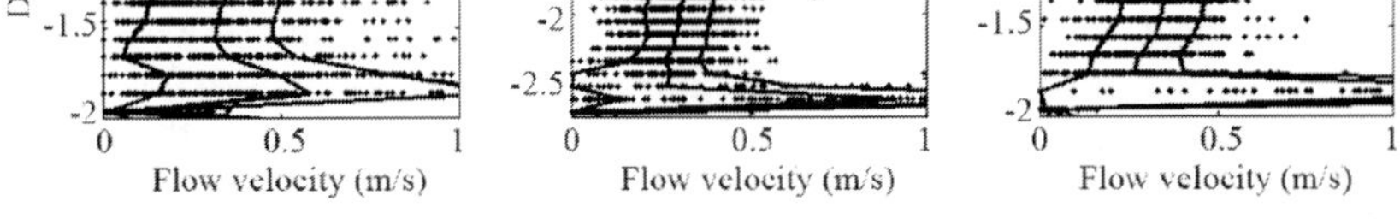

Figure 5. Velocity measurements at three different locations for case 1. The average velocity profile and 16- and 84-percentile boundaries are shown in each of the graphs. Stated above each graph, Us is the depth-averaged velocity in the surface layer and h the flow depth.

separately from flow in the surface layer U_s. Together, they give an estimate of the average flow velocity over the total depth U_T:

$$U_T = \frac{k}{h}U_r + \frac{h-k}{h}U_s \tag{1}$$

The depth-averaged flow velocity in the surface layer is represented by the power law

$$U_s = U_{r0}\left(\frac{h-k}{s}\right)^{\frac{2}{3}\left(1-(k/h)^5\right)} \tag{2}$$

where the power exponent reduces to a constant value of 2/3 if the total flow depth *h* is at least twice as large as vegetation height *k*. The representative separation between homogeneously distributed cylindrical vegetation elements vegetation (*s*), is calculated as [23]:

$$s = \frac{1}{\sqrt{m}} - D \tag{3}$$

The average flow velocity in between the vegetation (in the vegetation layer) is estimated as

$$U_r = U_{r0}\sqrt{\frac{h}{k}} \tag{4}$$

In Equations (2) and (4) the characteristic scaling velocity U_{r0} is given by

$$U_{r0} = \sqrt{\frac{2gi}{mDC_D}} \tag{5}$$

with *g* the gravitational acceleration, *i* the (energy) slope, m the number of vegetation elements per unit bed area, D the stem diameter and C_D the dimensionless drag coefficient. Using the earlier stated vegetation parameters in Section 2.2 the average flow velocities can be predicted using Equations (1)-(5). The only unknown parameter is the drag coefficient for which we adopt $C_D = 1.8$ [33].

Figure 6 shows the predicted flow velocities for the resistance layer, the surface layer, the entire flow depth and Manning's n based on a 20% standard deviation in the measured vegetation parameters and a 10% standard deviation in adopted C_D-value. Comparing the measured values for the surface velocity in Figure 5 ($U_s \approx 0.33$ - 0.38 m/s) with the predicted value in Figure 6 ($U_s \approx 0.38$ m/s), shows that the applied vegetation resistance model gives quite good results.

2.4. Discussion

For the considered case in the Green River we were only able to measure flow velocities in the surface layer, because the presence of vegetation distorted measurement near the bed (see Figure 5). Therefore, we cannot evaluate the predicted depth-averaged flow velocities of the vegetation layer U_r and of the total flow depth U_T (as given in Figure 6) against field measurements. However, using the predicted values for U_T, a Manning resistance parameter n can be calculated, which subsequently allows comparison with values cited in literature for grasslined channels. Manning's equation is given as:

$$U_T = \frac{R^{2/3}}{n}\sqrt{i} \tag{6}$$

which has the property that the resistance coefficient n is practically constant for hydraulically rough turbulent flow over a fixed bed [34]. Inserting the measured flow depth (at Location 3, see Figure 5) and the predicted values for U_T yields Manning resistance coefficients n = 0.045 +/– 0.007 $m^{-1/3}s$. This calculated n-value is consistent with the Manning values cited in literature for grass-lined floodplains. For "high grass" Chow [35] states a general value of n = 0.035, with a lower boundary of n=0.030 and an upper boundary of n = 0.050 $m^{-1/3}s$. However, as was mentioned in the introduction, n generally does not remain constant for flows over vegetation.

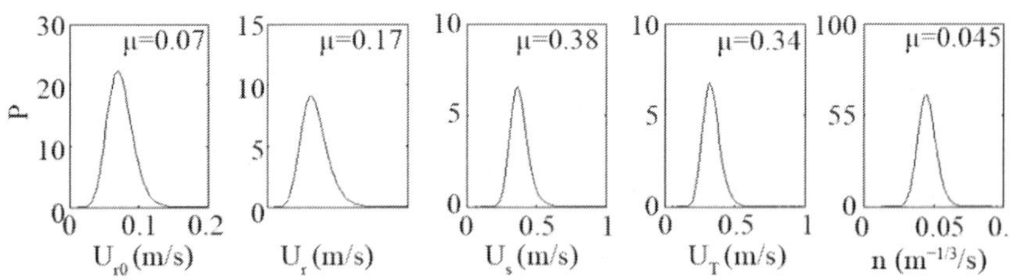

Figure 6. Probability densities of predicted flow velocities (μ = mean value) based on uncertainty in the drag coefficient and variations in measured vegetation characteristics.

Figure 7 shows how the model Equations (1)-(5) together with Manning's law, Equation (6), lead to a depthdependent Manning coefficient. It can be seen that at flow depths that are greater than the one considered, the n-value is expected to decrease. However, the projected changes for large flow depths appear to be quite insignificant, and also a constant Manning's n may be used. Shallower flows are expected to yield considerably stronger changes in n, with nearly doubled values for just submerged vegetation. Similar trends have been reported in empirical studies of vegetation resistance [1-5]. In particular, the study of Wilson and Horritt [4] shows that the Manning coefficient becomes practically constant if the flow depth is much larger than the vegetation height.

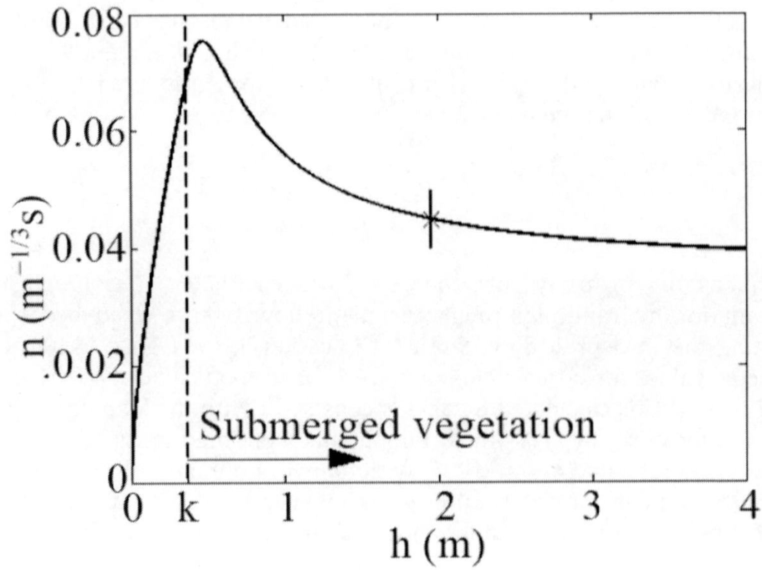

Figure 7. Manning's resistance coefficient n as predicted by the model Equations (1)-(5). The "x" symbol corresponds to the average conditions at Location 3 in the Green River.

We conclude that the model by Huthoff et al. [23] can be used to give a reliable estimate of the effective roughness under field conditions and that, qualitatively, predicted trends of the roughness model agree with trends cited in the literature.

3. CASE 2: VESSEL-BORNE FLOW MEASUREMENTS

3.1. Study Location

In 1998 a high-discharge event occurred on the Rhine River with a maximum recorded discharge of 9413 m^3/s (recorded on November 4th at station Lobith). During the high-discharge event, vessel-borne ADCP measurements were carried out along a cross-section of the River [36]. See Figure 8 for the research vessel and the ADCP device, and Figures 9 and 10 for the sampled river crosssection. Water flows from the east to the west, which in Figure 9(b) is from the lower right to the upper left. The flood-plain on the southern bank is clearly visible in Figure 9, as is the embankment that separates it from the main channel. Groynes are present on the northern bank and also further upstream on the southern bank. The flood-plain is covered with grassland.

Figure 8. The used research vessel for flow measurements in the Rhine River in November 1998 (left). The picture on the right shows the attached ADCP (pictures: aqua vision).

Figure 9. Aerial pictures of the study area. The white line shows the measurement transect.

3.2. Methodology and Results

Flow data were collected along a cross-section at km 863.9 in the Rhine on four consecutive days from November 3 to 6 and then again from November 9 to 11. The discharge peak occurred at November 4 and decreased quite rapidly thereafter. On November 9 the water level had dropped to a level that flow measurements were no longer possible in the floodplain. On each day, flow measurements were collected for multiple transits along the river cross-section (Figure **10**). The transits outline a narrow cross-sectional strip of the channel, for which the best fit straight-line was defined as the equivalent transect (dotted line in Figure **10**). In the subsequent analysis, all measurements are mapped onto the equivalent transect.

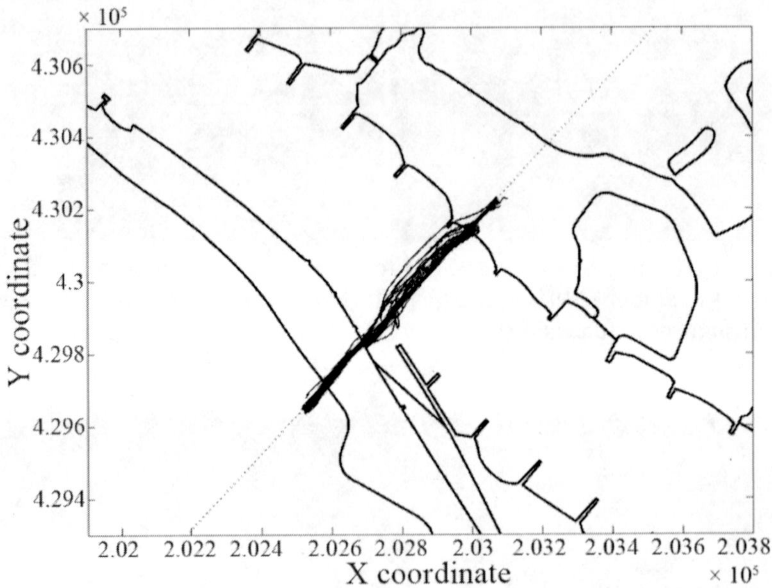

Figure 10. The measurement trajectories near rkm 863.9. The dotted line depicts the equivalent transect.

In Figure **11** an example is shown of one of the sampling runs. Flow velocities were measured below the research vessel at depths 25 cm apart with the highest measuring point 1.32 m below the water surface. Lateral sampling separations were on average 4 m, depending on the travelling speed of the vessel. A region of about 60 cm above the channel bed could not be reliably sampled for flow velocities, partly due to presence of obstructing objects (vegetation) or debris. The measurements were performed in bottom-track mode, correcting measurements for vessel movement and thus giving flow velocities with respect to the detected fixed bed. However, a bottom layer of sediment may be dragged along with the flow, which the ADCP interprets as a fixed bed and subsequently gives a bias towards underestimation of flow velocities. The measurements used here are corrected for this effect by comparison with independent GPS data of the research vessel. Details of this procedure are described in [36].

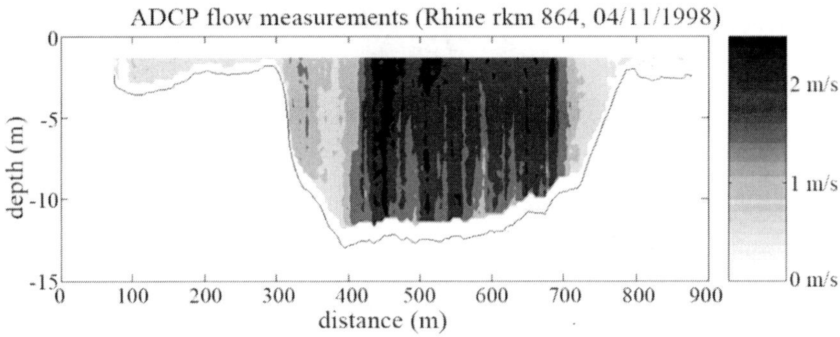

Figure 11. ADCP velocity measurements for one of the measurement runs shown in Figure 10.

Figure 11 shows that lower flow velocities are recorded in the shallower regions of the cross-section. However, a striking feature in the measurement results is that near the deepest part of the channel there appears to be a local minimum in flow velocities (near the lateral coordinate of 380 m). This feature is also observed in the remaining measurement runs (not shown here). The overview of the study area in Figure 9 may provide an explanation for this observation. Just upstream of the channel cross-section a small channel merges with the Rhine on its southern bank. The local dip in flow velocities is likely due to enhanced mixing in the wake of the confluence zone.

To be able to compare the measured flow velocities in the floodplains to model predictions of vegetation resistance, we depth-averaged the measured flow velocities. Next, we grouped all data collected on the same day, assuming that within this time frame no significant hydraulic changes occurred in the system. Figure **12** shows the resulting lateral profiles of depth-averaged hydraulic conditions as measured on November 4, 1998. Averaging of data is performed by taking the mean of all measurement points that fall within columns of 10 m wide. The bottom graph in Figure **12** shows the function $h^{2/3}/U$ set out against the lateral coordinate. Looking at Equation (6), this value represents the quantity $n/\sqrt{i}$. Therefore, if we assume that the downstream slope i is constant across the cross-section, then the bottom graph in Figure 12 reflects a measure for the effective hydraulic roughness. It thus appears that the hydraulic roughness of the main channel is nearly equal to that of the floodplain ($n/\sqrt{i} \approx 3.5$).

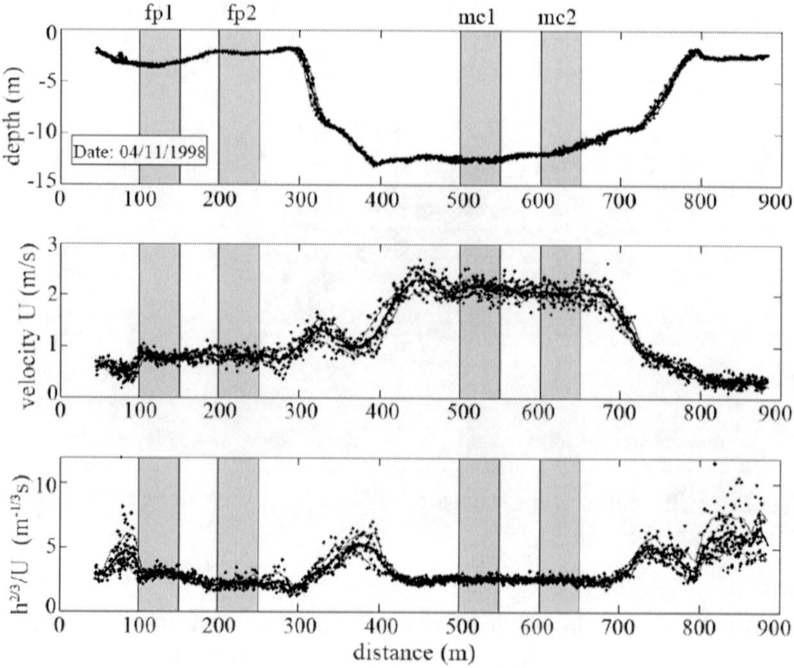

Figure 12. Measured depths (top graph), depth-averaged flow velocities (middle graph) and effective hydraulic resistance (bottom graph).

Next, the hydraulic characteristics (depth, flow velocity, effective flow resistance) are spatially-averaged over selected lateral regions, as marked in Figure 12 by the four vertical grey bars (each having a width of 50 m). Two regions are selected in the floodplain (fp1 and fp2) and also two regions are selected in the main channel: (mc1 and mc2). Figure 13 shows that the averaged hydraulic conditions in these four regions change during the measurement period. The change in flow depths in the floodplain and in the main channel clearly reflects passing of the peak of the flood wave. After the flood wave has passed, the effective hydraulic roughness in the main channel seemed to have increased. This observation is in line with the study by Wilbers [37], who showed that high-discharge events cause large bed forms on the channel bed, which increases the hydraulic roughness. Note that conditions in the floodplain were only measured during four days around the discharge peak. At lower discharges it was no longer possible to enter the floodplain for flow velocity sampling.

The depth-averaged flow velocities in the main channel could be obtained quite accurately, because the ADCP was able to sample most of the water column. However, due to the absence of flow measurements near the bed and near the water surface, in the floodplain typically only half of the water column could be sampled (see Figure **11**). To get a better representative flow velocity in the floodplains we therefore apply a correction procedure. Starting point of the correction procedure is the averaged velocity U_m that is based on the actual sampled data and represents flow in the layer $h_m < z < h - d_m$, (see Figure **14**). The highest measurement point is 1.32 m below the water surface and the lowest measurement point is approximately 0.6 m above the bed. Measurement depths were 25 cm apart, therefore the depth-averaging range of the ADCP is confined by a height from the bed $h_m = 0.6 - 0.25/2$ m and a depth below the water surface $d_m=1.32 - 0.25/2$ m. The steps in correcting the measured average velocity U_m is as follows:

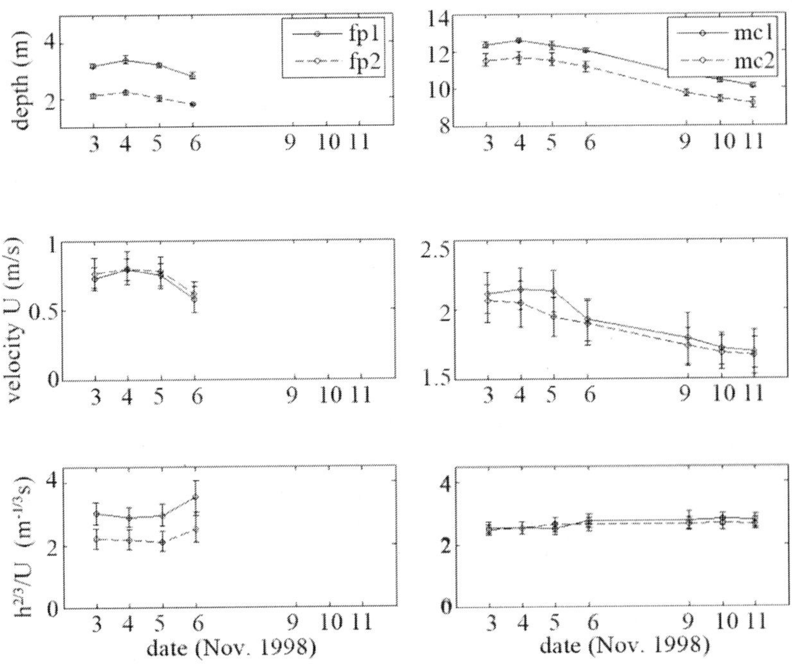

Figure 13. Time-dependency of the hydraulic characteristics within the averaged-regions in the floodplain (fp1 and fp2) and the main channel (mc1 and mc2, see Figure 12).

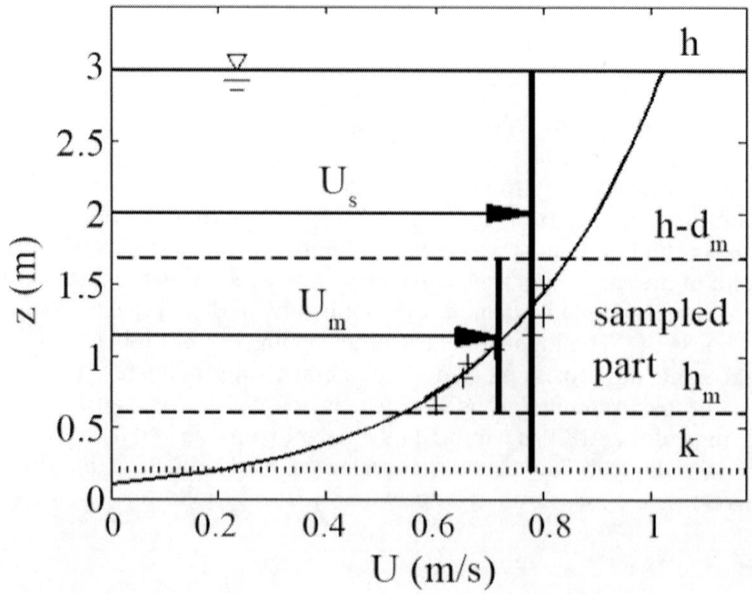

Figure 14. Sketch to illustrate the correction procedure of measured flow velocities.

1) We assume that the flow field above the vegetation follows a vertical logarithmic velocity profile, with a first guess for the corresponding (Nikuradse) roughness height of $k_N = 0.1$ m, which is a reasonable value for grass according to Van Velzen et al. [33].

2) The assumed profile is shifted vertically to fit the sampled flow velocity points (by minimizing least squares).

3) From the fitted profile, the average velocity in the surface layer U_s is determined, assuming an initial value of the vegetation layer height of k = 0.06 m.

Figure 14 illustrates the procedure for correcting flow velocities in the surface layer. Effectively, the correction-procedure described above corresponds to multiplying the measured average flow velocity U_m by a factor Φ:

$$U_s = \phi U_m \tag{7}$$

Where

$$\Phi = \frac{h-h_m-d_m}{h-k}\frac{\int_k^h \ln\frac{z}{k_N}dz}{\int_{h_m}^{h-d_m} \ln\frac{z}{k_N}dz} \tag{8}$$

In Figure 15(a) the results after correcting flow velocities are shown. As expected, the corrections are largest for flow measurements in the shallower region fp2.

3.3. Comparison to Model Predictions

We compare the corrected flow velocities in the surface layer (U_s), with values obtained with the simplified flow model proposed by Huthoff et al. [23]. If the depth of the surface layer (h – k) is more than three times the vegetation height, then the flow velocity in the surface layer is described by a simple scaling law (see Section 2.3):

$$U_s = U_{r0}\left(\frac{h-k}{s}\right)^{\frac{2}{3}} \tag{9}$$

This expression can be rearranged to

$$U_s^{3/2} = Ah - B \tag{10}$$

where $A = (U_{r0})^{3/2}/s$ and $B = Ak$. The bottom graph in Figure **15** shows the measured and corrected averaged velocities, $(U_m)^{3/2}$ and $(U_s)^{3/2}$ respectively, set out against the flow depth h. For each data set also a best fit straight line is shown. Following Equation (10), the ratio B/A yields an estimate of k, which reflects the height of the drag dominated flow layer (or: the vegetation height). **Table 1** gives an overview of these values.

The results in **Table 1** show that for location fp2 the effective height of the drag-dominated flow layer is approximately $k = 0.4$ m, if using the corrected flow velocities U_s. This value of 0.4 m is higher than expected for the floodplain under consideration (recall that we assumed 0.06 m in the correction procedure), but it is still acceptable for the type of vegetation that can be found in grassed floodplains. In contrast, for location fp1 an effective vegetation height of 1.8 m is found, which is unrealistically high given the nearly uniform vegetation characteristics in the floodplain meadow.

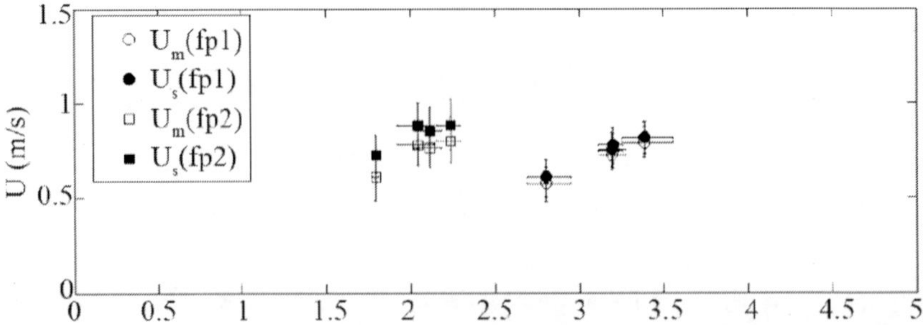

(a) Measured and corrected velocities.

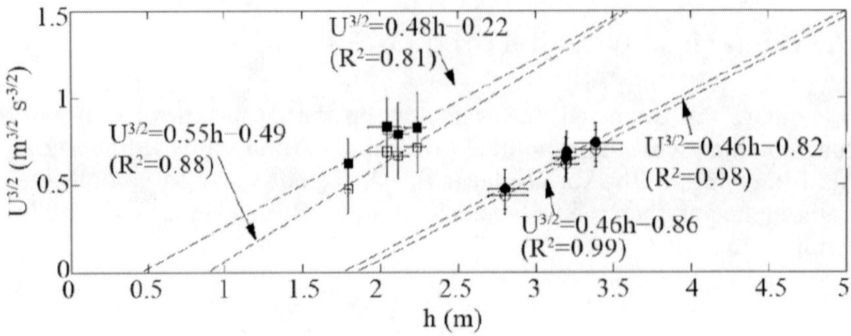

(b) Best fit for $U^{3/2} \sim h - k$.

Figure 15. (a) Depth-averaged velocity U vs flow depth for locations fp1 and fp2 (filled symbols: corrected velocities); (b) Regression relations of U3/2 vs depth.

Table 1. Equivalent vegetation heights (k) reproduced from flow measurements and Equation (10).

Location	U	A	B	$k = B/A$	R^2
fp1	U_m	0.46	0.86	1.9 m	0.99
fp1	U_s	0.46	0.82	1.8 m	0.98
fp2	U_m	0.55	0.49	0.9 m	0.88
fp2	U_s	0.48	0.22	0.4 m	0.81

A closer inspection of conditions in the floodplain reveals that the flow depth at location fp1 is about 1.3 m deeper than at location fp2 (Figure 13), while the average flow velocities at these locations are nearly the same. It appears that in the deeper part of the floodplain water does not flow freely. Figure 16 shows the floodplain shortly after a flooding event in 2007, leaving behind a clearly confined pool at location fp1. Therefore, the larger depth at fp1 is a local topographic effect. Consequently, at fp1 we cannot assume (quasi-) uniform flow and therefore we cannot estimate the hydraulic roughness based on the average flow velocity and local flow depth. However, fp2 is not as much affected by local topographical variations, so for this location we can compare measured data to prediction using model Equations (1)- (6). Using the corrected average flow velocities at fp2 (Figure 15(a)), a vegetation height of k = 0.4 m (**Table 1**) and by adopting a vegetation spacing of s = 1 cm for grassed floodplains (see Van Velzen et al. [33]), Equation (9) yields an approximate characteristic velocity of U_{r0} = 0.03 m/s. Next, if we use a channel slope of $i = 10^{-4}$, which is a representative value for the considered part of the Rhine River [38], model Equations (1)-(6) give the Manning values as shown in Figure 17. It can be seen that for shallow flows Manning values can be as high as n = 0.15, while for flow depths h > 1.5 m values converge towards a constant value in the range 0.02 - 0.025. Also, the results for an assumed vegetation height of k = 0.06 m are shown, indicating that for grass-lined floodplains the vegetation height has only a minor effect on overall flow resistance for relatively deep flows (i.e. if h > 2 m). The predicted values in Figure 17 are slightly lower than those stated in Chow [35], who gives n = 0.030 +/− 0.005 for short grass in floodplains.

3.4. Discussion

We studied two cases of flows in natural vegetated waterways to investigate the bed roughness properties and to compare these to predictions from the vegetation roughness model proposed by Huthoff et al. [23]. Because of incomplete data sampling we were forced to make assumptions on the general shape of the velocity profiles, in order to obtain representative depth-averaged flow velocities. Despite the uncertainties associated with the obtained flow conditions it was shown that local topographic variations can have important impacts on the flow field, and that these localized effects can easily be misinterpreted as local changes in hydraulic roughness. Therefore, field flow measurements should be treated with extreme care if used to determine hydraulic parameters such as Manning's roughness values.

Several studies have shown that remotely-sensed vegetation characteristics can be obtained with high enough accuracies to use them at as input for

Figure 16. A clearly confined pool indicating a local depresssion in the floodplain (picture: F. Huthoff).

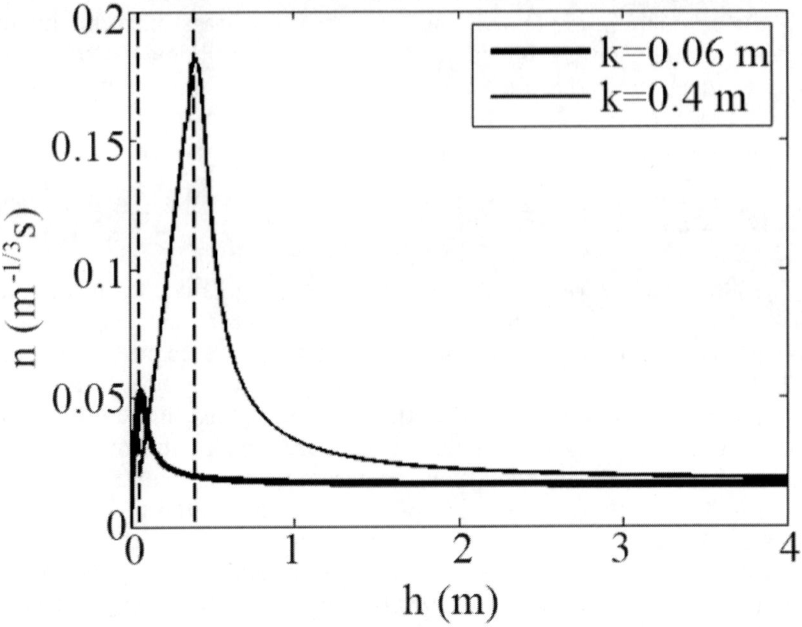

Figure 17. Manning's *n* vs depth according to model Equations (1)-(6) adopting a velocity of Ur_0 = 0.03 m/s in the vegetation layer and a vegetation spacing of *s* = 1 cm.

roughness parameterizations [32,39,40]. The vegetation roughness model proposed by Huthoff et al. [23] is based on such input parameters (such as vegetation height and vegetation spacing), which gives opportunities for advanced inundation modeling techniques where measured vegetation characteristics are translated into effective roughness values.

In the current study, data limitations did not allow for an accurate field validation of the considered roughness model. However, it was shown that the model equations give predictions that are in the correct range and also agree with roughness values stated in the literature. For grass-lined floodplains, the model equations suggest that constant Manning values may be appropriate for wellsubmerged vegetation (h > 2 m), which has traditionally been the preferred approach in inundation studies (e.g. [41,42]). In contrast, for situations where the vegetation is just submerged, the model equations predict significant variations in Manning's roughness values. The importance of these roughness variations for evolvement of floodplain inundations should be the topic of more focused research in the future.

4. CONCLUSION

In the case studies considered here, the waterways with submerged vegetation included predominantly grass species, having average stem heights that are easily an order of magnitude smaller than the flow depths. The vegetative roughness model proposed in Huthoff et al. [23] is consistent with measured flow velocities over such vegetation, predicting nearly constant Manning roughness values for well-submerged vegetation. In contrast, model equations predict that constant Manning's roughness values are no longer appropriate when vegetation is just submerged. Future research should focus on the importance of vegetation roughness-changes under such conditions. Further, it was shown that topographical variations of the bed may have major impacts on local flow velocities, underlining that extreme care should be taken when using field data to estimate floodplain roughness coefficients.

5. ACKNOWLEDGEMENTS

Fredrik Huthoff, Menno W. Straatsma, Denie C. M. Augustijn, Suzanne J. M. H. Hulscher (Netherlands). We gratefully thank Aqua Vision for letting us use their data and Anne Wijbenga (HKV Consultants) for his valuable comments. We appreciate the help from Ivo Thonon and Chris Roosendaal (both Utrecht University), who assisted during the measurement campaign in Driel, and the help from Abe Klaas De Jong (HKV Consultants) and Blanca Pérez Lapeña (University of Twente), who assisted in producing the figures.

REFERENCES

1. H. L. Cook and F. B. Campbell, "Characteristics of Some Meadow Strip Vegetation," Agricultural Engineering, Vol. 20, No. 9, 1939, pp. 345-348.
2. US Soil Conservation Service, "Handbook of Channel Design for Soil and Water Conservation," United States Department of Agriculture, Annapolis, 1947.
3. J. E. P. Green and J. E. Garton, "Vegetation Lined Channel Design Procedures," Transactions of the American Society of Agricultural Engineers, Vol. 26, No. 2, 1983, pp. 437-439.
4. C. A. M. E. Wilson and M. S. Horritt, "Measuring the Flow Resistance of Submerged Grass," Hydrological Processes, Vol. 16, No. 13, 2002, pp. 2589-2598.
5. R. Garcia Diaz, "Analysis of Manning Coefficient for Small-Depth Flows on Vegetated Beds," Hydrological Processes, Vol. 19, No. 16, 2005, pp. 3221-3233.
6. F. Lightbody and H. M. Nepf, "Prediction of Velocity Profiles and Longitudinal Dispersion in Emergent Salt Marsh Vegetation," Limnology and Oceanography, Vol. 51, No. 1, 2006, pp. 218-228.
7. J. C. Green, "Comparison of Blockage Factors in Modelling the Resistance of Channels Containing Submerged Macrophytes," River Research and Applications, Vol. 21, No. 6, 2005, pp. 671-686.
8. T. Sukhodolova, A. Sukhodolov and C. Engelhardt, "A Study of Turbulent Flow Structure in a Partly Vegetated River Reach," In: G. Carravetta and D. Morte, Eds., River Flow, Vol. 1, 2004, pp. 469-478.
9. J. K. Lee, L. C. Roig, H. L. Jenter and H. M. Visser, "Drag Coefficients for Modeling Flow through Emergent Vegetation in the Florida Everglades," Ecological Engineering, Vol. 22, No. 4-5, 2004, pp. 237-248.
10. J. Järvelä, "Flow Resistance of Flexible and Stiff Vegetation: A Flume Study with Natural Plants," Journal of Hydrology, Vol. 269, No. 1-2, 2002, pp. 44-54.
11. Z. Shi and J. M. R. Hughes, "Laboratory Flume Studies of Microflow Environments of Aquatic Plants," Hydrological Processes, Vol. 16, No. 16, 2002, pp. 3279-3289.
12. C. A. M. E. Wilson, T. Stoesser, P. D. Bates and A. Batemann-Pinzen, "Open Channel Flow through Different Forms of Submerged Flexible Vegetation," Journal of Hydraulic Engineering, Vol. 129, No. 11, 2003, pp. 847- 853.
13. Armanini, M. Righetti and P. Grisenti, "Direct Measurement of Vegetation Resistance in Prototype Scale," Journal of Hydraulic Research, Vol. 43, No. 5, 2005, pp. 481-487.
14. Murota, T. Fukuhara and M. Sato, "Turbulence Structure in Vegetated Open Channel Flow," Journal of Hydroscience and Hydraulic Engineering, Vol. 2, No. 1, 1984, pp. 47-61.
15. T. Tsujimoto, T. Kitamura and T. Okada, "Turbulent Structure of Flow over Rigid Vegetation-Covered Bed in Open Channels," KHL Progressive Report 2, Kanazawa University, Kanazawa, 1991.
16. D. Klopstra, H. J. Barneveld, J. M. van Noortwijk and E. H. van Velzen, "Analytical Model for Hydraulic Roughness of Submerged Vegetation," The 27th International IAHR Conference, San Fransisco, 10-15 August 1997, pp. 775-780.
17. D. Velasco, A. Bateman and V. De Medina, "A New Integrated Hydromechanical Model Applied to Flexible Vegetation in Riverbeds," In: G. Parker and M. Garcia, Eds., River, Coastal and Estuarine Morphodynamics, 2005, pp. 217-227.

18. D. G. Meijer and E. H. van Velzen, "Prototype-Scale Flume Experiments on Hydraulic Roughness of Submerged Vegetation," The 28th International IAHR Conference, Graz, 22-27 August 1998, Article ID: D108.
19. M. G. Khublaryan, A. P. Frolov and V. N. Zyryanov, "Modeling Water Flow in the Presence of Higher Vegetation," Water Resources, Vol. 31, No. 6, 2004, pp. 668- 674.
20. Y. Shimizu and T. Tsujimoto, "Numerical Analysis of Turbulent Open-Channel Flow over Vegetation Layer Using a k-ε Turbulence Model," Journal of Hydroscience and Hydraulic Engineering, Vol. 11, No. 2, 1994, pp. 57- 67.
21. F. Lopez and M. H. Garcia, "Mean Flow and Turbulence Structure of Open Channel Flow through Non-Emergent Vegetation," Journal of Hydraulic Engineering, Vol. 127, No. 5, 2001, pp. 392-402. doi:10.1061/(ASCE)0733-9429(2001)127:5(392)
22. Defina and A. C. Bixio, "Mean Flow and Turbulence in Vegetated Open Channel Flow," Water Resources Research, Vol. 41, No. 7, 2005, Article ID: W07006.
23. F. Huthoff, D. C. M. Augustijn and S. J. M. H. Hulscher, "Analytical Solution of the Depth-Averaged Flow Velocity in Case of Submerged Rigid Cylindrical Vegetation," Water Resources Research, Vol. 43, 2007, Article ID: W06413.
24. N. S. Cheng, "Representative Roughness Height of Submerged Vegetation," Water Resources Research, Vol. 47, No. 8, 2011, Article ID: W08517.
25. Fischenich and S. Dudley, "Determining Drag Coefficients and Area for Vegetation," Ecosytem Management & Restoration Research Program, US Army Corps of Engineers, Washington DC, 2000.
26. F.-C. Wu, H. W. Shen and Y.-J. Chou, "Variation of Roughness Coefficients for Unsubmerged and Submerged Vegetation," Journal of Hydraulic Engineering, Vol. 125, No. 9, 1999, pp. 934-942.
27. Sumner, J. L. Heseltine and O. J. P. Dansereau, "Wake Structure of a Finite Circular Cylinder of Small Aspect Ratio," Experiments in Fluids, Vol. 37, No. 5, 2004, pp. 720-730.
28. H. M. Nepf, "Drag, Turbulence, and Diffusion in Flow through Emergent Vegetation," Water Resources Research, Vol. 35, No. 2, 1999, pp. 479-489.
29. M. R. Raupach, "Drag and Drag Partition on Rough Surfaces," Boundary-Layer Meteorology, Vol. 60, No. 4, 1992, pp. 375-395.
30. M. Fathi-Maghadam and N. Kouwen, "Nonrigid, NonSubmerged, Vegetative Roughness on Floodplains," Journal of Hydraulic Engineering, Vol. 123, No. 1, 1997, pp. 51-57.
31. J. Järvelä, "Determination of Flow Resistance Caused by Non-Submerged Woody Vegetation," Journal of River Basin Management, Vol. 2, No. 1, 2004, pp. 61-70.
32. M. Straatsma, "3D Float Tracking: In Situ Floodplain Roughness Estimation," Hydrological Processes, Vol. 23, No. 2, 2009, pp. 201-212.
33. H. Van Velzen, P. Jesse, P. Cornelissen and H. Coops, "Stromingsweerstand Vegetatie in Uiterwaarden," Report RIZA, Arnhem, 2003. [Citation Time(s):3]
34. B. C. Yen, "Open Channel Flow Resistance," Journal of Hydraulic Engineering, Vol. 128, No. 1, 2002, pp. 20-39. doi:10.1061/(ASCE)0733-9429(2002)128:1(20)
35. V. T. Chow, "Open-Channel Hydraulics," McGraw-Hill, New York, 1959.
36. J. M. Eij, "Verwerking van ADCP Stromingsmetingen rond de Pannerdensche Kop tijdens Hoogwater in November 1998," Technical Report AV DOC 040211, Aqua Vision, Utrecht, 2004.
37. W. E. Wilbers, "The Development and Hydraulic Roughness of Subaqueous Dunes," Ph.D. Thesis, University of Utrecht, Utrecht, 2004.
38. P. Y. Julien, G. J. Klaassen, W. B. M. Ten Brinke and A. W. E. Wilbers, "Case Study: Bed Resistance of Rhine River during 1998 Flood," Journal of Hydraulic Engineering, Vol. 128, No. 12, 2002, pp. 1042-1050.
39. C. Mason, D. M. Cobby, M. S. Horritt and P. D. Bates, "Floodplain Friction Parameterization in Two-Dimensional River Flood Models Using Vegetation Heights Derived from Airborne Scanning Laser Altimetry," Hydrological Processes, Vol. 17, No. 9, 2003, pp. 1711-1732.

40. M. W. Straatsma and M.J. Baptist, "Floodplain Roughness Parameterization Using Airborne Laser Scanning and Spectral Remote Sensing," Remote Sensing of Environment, Vol. 112, No. 3, 2008, pp. 1062-1080.
41. P. Nicholas and C. A. Mitchell, "Numerical Simulation of Overbank Processes in Topographically Complex Floodplain Environments," Hydrological Processes, Vol. 17, No. 4, 2003, pp. 727-746.
42. V. Tayefi, S. N. Lane, R. J. Hardy and D. Yu, "A Comparison of Oneand Two-Dimensional Approaches to Modelling Flood Inundation over Complex Upland Floodplains," Hydrological Processes, Vol. 21, No. 23, 2007, pp. 3190-3202.

CHAPTER 2

Inactive Supply Wells as Conduits for Flow and Contaminant Migration: Conditions of Occurrence and Suggestions for Management

Robert M. Gailey[1]

[1]Department of Civil and Environmental Engineering, University ofCalifornia Davis, 2001 Ghausi Hall, 1 Shields Avenue,Davis, CA 95616, USA

ABSTRACT

Water supply wells can act as conduits for vertical flow and contaminant migration between water-bearing strata under common hydrogeologic and well construction conditions. While recognized by some for decades, there is little published data on the magnitude of flows and extent of resulting water quality impacts. Consequently, the issue may not be acknowledged widely enough and the need for better management persists. This is especially true for unconsolidated alluvial groundwater basins that are hydrologically stressed by agricultural activities. Theoretical and practical considerations indicate that significant water volumes can migrate vertically through wells. The flow is often downward, with shallow groundwater, usually poorer in quality, migrating through conduit wells to degrade deeper water quality. Field data from locations in California, USA, are presented in combination with modeling results to illustrate both the prevalence of conditions conducive to intraborehole flow and the resulting impacts to water quality. Suggestions for management of planned wells include better enforcement of current regulations and more detailed consideration of hydrogeologic conditions during design and installation. A potentially greater management challenge is presented by the large number of existing wells. Monitoring for evidence of conduit flow and solute transport in areas of high well density is recommended to identify wells that pose greater risks to water quality. Conduit wells that are discovered may be addressed through approaches that include structural modification and changes in operations.

Keywords: Wellbore leakage Groundwater flow Contamination Corrective measures

INTRODUCTION

It has long been known that vertical flow and contaminant migration can occur through inactive water supply wells when they interconnect water-bearing zones with different hydraulic heads and water qualities—Johnson (1961); California Department of Water Resources (CADWR 1962); Davis et al. (1964). Observations of flow include (1) the sound of water cascading down well screens located above standing water levels (Davis et al. 1964) and (2) downhole videos showing particles entrained in water moving swiftly within well casings and across well screens (observations and consulting reports by field hydrogeologists including the author) as well as measurements of vertical flows through casings (McCombs and Fiedler 1927; Davis et al. 1964). Evidence of contaminant migration includes wells that must be pumped to waste after periods of inactivity in order to reestablish acceptable water quality (Mayo 2010). Such flow and transport can also occur in abandoned supply wells (Gass et al. 1977) as well as improperly destroyed wells (CADWR 1981). While some research and publication on intraborehole flow and transport has occurred, relatively little of the available field data has been documented beyond private reports to well owners and only a few detailed case studies support research findings. Consequently, the prevalence of groundwater supply wells acting as vertical conduits and the extent to which water quality impacts stem from this phenomenon may not be widely understood.

Given the significant reliance on groundwater both globally (Wada et al. 2012) and in specific areas such as California (CADWR 2014), effective resource management should include an understanding of the threat to water quality posed by wells that act as conduits. If the threat is appreciated at a sufficiently high regulatory level, approaches to minimize groundwater flow and the spread of contaminants through inactive, abandoned and improperly destroyed wells may be further developed. This work (1) quantifies the effects of intraborehole flow and transport through inactive wells in unconsolidated alluvial groundwater basins and (2) suggests management approaches for reducing impacts. As discussed in the following section, these contributions will add to a developing body of knowledge regarding wells that act as conduits with results from field investigation for several wells in California, USA (Fig. 1).

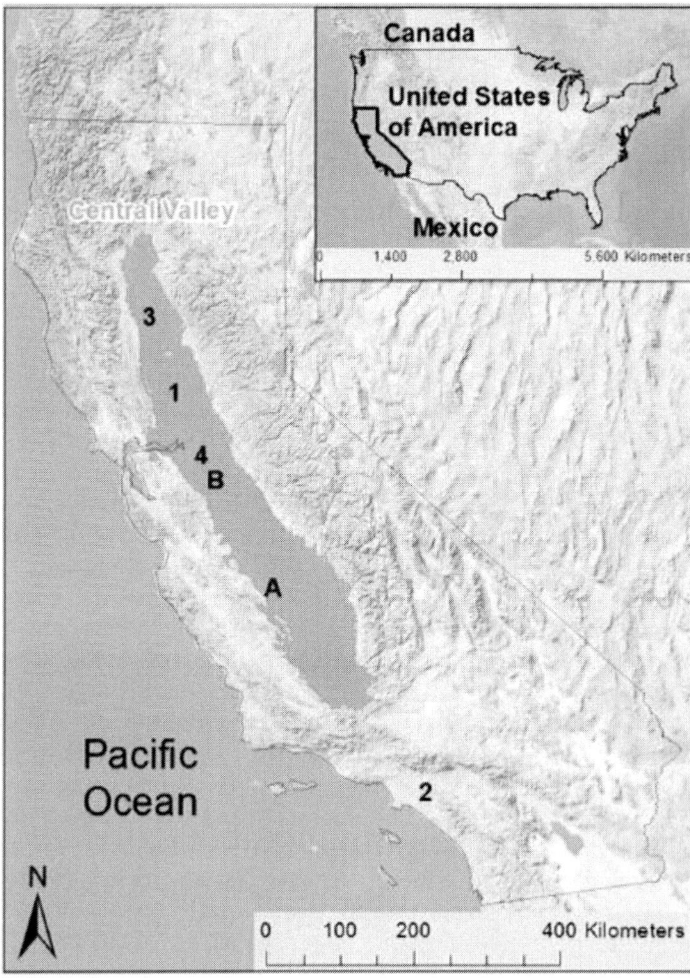

Figure. 1 Study locations in California, USA. *Black letters* indicate approximate locations of field sites cited in the literature review. *Black numbers* indicate approximate locations of case examples

Previous research

Table 1 summarizes research on water supply wells acting as vertical conduits. Both flow and contaminant transport are addressed in most cases. Many contributions are theoretical modeling studies. Some include modeling and field data while others evaluate field data without modeling. The MODFLOW Multi-Node Well package (Halford and Hanson 2002; Konikow et al. 2009) was used for many of the studies that model intraborehole flow (Hanson et al. 2004; Konikow and Hornberger 2006; Zinn and Konikow 2007; Clark et al. 2008; Johnson et al. 2011; Yager and Heywood 2014) although some researchers represented flow through the inside of the well casing with an equivalent porous medium (Lacombe et al. 1995; Mejia et al. 2012). Some works consider a small spatial scale (the well and immediate vicinity), while others focus on larger areas (wellfields and regions). The locations of most field examples are in the USA. The hydrogeology consists of sedimentary basins in all but one case and unconsolidated sediments in all but four cases; of 27 citations, there are 16 unique field sites.

Theoretical considerations of intraborehole flow include Sokol (1963) who developed an analytical representation of steady-state flow between multiple confined aquifers. Silliman and Higgins (1990) later developed an analytical approach for representing steady-state flow between two aquifers that included head loss across the well screens and allowed the shallower water-bearing zone to be unconfined. Lacombe et al. (1995) developed a numerical approach to simulate transient flow and solute transport across an aquitard via a leaky borehole under a variety of conditions. The authors concluded that rapid migration and widespread distribution of contamination are possible and the potential impacts of leaky boreholes on groundwater quality should be further evaluated. Konikow and Hornberger (2006) addressed flow and solute transport through hypothetical wells to consider potential effects on groundwater systems. Their results indicated that, under some conditions, intraborehole flow can occur even when the well is pumped. Zinn and Konikow (2007) explored the effects of intraborehole flow on groundwater age in a hypothetical regional system. They demonstrated that the location of a conduit well in a flow system can determine direction of flow (upwards or downwards) and the depth where age changes as a result of the flow. Changes in pumping stresses within the groundwater system also affected direction and magnitude of flow in conduit wells as well as the depth of impact by transport. Johnson et al. (2011) reinforced the conclusions of others noting that (1) proximity to pumping wells as well as seasonality of pumping influenced the magnitude of vertical flows and (2) bypassing transport through both shallow water-bearing zones and lower hydraulic conductivity sediments reduced protection to deeper groundwater since less solute attenuation occurred.

Table 1 Research contributions regarding water supply wells acting as vertical conduits for flow and contaminant migration

Reference	Scope		Scale		Processes		Setting	
	Theoretical	Field data	Well	Regional	Flow	Transport	Location	Geology
Bexfield et al. 2012	-	x	x	x	x	x	Albuquerque, New Mexico	Unconsol. alluvial
Bexfield and Jurgens 2014	-	x	x	x	x	x	Modesto, California, USA	Unconsol. alluvial
	-	x	x	x	x	x	Albuquerque, New Mexico	Unconsol. alluvial
Clark et al. 2008	-	x	x	x	x	x	York, Nebraska, USA	Unconsol. alluvial
Corcho Alvarado et al. 2009	-	x	x	-	x	x	Paris Basin, France	Consol. sedimentary
Davis et al. 1964	-	x	x	-	x	-	San Joaquin Valley, California, USA	Unconsol. alluvial
Dragon et al. 2009	-	x	x	x	x	x	Wielkopolska, Poland	Glacio-fluvial and fluvial
Hanson et al. 2004	-	x	-	x	x	-	Santa Clara Valley, California, USA	Unconsol. alluvial
Hart et al. 2006	-	x	-	x	x	-	Southeastern Wisconsin, USA	Consol. sedimentary
Jiménez-Martínez et al. 2011	-	x	-	x	x	x	Campo de Cartagena, Spain	Consol. sedimentary
Johnson et al. 2011	x	-	x	x	x	-	-	Multi-layered porous medium
Jurgens et al. 2008	-	x	x	x	x	x	Modesto, California, USA	Unconsol. alluvial
Jurgens et al. 2014	-	x	x	x	x	x	Albuquerque, New Mexico	Unconsol. alluvial
Konikow and Hornberger 2006	x	-	x	x	x	x	-	Multi-layered porous medium
Lacombe et al. 1995	x	-	x	x	x	x	-	Multi-layered porous medium
Landon et al. 2008	-	x	x	x	x	x	York, Nebraska, USA	Unconsol. alluvial
Landon et al. 2009	-	x	x	x	x	x	Modesto, California, USA	Unconsol. alluvial
	-	x	x	x	x	x	York, Nebraska, USA	Unconsol. alluvial
Lu et al. 2008	-	x	-	x	x	x	Chianan Plain, Taiwan	Unconsol. near-shore
Mayo 2010	-	x	x	-	x	x	San Luis Valley, Colorado, USA	Unconsol. Lacustrine and fluvial
McCombs and Fiedler 1927	-	x	x	-	x	-	Oahu, Hawaii, USA	Vesicular basalt
	-	x	x	-	x	-	Pecos Valley, New Mexico, USA	Consol. sedimentary
Mejia et al. 2012	x	-	x	x	x	x	-	Multi-layered porous medium
Sokol 1963	x	-	x	-	x	-	-	Multi-layered porous medium
Viers et al. 2012	x	x	-	x	x	x	San Joaquin Valley, California, USA	Unconsol. alluvial
	x	x	-	x	x	x	Salinas Valley, California, USA	Unconsol. alluvial
Williamson et al. 1989	-	x	-	x	x	-	Central Valley, California, USA	Unconsol. alluvial
Yager and Heywood 2014	-	x	x	-	x	x	Modesto, California, USA	Unconsol. alluvial
	-	x	x	-	x	x	Albuquerque, New Mexico	Unconsol. alluvial
Zinn and Konikow 2007	x	-	x	x	x	x	-	Multi-layered porous medium

Early field research documented flow at the well scale. McCombs and Fiedler (1927) may provide the earliest documented measurements of intraborehole flows. The most extensive dataset on vertical flows through inactive water supply wells may be by Davis et al. (1964). They presented ambient flow (well not pumping) data, collected as a profile along the well length with an impeller flowmeter, from 11 idle agricultural supply wells located in the western San Joaquin Valley of California (see A on Fig. 1 for location and Table 2 for data). The wells penetrated the deeper water-bearing zone beneath a regional aquitard (Corcoran Clay). Flow measurements ranged from 48 to 478 gal/min, or 187–1,852 l/min, at depths within the range of 485–1,350 ft below ground surface (ft bgs), or 148–411 m below ground surface (m bgs). Flow profiles increased with depth in the wells as groundwater entered the wells, largely through holes in the casing, and decreased as water exited the wells through the screen perforations. As indicated by the Reynolds numbers, turbulent flow occurred in the wells. The authors noted that the observed intraborehole flow rates were consistent with intensive groundwater pumping for agricultural irrigation that occurred throughout the groundwater system for years before the data were collected. In this area, pumping from depth and return flows from irrigation at the ground surface created vertical head differences as great as 300 ft (91 m) that drove flow down through the wells.

Table 2 Flow profiling conducted by Davis et al. 1964. Kinematic viscosity used to calculate *Reynolds numbers*: 1.1092 mm²/s at 16 °C

Well number	Screened interval	Conduit flow					Well diameter		Flow velocity	Reynolds number[a]
		Depth		Cumulative		Incremental				
	(bgs)	(ft bgs)	(m bgs)	(cfs)	(l/min)	(l/min)	(in)	(cm)	(m/s)	(–)
14/12-12 N1	558–1,709 ft	550	168	0.00	0	0	12.75	32	0.00	0.0E + 00
	(170–521 m)	600	183	0.20	340	340	12.75	32	0.07	2.0E + 04
		650	198	0.23	391	51	12.75	32	0.08	2.3E + 04
		700	213	0.13	221	−170	12.75	32	0.04	1.3E + 04
		800	244	0.16	272	51	12.75	32	0.06	1.6E + 04
		900	274	0.12	204	−68	10.75	27	0.06	1.4E + 04
		1,000	305	0.19	323	119	10.75	27	0.09	2.3E + 04
		1,100	335	0.16	272	−51	10.75	27	0.08	1.9E + 04
		1,200	366	0.16	272	0	10.75	27	0.08	1.9E + 04
		1,300	396	0.11	187	−85	10.75	27	0.05	1.3E + 04
14/13-17 E1	Unknown	600	183	0.00	0	0	18.00	46	0.00	0.0E + 00
		700	213	0.00	0	0	18.00	46	0.00	0.0E + 00
		700	213	0.00	0	0	18.00	46	0.00	0.0E + 00
		734	224	0.12	204	204	18.00	46	0.02	8.5E + 03
		750	229	0.13	221	17	18.00	46	0.02	9.2E + 03
		760	232	0.12	204	-17	12.75	32	0.04	1.2E + 04
		800	244	0.15	255	51	12.75	32	0.05	1.5E + 04
		900	274	0.19	323	68	12.75	32	0.07	1.9E + 04
		1,000	305	0.19	323	0	12.75	32	0.07	1.9E + 04
		1,100	335	0.00	0	−323	12.75	32	0.00	0.0E + 00
14/13-29 Q1	621–1,803 ft	800	244	0.00	0	0	12.75	32	0.00	0.0E + 00
	(189–550 m)	900	274	0.12	204	204	12.75	32	0.04	1.2E + 04
		1,000	305	0.14	238	34	10.75	27	0.07	1.7E + 04
		1,050	320	0.18	306	68	10.75	27	0.09	2.1E + 04
15/13-8 N1	639–1,798 ft	650	198	0.00	0	0	12.75	32	0.00	0.0E + 00
	(195–548 m)	675	206	0.49	832	832	12.75	32	0.17	4.9E + 04
		707	215	0.44	747	−85	12.75	32	0.15	4.4E + 04
		750	229	0.44	747	0	12.75	32	0.15	4.4E + 04
		800	244	0.47	798	51	12.75	32	0.16	4.7E + 04

		1,000	305	0.47	798	0	10.75	27	0.23	5.6E + 04
16/15-26 N3	Unknown	700	213	0.00	0	0	12.75	32	0.00	0.0E + 00
		800	244	0.00	0	0	12.75	32	0.00	0.0E + 00
		910	277	0.19	323	323	12.75	32	0.07	1.9E + 04
		1,000	305	0.26	442	119	12.75	32	0.09	2.6E + 04
		1,100	335	0.00	0	−442	12.75	32	0.00	0.0E + 00
17/15-27 K1	912–2,130 ft (278–649 m)	925	282	0.00	0	0	12.75	32	0.00	0.0E + 00
		975	297	0.18	306	306	12.75	32	0.06	1.8E + 04
		1,025	312	0.48	815	510	12.75	32	0.17	4.8E + 04
		1,100	335	0.48	815	0	10.75	27	0.23	5.7E + 04
		1,200	366	0.48	815	0	10.75	27	0.23	5.7E + 04
		1,300	396	0.48	815	0	10.75	27	0.23	5.7E + 04
17/16-4 E1	416–1,580 ft (127–482 m)	475	145	0.00	0	0	10.75	27	0.00	0.0E + 00
		485	148	0.54	917	917	10.75	27	0.26	6.4E + 04
		500	152	0.36	612	−306	10.75	27	0.17	4.3E + 04
		525	160	0.46	781	170	10.75	27	0.22	5.5E + 04
		549	167	1.09	1852	1070	10.75	27	0.53	1.3E + 05
		600	183	0.96	1631	−221	10.75	27	0.46	1.1E + 05
17/16-8 L1	553–1,800 ft (169–549 m)	550	168	0.00	0	0	12.75	32	0.00	0.0E + 00
		650	198	0.54	917	917	12.75	32	0.19	5.4E + 04
		700	213	0.14	238	−679	12.75	32	0.05	1.4E + 04
		800	244	0.13	221	−17	12.75	32	0.04	1.3E + 04
		900	274	0.27	459	238	12.75	32	0.09	2.7E + 04
		1,000	305	0.20	340	−119	12.75	32	0.07	2.0E + 04
		1,100	335	0.16	272	−68	12.75	32	0.06	1.6E + 04
		1,200	366	0.00	0	−272	10.75	27	0.00	0.0E + 00
19/17-19 D1	721–2,063 ft (220–629 m)	700	213	0.00	0	0	12.75	32	0.00	0.0E + 00
		800	244	0.12	204	204	12.75	32	0.04	1.2E + 04
		900	274	0.12	204	0	12.75	32	0.04	1.2E + 04
		1,000	305	0.00	0	−204	12.75	32	0.00	0.0E + 00
19/17-19 P1	665–2,030 ft (203–619 m)	400	122	0.00	0	0	16.00	41	0.00	0.0E + 00
		450	137	0.51	866	866	16.00	41	0.11	4.1E + 04
		700	213	0.21	357	−510	12.75	32	0.07	2.1E + 04
		900	274	0.16	272	−85	12.75	32	0.06	1.6E + 04
		1,100	335	0.16	272	0	10.75	27	0.08	1.9E + 04
19/18-33 N2	632–2,025 ft (193–617 m)	575	175	0.00	0	0	16.00	41	0.00	0.0E + 00
		600	183	0.00	0	0	16.00	41	0.00	0.0E + 00
		625	191	0.42	713	713	16.00	41	0.09	3.4E + 04
		650	198	0.47	798	85	12.75	32	0.16	4.7E + 04
		750	229	0.40	679	−119	12.75	32	0.14	4.0E + 04
		850	259	0.38	646	−34	12.75	32	0.13	3.8E + 04
		950	290	0.38	646	0	12.75	32	0.13	3.8E + 04
		1,050	320	0.43	730	85	12.75	32	0.15	4.3E + 04
		1,150	351	0.39	662	−68	12.75	32	0.13	3.9E + 04
		1,240	378	0.38	646	−17	12.75	32	0.13	3.8E + 04
		1,350	411	0.37	629	−17	12.75	32	0.13	3.7E + 04

Data for *Screened interval*, *Conduit flow* (including *Depth* and *Cumulative*) and *Well diameter* are as presented by Davis et al. (1964). Data for *Incremental* conduit flows, *Flow velocity*, *Reynolds number* and entries in columns with headings that include SI units are calculated from those by the author

ft bgs feet below ground surface; *m bgs* meters below ground surface; *cfs* cubic feet per second; *L/min* liters per minute; *in* inches; *cm* centimeters; *m/s* meters per second

[a] Reynolds number values greater than 2.1E + 03 in pipes indicate departures from laminar flow (Vennard and Street 1982)

More recent research documented and evaluated well-scale transport occurring in the field. Corcho Alvarado et al. (2009) analyzed vertical profile data for tritium activity and temperature from an inactive long-screened supply well in an area where downward vertical hydraulic gradients were anticipated and found that the results were consistent with intraborehole transport. Landon et al. (2009) concluded that a long-screened public supply well in Modesto, California (B on Fig. 1) acted as a conduit partly based on (1) local downward vertical hydraulic gradients and (2) correlation between periods of pump inactivity and concentration increases in deeper water-bearing zones of constituents typically found in shallow water-bearing zones. Bexfield et al. (2012), Bexfield and Jurgens (2014), Jurgens et al. (2008) and (2014) documented vertical groundwater flow and contaminant migration through municipal supply wells in two different parts of USA (Modesto, California and Albuquerque, New Mexico) during periods of nonoperation. Ambient flow was downwards in one well and upwards in the other. Effects on water quality, including elevated arsenic, nitrate and uranium concentrations, changed seasonally with vertical hydraulic gradients. The authors indicated the potential for similar occurrences in other alluvial basins with stratified water quality and vertical hydraulic gradients. Yager and Heywood (2014) modeled these two wells and showed that flow and contaminant migration during nonoperation degraded water quality. They showed that pumping during the off-season could reduce the undesired flows and water quality degradation but noted that the costs of the additional pumping should be weighed against costs of other options such as structurally modifying the wells to prevent connection of separate water-bearing strata.

Others have evaluated the effects of flows through inactive wells at the regional scale. For the Central Valley of California, Williamson et al. (1989) estimated through numerical modeling that, in areas where the density of multi-aquifer wells equaled or exceeded one per square mile, the effective vertical hydraulic conductivity of confining layers in the groundwater system could increase by a factor of seven as a result of flow through well casings. Similarly, while modeling the western portion of the San Joaquin Valley, California, Phillips and Belitz (1991) and Belitz et al. (1993) found that the calibrated value for vertical hydraulic conductivity for the regional aquitard was higher than expected from permeameter data and reasoned that flow through the many multi-aquifer wells in the study area was the cause. Bertoldi et al. (1991) and Faunt (2009) provide additional support for these findings. Hanson et al. (2004) modeled groundwater flow in the Santa Clara Valley of California and estimated that approximately 19% of flow from shallower to deeper strata occurred through wells. Hart et al. (2006) investigated potential flow mechanisms responsible for a discrepancy between core-scale and regional estimates of vertical conductivity for an aquitard in Wisconsin, USA. Using a combination of numerical and analytical modeling, they found that flow through a single well could be approximately 700 l/min and a spacing of one well every 10 km would be sufficient to account for the difference in flow implied by the different estimates of hydraulic conductivity. For a public supply well near York, Nebraska, a combination of field and modeling studies indicated that approximately 25% of the water that migrated from shallow to deeper groundwater crossed the aquitard by flowing directly through wells or boreholes that connected water-bearing strata separated by the aquitard (Clark et al. 2008, Landon et al. 2008, 2009). Viers et al. (2012) and Mejia et al. (2012) evaluated the potential for nitrate migration from shallow groundwater to deeper strata over broad geographic areas using algebraic and numerical modeling methods, respecitvely. The potential effects of intraborehole flow and transport on a regional scale have also been considered in Poland (Dragon et al. 2009), Spain (Jiménez-Martínez et al. 2011) and Taiwan (Lu et al. 2008).

There are three significant limitations of the currently available literature. First, only three studies present actual measurements of intraborehole flow. McCombs and Fiedler (1927) measured a maximum flow equivalent to 520 gal/min (1970 l/min). Davis et al. (1964) measured a maximum flow rate of 478 gal/min (1,852 l/min). Bexfield et al. (2012) measured a maximum flow rate of 204 gal/min (772 l/min). Second, only one study presents information regarding the extent of water quality impact. Mayo (2010) reported that, after a well had been out of operation for 71 days, reversing the water quality impact from migration of total dissolved solids required running the well at its operational pumping rate of 175 gal/min (660 l/min) for approximately 15 days (approximately 3.8×10^6 gal or 14.3×10^6 l extracted). Third, only one study suggests management approaches. Yager and Heywood (2014) suggested off-season pumping as a possible approach for limiting water quality impacts for specific wells. This current work provides additional field information on intraborehole flow measurements and the extent of water quality impacts as well as suggests several management approaches.

Overview of flow and transport processes

Conditions when intraborehole flows occur

Figure 2a depicts a sedimentary groundwater basin with multiple permeable water-bearing zones common in many parts of the world. Vertical head gradients, both upward and downward, will be present in basins as a natural consequence of recharge and discharge (Fetter 1994; Phillips et al. 2007). Downward gradients also can result from hydraulic stresses induced by human activity such as groundwater extraction from deeper strata and deep percolation associated with crop irrigation

(Fig. 2b). In alluvial aquifers and other stratified sedimentary groundwater systems, pronounced bulk anisotropy of hydraulic conductivity ($K_{vertical}/K_{horizontal}$), perhaps on the order of 10^{-3}–10^{-4}, often limits vertical flow and vertical hydraulic gradients can exceed horizontal gradients by orders of magnitude (Fogg 1986). For conditions that exist in agriculturally stressed alluvial groundwater systems, horizontal and vertical hydraulic gradients can be on the order of 10^{-3} and 10^{-1}, respectively (see field examples Nos. 2 and 4 in the following). When an interconnecting vertical pathway exists, the vertical gradients are sustained regionally by the previously referenced hydraulic stresses and drive vertical flows on a continual basis.

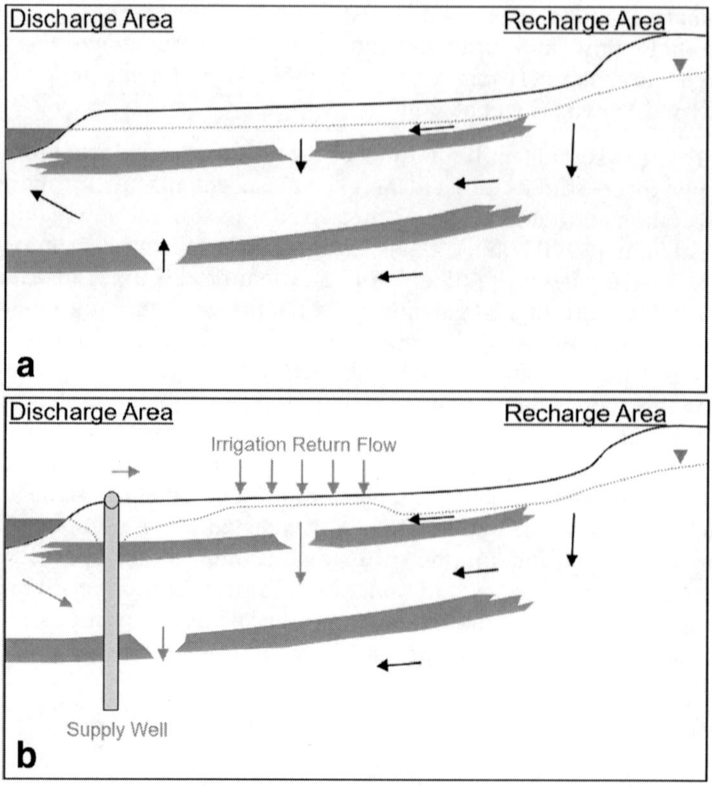

Figure. 2 Simplified depiction of unconsolidated sedimentary groundwater basin: **a** natural system and **b** changes in groundwater stresses and flows from human activities. *White areas* indicate water-bearing zones. *Gray areas* indicate aquitards. Spatial variation in hydraulic conductivity within hydrostratigraphic units not shown. *Black arrows* indicate direction of groundwater flow. *Red arrows* indicate changes induced by human activities. *Blue dotted line and triangle* indicate the water table

Interconnecting vertical pathways may exist naturally within the sedimentary structure (gaps in aquitards from erosion before burial or discontinuities in silts and clays deposited by spatially variable processes) or artificially as a result of well construction. Figure 3 shows a range of common well designs and preferential pathways for groundwater flow when vertical gradients are present. A properly designed and maintained well that provides little potential for vertical flow is depicted at the left of Fig. 3 and serves as a basis for comparison. This well consists of (1) a well casing and screen installed within a borehole, (2) a cement seal filling the annular space between the borehole wall and the well casing that extends from ground surface down into the aquitard, (3) a gravel pack filling the annular space that extends from above the top to below the bottom of the screen and (4) a gravel fill tube for replenishing the gravel pack in the event of any settlement resulting from operations. Common elements that create the potential for wells to act as vertical conduits include long screens, multiple shorter screens separated over long vertical distances and long gravel packs (Santi et al. 2006). Another common well condition leading to a well becoming a vertical conduit is localized failure of the casing or other steel component (tool port or gravel fill tube) from corrosion or failure of a welded joint. Such a failure allows hydraulic communication between shallower and deeper strata separated by relatively long distances through the interior of the well. The term "long" is used within the context of the hydrogeology at a particular well site. Well elements are considered long if they cross hydraulic barriers created by finer-grained strata where differences in hydraulic head occur (see Fig. 3). Conditions that result in flow through the well casing instead of the gravel pack will generally result in higher velocities and larger volumes of water transferred between strata because, as discussed in the following, resistance to flow will be many orders of magnitude less inside the casing.

Consistent with Darcy's Law, the potential magnitude of intraborehole flow depends upon both resistance along the flow path and the hydraulic gradient that drives the flow. Typical vertical hydraulic conductivity values for sediments range from approximately 10^{-7} to 10 cm/s (Freeze and Cherry 1979; Fetter 1994), while effective pipe conductivities for well casings or screens and gravel fill tubes are much higher at approximately 10^4 to 10^6 cm/s. Pipe conductivities may be calculated using the Hagen-Poiseuille equation for assumed laminar flow conditions: $K_{\text{well}} = r^2 \rho g / 8\,\mu$, where r is the inner radius of the well casing or screen, ρ is the fluid density, g is the gravitational acceleration and μ is the fluid viscosity (Reilly et al. 1989; Lacombe et al. 1995). Because the hydraulic conductivity values for well components are orders of magnitude greater than those for sediments, it is energetically more favorable for water to flow through the well, either the gravel-filled borehole annulus or the open casing, than across the lower hydraulic conductivity strata. The range in vertical hydraulic gradients are discussed in the field examples in the following, but experience suggests that they can be at least as high as −0.8 (see field example No. 2 in the following). As indicated previously and further discussed in the following field examples, flow rates can range up to thousands of liters per minute.

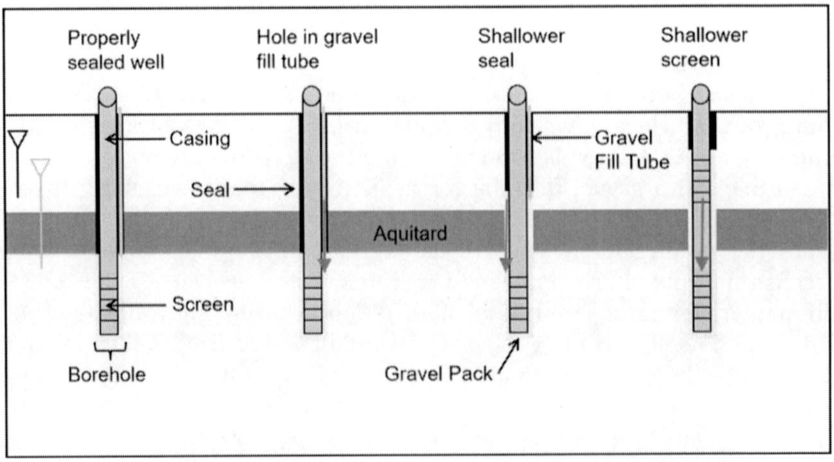

Figure. 3 Range of well conditions. *Red arrows* indicate potential vertical flow pathways. *Blue symbols* indicate piezometric levels at different depths in the groundwater system

Based on field data, Mayo (2010) found that vertical head differences of 0.1 m could be significant for intraborehole flow where the higher-head stratum is contaminated; however, even smaller head differences can drive intraborehole flows that are quite high since the effective hydraulic conductivity for a well is many orders of magnitude greater than for a natural porous medium. A head difference of as little as 0.01 m over a vertical distance of 10 m would create a vertical gradient on the same order of magnitude as typically observed for horizontal gradients, but the flow through a conduit well would be many orders of magnitude greater than through the natural system given the higher hydraulic conductivity of the well. Because head differences of 0.01 m are not easily measured in the field, many intraborehole flows may not be apparent unless measurements of flow in the wellbore are made or water quality data are collected over time.

Transport impacts from intraborehole flows

Areas where flow through wells is investigated are often located in groundwater systems significantly affected by human activities that create or enhance downward gradients. In these areas, shallow water that is also often affected by activities at ground surface and poorer in quality (i.e., nitrate, pesticides and other organic compounds from anthropogenic sources; Burow et al. 2007) migrates downward. Migration through wells bypasses lower hydraulic conductivity sediments and reduces water quality protections afforded by (1) longer travel times of perhaps 100s–1,000s of years and (2) beneficial chemical reactions such as denitrification (McMahon et al. 2008). Deeper water quality is degraded in these cases; however, shallow groundwater quality impacts may also occur where flow is upward through wells and deeper strata contain naturally occurring undesirable constituents such as arsenic, chloride or hexavalent chromium (Izbicki et al. 2005a, b, 2006, 2008; Izbicki et al. 2015).

In addition to problems from well construction, well management practices contribute to water quality impacts from vertical flows through wells. Extended well inactivity (months to years) allows ongoing flows to accumulate larger volumes of water and masses of dissolved constituents moving between strata. Irrigation and some municipal supply wells are commonly left inactive each year during the winter when demands decrease. Common naturally occurring constituents that can be mobilized include the inorganic constituents mentioned in the previous as well as dissolved gasses such as oxygen. Anthropogenic constituents that can be mobilized include nitrate, volatile organic compounds, total dissolved solids including chloride, fumigants, pesticides and uranium (if mobilized by changing redox conditions with agricultural irrigation, see Jurgens et al. 2010). When intraborehole flow occurs, contaminated water displaces better quality water at depth when the well is inactive (Fig. 4a). As depicted in Fig. 4b, the quality of produced water then includes a less favorable mix upon restarting the well (Bexfield and Jurgens 2014).

The hydraulic performance of a well can also be reduced by conduit flow and transport. Mixing of geochemically dissimilar waters from different strata often causes precipitation of inorganic chemical species that clog well screens (Houben and Treskatis 2007; Van Beek 2012). Oxygenation of deeper waters can spur bacterial growth that also clogs well screens (Mansuy 1999). In both cases, suspended sediment from the aquifer material that is often entrained in the flow can become adhered to well components by the inorganic precipitates and bacterial growth resulting in greater volume of clogging material and exacerbating hydraulic performance problems (Houben and Treskatis 2007). The well pump can become fouled and inefficient through the same processes (Houben and Treskatis 2007; Smith and Comeskey 2010).

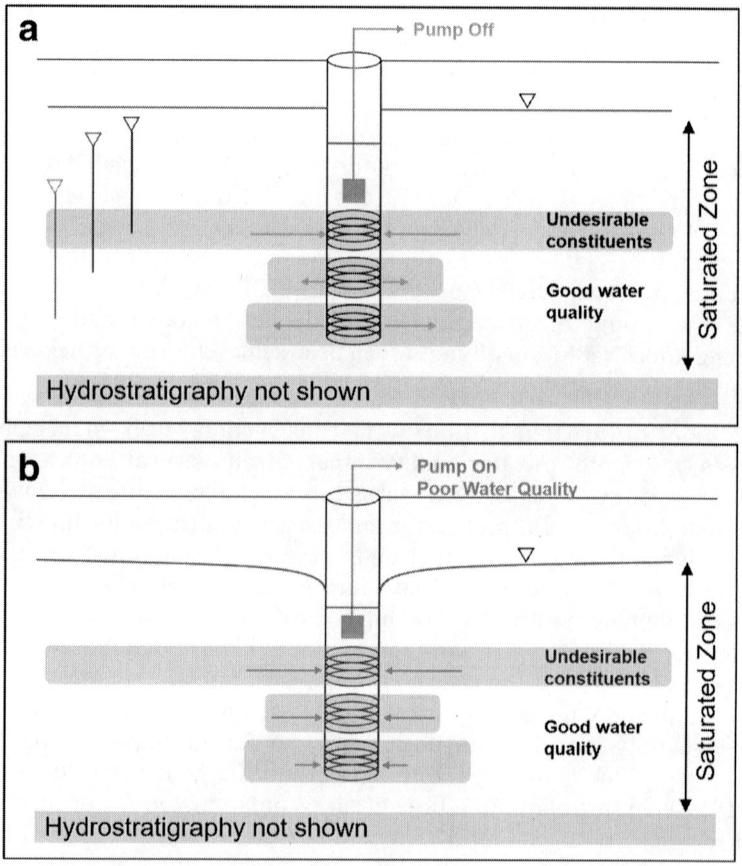

Figure. 4 Schematic of well that acts as a conduit: **a** vertical flow and contaminant migration through an inactive well (*Blue symbols* indicate piezometric levels at different depths in the groundwater system) and **b** water quality degradation during pumping after contaminant migration through conduit. Lengths of *arrows* indicate flow magnitudes in a heterogeneous hydraulic conductivity field

It should be noted that monitoring wells can also act as vertical conduits depending upon factors including length of well screen and location within the flow field. Such occurrences can be important for contaminated site cleanup and research has focused on understanding concentration data collected from monitoring wells (Reilly et al. 1989; Church and

Granato 1996; Elci et al. 2001; Ma et al. 2011; McMillan et al. 2014; Vitale and Robbins 2016). However, the differences in construction between water supply wells and monitoring wells, primarily casing diameter and length of well screen, create the potential for greater flows through supply wells where there is less resistance to flow as well as greater head difference between shallow and deep strata. In general, the flow magnitudes and migration distances, both vertical and horizontal, are expected to be greater for water supply wells than monitoring wells. Vertical flow through monitoring wells is not considered in this work.

METHODS OF INVESTIGATION

Field work

The general locations of wells investigated in the field are indicated on Fig. 1. Field examples Nos. 1 and 2 present intraborehole flow measurements, while field examples Nos. 3 and 4 address water-quality-impact evaluations. For field example No. 1, intraborehole flow under non-pumping conditions was measured on a project conducted by the author using a dye tracer method derived from the method of Izbicki et al. (1999). Similarly, field example No. 2 involved measuring intraborehole flow under non-pumping conditions on a project conducted by others (see discussion of field example below) using an impeller flowmeter (Welenco 1996; Keys 1997). The previously referenced measurement methods entail recording dye travel times and impeller rotation rates at different points along well screens to estimate intraborehole flows. For field example No. 3, water quality impact in the vicinity of a well that acted as a vertical conduit was evaluated on a project conducted by the author through (1) pumping the well for an extended period while monitoring nitrate concentrations in the discharge and (2) profiling flow into the well along the well screen with a dye tracer and obtaining depth-specific water samples for analysis using the previously referenced method of Izbicki et al. (1999). Finally, field example No. 4 entailed a hydrogeologic investigation on a project conducted by the author. Methods used included flow and concentration profiling along the well screen of a municipal well using the previously referenced method of Izbicki et al. (1999) as well as subsurface characterization by routine methods (monitoring well installation, water sampling and performing a pumping test).

Management analysis

Well-specific approaches

Options for managing intraborehole flow and transport on an individual well basis were explored through modeling a hypothetical two-aquifer system penetrated by a single well as in Fig. 3. The model was implemented as a half-space similar to Konikow and Hornberger (2006) using the MODFLOW-2000 (Harbaugh et al. 2000) and MT3DMS (Zheng and Wang 1999) modeling codes. Horizontal finite difference grid dimensions were 10 × 10 ft (3 × 3 m) in an area surrounding the simulated well (Fig. 5). Grid refinement to 0.25 × 0.25 ft (8 × 8 cm) occurred for the well itself and grid coarsening was implemented towards the domain boundaries. Vertical grid dimensions were 10 ft (3 m) in the shallow aquifer and the intervening aquitard and 5 ft (1.5 m) in the deep aquifer. The shallow and deep aquifers were simulated as 100-ft (30 m) thick sands with horizontal hydraulic conductivity of 2.8×10^{1} ft/day (10^{-2} cm/s) and vertical hydraulic conductivity of 2.8 ft/day (10^{-3} cm/s). The intervening aquitard was simulated as a 50-ft (15 m) thick clay with horizontal hydraulic conductivity of 2.8×10^{-2} ft/day (10^{-5} cm/s) and vertical hydraulic conductivity of 2.8×10^{-3} ft/day (10^{-6} cm/s). A spatially uniform storage coefficient value of 10^{-4} was used. Recharge was made horizontally uniform at a value of 3×10^{-3} ft/day (9×10^{-4} m/day). Specified-head boundary conditions that remained constant over time were set on the (1) up-flow and down-flow edges of the model grid in the aquifers to establish a horizontal hydraulic gradient of 0.001 and (2) bottom of the model grid to establish a downward vertical hydraulic gradient. The combination of the head boundary conditions and recharge resulted in a head difference across the aquitard of 7.5 ft (2.3 m) in the absence of the conduit well, which corresponded to a vertical gradient of −0.15. The concentration of a conservative solute was set at 100 mg/l in the shallow aquifer and 10 mg/l in the deep aquifer. Advective solute transport was simulated using a porosity of 0.3 and implemented using the method of characteristics.

The 12-in (30 cm) diameter well casing and screen were simulated within a 24-in (61 cm) diameter borehole. An effective daily average pumping rate of 125 gal/min (473 l/min), equivalent to 500 gal/min (1,893 l/min) for 6 h/day, was applied; however, only half of this value was used since the model was implemented as a half-space. The well consisted of (1) a 50 ft (15 m) long well seal from 0 to 50 ft bgs (0–15 m bgs), (2) a 50 ft (15 m) long well screen in the shallow aquifer placed immediately above the aquitard from 50 to 100 ft bgs (15–30 m bgs), (3) a 50 ft (15 m) long well screen in the deep aquifer placed immediately below the aquitard from 150 to 200 ft bgs (45–60 m bgs) and (4) one continuous gravel pack from 50 to 205 ft bgs (15–60 m bgs). The well was simulated with an assemblage of model grid cells (Horn and Harter 2009; Houben and Hauschild 2011). To simulate the well seal, inactive cells were placed at grid locations that corresponded to the borehole annulus and well casing from 0 to 50 ft bgs (0–15 m bgs). The gravel pack was simulated with horizontal and vertical hydraulic conductivities set at 2.8×10^3 ft/day, or 1 cm/s, generally consistent with Houben and Hauschild (2011). To simulate flow in the well, a value one hundred times that of the gravel pack (2.8×10^5 ft/day, or 100 cm/s) was used for cells at grid locations that corresponded to the inside of the well. This value was, however, lower than that predicted for laminar flow conditions by Hagen-Poiseuille (2.5×10^6 cm/s) in order to simulate enhanced head loss within the well during pumping as a result of turbulence. This approach was reasonable given that the Reynolds number for the specified casing diameter and pumping rate would be well beyond the upper range for laminar flow (1.2×10^5 compared to 2.1×10^3; Vennard and Street 1982).

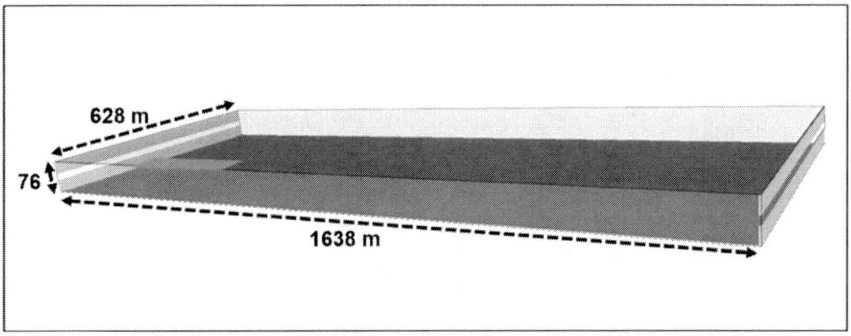

Figure. 5 Groundwater model configuration. *Arrows* indicate model domain dimensions. The *gray area* indicates the location of the higher density finite difference grid. The *red line* indicates the conduit well location. *Orange areas* indicate specified head boundaries

Simulating the unmanaged case entailed modeling transient flow and transport with a pumping schedule as follows: inactivity from day 0 through 180, pumping from day 181 through 360, inactivity from day 361 through 540, pumping from day 541 through 720. Consistent with typical pump settings for water supply wells, the pump intake location was simulated as located in the well casing above the shallower screen. Management options were evaluated by changing the pumping schedule and pump intake depth.

Representing the well as an equivalent porous medium, rather than using the MODFLOW Multi-Node Well Package (Konikow et al. 2009), was consistent with previous approaches by some researchers (Lacombe et al. 1995; Halford 2009; Horn and Harter 2009; Halford et al. 2010; Houben and Hauschild 2011; Mejia et al. 2012; McMillan et al. 2014) and allowed simulation of head loss as water flowed vertically inside the well during pumping. Because the MODFLOW Multi-Node Well Package assumes no head loss inside the well, it may over-allocate pumping flows along the well screen for portions of the screen located farthest from the pump. In such cases, variation in the areal extent of the groundwater capture zone with depth and the potential migration of solutes beyond the capture zone would not be properly simulated. One of the management approaches addressed in this work, changing the pump depth to adjust the capture zone extent with depth, required the simulation of head loss inside the well; therefore, the Multi-Node Well Package was not used. Since this work also simulates intraborehole flows under non-pumping conditions, the selected approach may underestimate those flows because of the lower hydraulic conductivity used inside the casing. As a result, expected volumes transferred between aquifers could be larger and water quality impacts could be greater than those simulated.

Regional approaches

Options for managing intraborehole flow and transport by focusing limited resources on a regional basis were explored using geographical analysis of well construction log data. The northern portion of California's Central Valley (Fig. 1) was used for demonstration based on data availability but the concepts apply more broadly. The number and diameter of wells in the study area were tabulated on a 1-mi (1.6-km) grid spacing, then the increase in vertical hydraulic conductivity was estimated using the method presented in the Appendix. When combined with a threshold for factor increase in natural vertical hydraulic conductivity, the results can be used to target areas of potential high impact for investigation.

RESULTS AND DISCUSSION OF FIELD INVESTIGATIONS

Measurements of intraborehole flow

Field example No. 1: upward flow in a regional discharge area

A 14-in (36 cm) diameter well near the Sacramento River (see No. 1 on Fig. 1) was investigated to determine the cause of water quality impact. Part of the investigation involved measuring intraborehole flow. Figure 6a presents a cumulative flow profile under ambient conditions during the winter. (Cumulative flows are measured inside the well along its length.) Upward flow occurred over the entire 229-ft (70 m) screened interval. Upward hydraulic gradients and groundwater flow are common in areas where the saturated zone is in contact with the river and seasonal pumping stresses do not prevent natural groundwater discharge. The maximum flow along the well was approximately 35 gal/min (132 l/min).

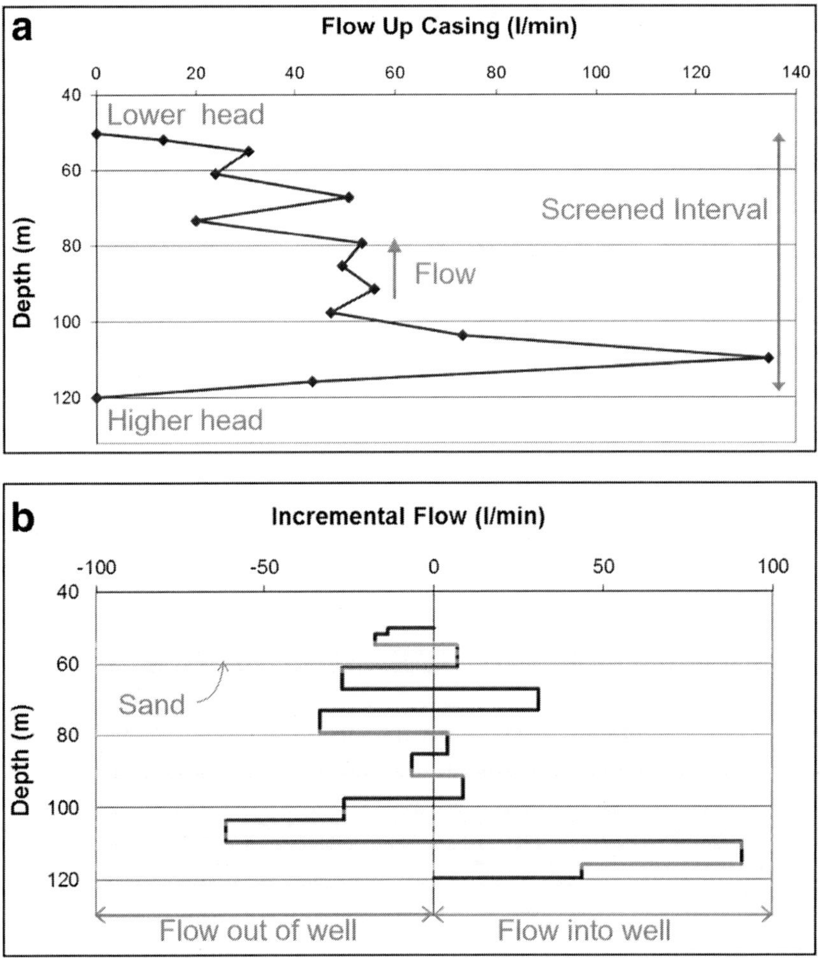

Figure. 6 Field example No. 1: **a** cumulative flow profile conducted under ambient (nonpumping) conditions; **b** incremental flow profile derived from the cumulative flow profile

Figure 6b presents the incremental flow profile for the data presented in Fig. 6a as well as the depths and thicknesses of sand layers identified in the lithologic log that was recorded when the wellbore was drilled. (Incremental flows are flows into and out of the well calculated as differences between vertically adjacent flow measurements inside the well with upstream subtracted from downstream.) Depths of the larger inflows and outflows generally coincided with the sand layers. Depths where flows occurred without indications of sand could result from inaccuracies in the lithologic log (sand actually present) or interaction of flows inside the well with the gravel pack. Even though the entire screened interval for this well was within the same water-bearing zone, stratification of the alluvial sediments created conditions conducive to the long-screened well acting as vertical conduit for flow (development of vertical hydraulic gradients and a preferential flow path through the well as a result of the contrast in hydraulic conductivity between the fine-grained strata contained in the alluvial sequence and the open well structure).

Field example No. 2: downward flow in a heavily pumped basin

Figure 7a presents a schematic for an experimental well installed in southern California by the Orange County Water District (OCWD) to evaluate the possibility of conveying recharge water to depth through wells (see No. 2 on Fig. 1). A significant amount of water enters the groundwater system as recharge to the shallow aquifer through artificial ponds and exits the system through water supply wells pumping from the deeper Principal Aquifer. However, an aquitard present from approximately 185–230 ft bgs (56–70 m bgs) impedes the flow of water between aquifers. The well was constructed of 8-in (20 cm) diameter casing and wire-wrapped screen. There was one 25-ft (8 m) screened interval in the shallow aquifer and three screened intervals with a combined length of 49 ft (15 m) in the Principal Aquifer.

Figure 7b presents results of 24 flow measurements collected by OCWD with an impeller flowmeter between the shallowest and deeper screened intervals over 47 months. Flows ranged between approximately 80 and 360 gal/min (300 and 1,360 l/min). Based upon measurements from shallow and deep short-screened monitoring wells located approximately 38 ft (12 m) away from the experimental well, the head difference across the aquitard varied between approximately 14 and 38 ft (4 and 12 m; Fig. 7b). Using these head differences and the aquitard thickness, the vertical hydraulic gradient driving flow down the well was calculated as ranging from approximately −0.3 to −0.8. The flow measurements indicate that approximately 350×10^6 gal (1.3×10^9 l) of water flowed down the well over the 47-month monitoring period.

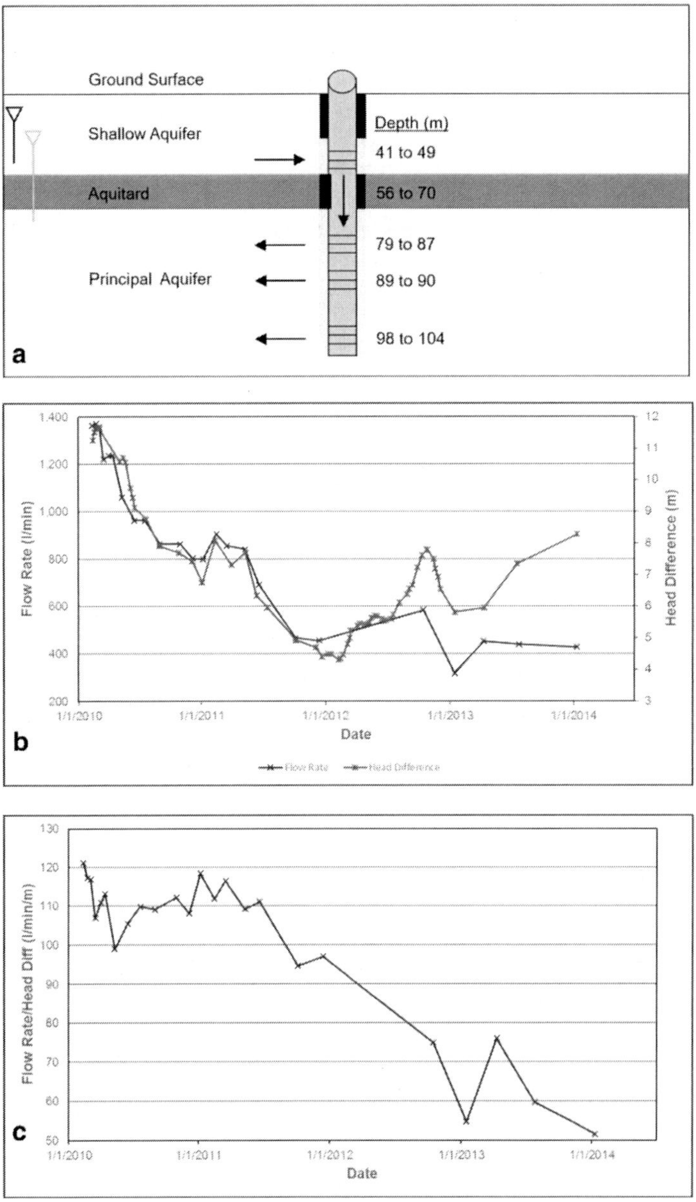

Figure. 7 Field example No. 2: **a** schematic of experimental transfer well (*Blue symbols* indicate piezometric levels at different depths in the groundwater system), **b** flows in well under ambient conditions and head differences driving flows, and **c** change in flow response to driving head difference over time

Figure 7c plots the decrease over time in rate of flow divided by the observed head difference across the aquitard. This trend is taken as an indication that the deeper well screens, gravel pack and, possibly, the sediments in the near-well environment began to clog. The causes of clogging were not investigated by OCWD but could have included several processes discussed in the previous (accumulation of suspended sediments as well as development of inorganic precipitation and/or bacterial growth resulting from mixing of geochemically different waters).

Extent of water quality impact

Field example No. 3: water quality degradation at the well scale

A 16-in (41 cm) diameter well in the Sacramento Valley was investigated to determine the cause of nitrate contamination (see No. 3 on Fig. 1). The lithologic log for the well and other available information regarding the local geology indicated that the sediments in the groundwater basin comprised a stratified sequence of finer and coarser materials. Significant groundwater pumping and agricultural irrigation occurred in the area and downward hydraulic gradients had been measured in the greater vicinity of the well. The seven screened intervals for the well spanned a significant vertical distance (451 ft, or 137 m; Fig. 8a) and the well had not been pumped for over 1 year. Together, these conditions created the potential for the well to act as a vertical conduit for downward flow and migration of shallow contaminants to depths where better quality water was present.

Figure 8a presents incremental nitrate water-quality profiles for the well collected under pumping conditions at different times as the well was continuously purged for an extended period. The later-time profiles indicate that nitrate concentrations decreased with depth. This is consistent with (1) the constituent source being at the ground surface, (2) concentration attenuation with distance from a solute source and (3) increased application rates over time along with long transport times to the deeper part of the groundwater system. Figure 8b presents the wellhead concentrations over a 76-day period when the well was pumped continuously at 600 gal/min (2,270 l/min). The initial concentration from the pumping period (slightly less than 24 mg/l on Fig. 8b) was consistent with a mix of concentrations from the two shallowest depths sampled for the profile (21 and 24 mg/l on Fig. 8a). These observations are consistent with the expectation that the well acts as a vertical conduit for downward flow and contaminant migration during periods of no pumping. Referring back to Fig. 4a, groundwater that contained nitrate concentrations from the shallower strata migrated down through the well, displaced water at depth that contained lower concentrations, and created a zone near the well containing relatively uniform water quality along the entire well

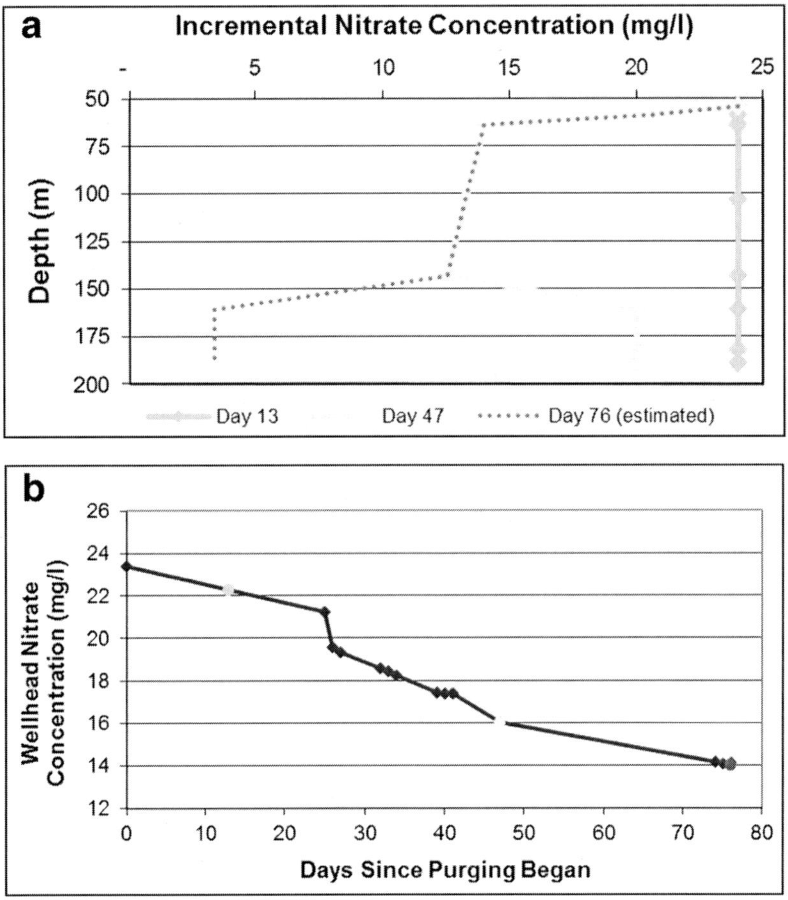

Figure. 8 Field example No. 3: **a** Evolution of an incremental nitrate concentration profile with extended purging (A cumulative water quality profile consists of concentrations associated with samples collected from inside the well, while an incremental water quality profile is the result of differencing the flow-weighted cumulative concentration profile as described by Izbicki et al. 1999. The incremental water-quality profile indicates the depths and concentrations of dissolved constituents in the strata adjacent to the well screen. *Day 76* concentration profile at depths greater than 150 m estimated based upon day 76 wellhead concentration, shallower portion of *Day 47* concentration profile and the incremental flow profile.); **b** Nitrate concentration decrease at wellhead during extended period of well pumping. *Colored diamonds* (**a**) indicate the incremental concentrations. *Colored circles* (**b**) correspond to profiles presented (**a**). The applicable drinking-water quality standard is 45 mg/l

screen. Consistent with Fig. 4b, water quality at the beginning of pumping reflected shallow water quality since the water located along the entire length of the well screen originated in the shallower strata. As pumping progressed, the higher concentration water that had migrated to depth was removed from the deeper strata and the well ultimately pumped from the heterogeneous distribution of water quality representative of conditions in the greater vicinity of the well (day 76 on Fig. 8a).

Dynamic (pumping) flow and water quality profiles for the well at 13 and 47 days into the 76-day pumping period document the well purging process. The incremental water quality profiles for these data collection events are presented on Fig. 8a. After 13 days of pumping, the water quality profile was homogeneous with a single concentration along the well screen that was consistent with shallow water quality. At this point, with a pumping rate of 600 gal/min (2,270 l/min), over 11×10^6 gal (42×10^6 l) of water containing a relatively high nitrate concentration had been extracted. Clearly, a larger volume of water had migrated down the well. After 47 days of pumping, concentrations decreased with depth, as would be expected once the well was purged, for all but the deepest portion of the water quality profile. The effects of shallow water migrating to depth through the well had been reversed in the sandier strata. Only the deeper strata containing clay, as indicated by the lithologic log, retained poorer quality water that had migrated from the shallower strata. After another 29 days, the effects of contaminated water flowing down the well while it was inactive appeared to have been fully reversed. These data clearly illustrate two points: (1) purging a well after long periods of inactivity can require weeks to months and (2) adequate purging of water supply wells before concentration profiling is necessary to obtain accurate information.

A total of more than 65×10^6 gal (246×10^6 l) of water was pumped to reverse the effects of the well acting as a vertical conduit. In this case, the vertical flow did not create a water quality problem because the nitrate concentration in the migrating water (21–24 mg/l) was below the drinking water standard (45 mg/l) and no action was required. However, shallower groundwater concentrations of nitrate, and possibly other constituents, exceed drinking water standards in many areas. The final example considers such a case.

Field example No. 4: water quality degradation at the wellfield scale

The wells considered in this example were near the Sacramento-San Joaquin River Delta (see No. 4 on Fig. 1). Groundwater pumping and agricultural irrigation are prevalent in the area. Lithologic logs for nearby wells plus other information on the local geology indicated a stratified sequence of finer and coarser sediments within the groundwater system.

Consistent with these conditions, a downward vertical hydraulic gradient had been documented locally. A pressure transducer survey was conducted for 13 months using monitoring wells with 20-ft (6 m) screens. Two shallow and two deep wells were installed to approximately 35 (shallow) and 180 (deep) ft bgs (11 and 55 m bgs) and equipped with pressure transducers. The data collected showed downward vertical hydraulic gradients throughout the year with seasonal variation that coincided with pumping. Gradients were between 0 and −0.01 during the winter and roughly doubled to between −0.01 and −0.02 during the irrigation season. A 30-day pumping test on a nearby municipal well screened in the deeper portion of the groundwater system at approximately 2,000 gal/min (7,570 l/min) increased the gradients to between −0.03 and −0.04 at a distance of approximately 1,000 ft (305 m) from the pumping well. Horizontal hydraulic gradients varied somewhat from shallow to deep and between seasons but were less than or equal to 0.001.

An 8-5/8-in (22 cm) diameter industrial supply well constructed in 1974 to a depth of approximately 200 ft bgs (61 m bgs) had an annular seal that extended to the minimum regulatory standard of 50 ft bgs (15 m bgs) and a gravel pack that extended from approximately 150 to 200 ft bgs (46–61 m bgs). It was not clear what type of material was used to fill the annular space from 50 to 150 ft bgs (15–46 m bgs). Possibilities included gravel pack as well as a collapse of the borehole and infilling with the surrounding mix of sediments. A 20-ft (6 m) screen was present at the bottom of the well. This well was on the grounds of a fertilizer formulation facility where shallow groundwater was highly contaminated with nitrate. Proximity of the industrial supply well to shallow groundwater contamination and the short annular seal created potential for nitrate migration to depths with better water quality. Decades after the industrial supply well was constructed, the afore-referenced municipal supply well capable of producing approximately 2,000 gal/min (7,570 l/min) was installed approximately 1,100 ft (335 m) from the industrial supply well. The proximity of the high-capacity municipal well increased downward vertical gradients in the area during pumping operations.

The higher nitrate concentrations in shallow groundwater near the industrial supply well ranged from approximately 1,000 to 4,000 mg/l nitrate (compared to a drinking water standard of 45 mg/l). The industrial well was pumped only intermittently and seasonal variations in nitrate concentration at the wellhead ranged from approximately 20 to 900 mg/l. Ten years of water quality data indicated that the highest concentrations occurred in the well during summer when downward gradients were the greatest. Investigation of the groundwater nitrate distribution, as well as groundwater age dating (Singleton et al. 2010), indicated impacts to deeper water quality (Fig. 9). The concentration at the industrial well (917 mg/l on Fig. 9) was higher than in surrounding wells screened at similar depths by a factor of ten or more. Also, the vertical distribution of nitrate in the municipal supply well did not decrease with depth, as was the case for example No. 3 above, which would be expected for a constituent where the source is at ground surface. Furthermore, the groundwater age distribution indicated younger water at depth near the industrial supply well (average of 27 years on Fig. 9) relative to the surrounding groundwater by a factor of as much as 100%. These observations indicated that the industrial supply well acted as a vertical conduit for downward flow and nitrate migration and that the impact to groundwater quality extended more than 1,000 ft (300 m) from the well.

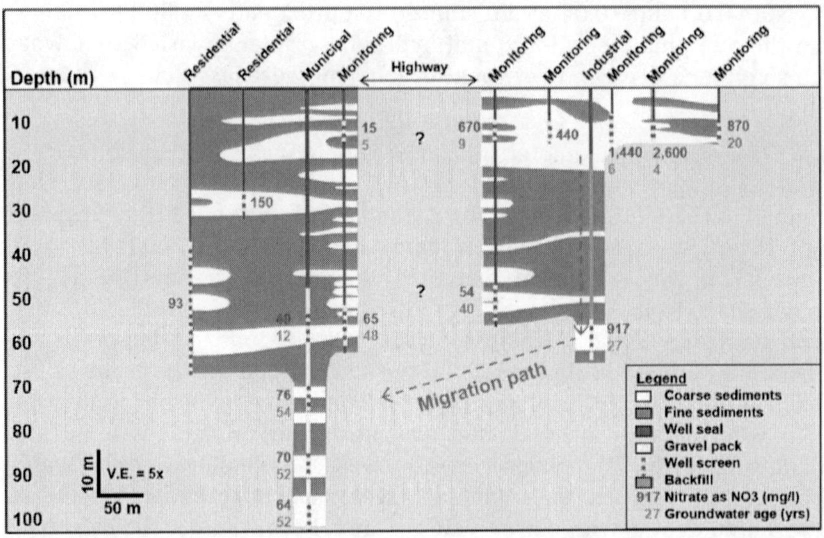

Figure. 9 Field example No. 3: Nitrate concentrations and groundwater age in the vicinity of an industrial supply well acting as a vertical conduit. Data from Singleton et al. (2010) and this study

RESULTS AND DISCUSSION OF MANAGEMENT ANALYSIS

Well-specific approaches

The volume of water transferred between strata through a well that acts as a conduit depends upon magnitude and duration of flow. A sustained flow of only 1 gal/min (4 l/min) becomes 1,440 gal (5,450 l) in a day and over

260 × 10^3 gal (992 × 10^3 l) in a 6-month period of inactivity (a representative duration of inactivity for wells used seasonally). Figure 10a–d depicts simulated results of flow and contaminant migration down a typical well that is pumped only seasonally in a layered sand and clay system.

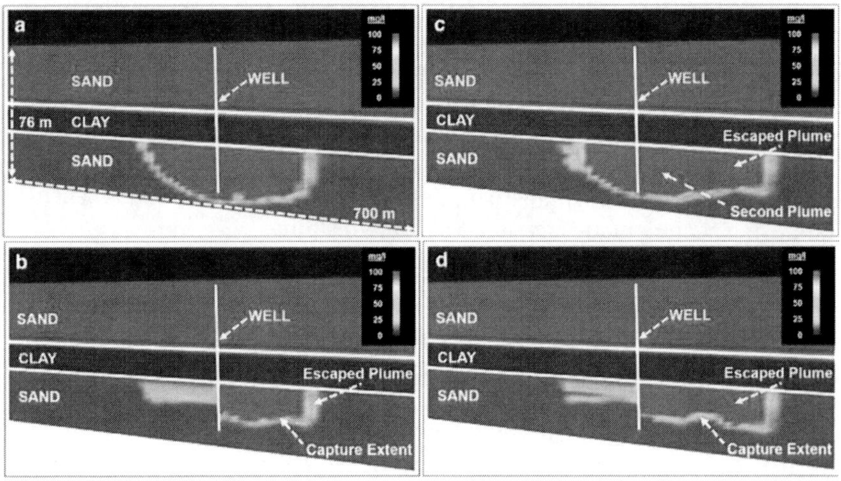

Figure. 10 Simulated flow and transport in a conduit well: **a** contaminant migration to the deeper zone while the well is idle for 180 days (days 1–180); **b** pumping the well for 180 days (days 181–360); **c** a second 180-day period when the well is idle (days 361–540); **d** a second 180-day period of pumping the well (days 541–720)

Over the 24-month simulation period, inactive and pumping periods of 6 months each occur in a series. Approximately 4.3 × 10^6 gal (16 × 10^6 l) of groundwater is redistributed from the shallow to deep aquifer during each inactive period and approximately 33 × 10^6 gal (124 × 10^6 l) is pumped from the groundwater system during each active period. Only 29% of the volume extracted comes from the deep aquifer because the well pump is positioned above the shallower screen, head loss occurs in the well casing and there is a downward hydraulic gradient. Moreover, the vast majority of flow to the well in the deep aquifer is delivered from the up flow direction by regional flow. Only 6% of the volume extracted comes from the down flow direction in the deep aquifer, which limits the down flow capture zone extent and the regional flow system transports some of the water transferred during inactive periods away from the well. This uncaptured water transfers contaminant mass to the deep aquifer where it mixes with better quality water and degrades the resource.

To prevent such contamination, local regulatory requirements and industry standards for well construction should be followed when constructing new wells (i.e., CADWR 1991, 2003). Because regulatory criteria often provide only minimum requirements and are prepared in the absence of site-specific details, more stringent design criteria such as shorter screened intervals and gravel packs, may be warranted in some cases. The hydrogeology for each new well site should be considered during design development. Key information to consider includes sediment stratigraphy and depositional environment, variations in vertical hydraulic gradient over the year, water quality near the well site and potential changes in conditions over the life of the well (well condition, aquifer stresses and water quality). In addition, best methods should be applied when abandoning test holes since a test hole also can act as a vertical conduit. This includes isolating vertically separate water-bearing strata using sealing materials that will remain stable.

Where significant water quality impacts occur from existing wells acting as vertical conduits, operational and/or structural changes should be considered. Operational changes could include implementing intermittent pumping during periods of low demand to control migration through a well. Comparison of Fig. 11a,b with Fig. 10b,d demonstrates the improvement in water quality from more frequent pumping, by a factor of two in this case, to limit contaminated water migration out of the well capture zone. This approach would only be possible when at least some water demand exists during the off-season and rotation of pumping among wells is possible. Structural changes could include changing the depth of the pump intake (moving the pump deeper in the case of water flowing downward through a well). Comparison of Fig. 11c,d with Fig. 10b,d conceptually demonstrates the improvement in water quality from moving the pump deeper in the well to reduce the amount of contaminant mass that escapes the well when pumping. Approximately 48% of the extracted volume comes from the deep aquifer, as opposed to 29% when the pump was higher in the well, and much less of the plume escapes. Other structural changes could include installing patches or packers to prevent contaminated water from entering and migrating through the well and injecting cement into the gravel pack within some intervals to prevent migration along the outside of the well. Finally, destroying wells should include application of methods that target the migration pathways of concern and collection of sufficient confirmatory information so that wells do not act as conduits for vertical contaminant migration after their useful lives have ended—for example, where a long gravel pack exists above the screen outside the well casing, it may be appropriate to perforate the casing to allow injection of sealing material.

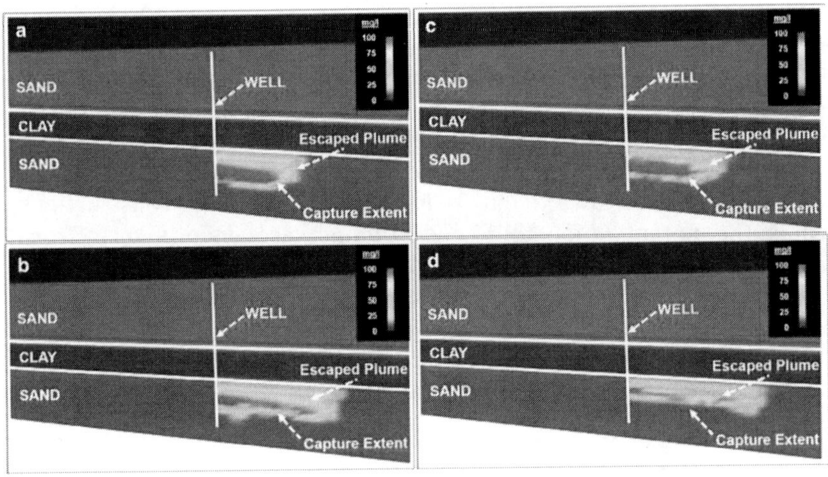

Figure. 11 Simulated flow and transport in a modified well: **a** resting and pumping the well every 90 days over a period of 360 days; **b** resting and pumping the well every 90 days over a period of 720 days; **c** moving pump deeper in well so that it is adjacent to the deeper water-bearing unit then resting and pumping the well every 180 days over a period of 360 days; **d** moving pump deeper in well and pumping the well every 180 days over a period of 720 days. The results for 360 and 720 days should be compared with parts b and d, respectively, of Fig. 10. All other details of the simulations are as described in section 'Well-specific approaches'

Regional approaches

Considering the great number of wells that exist in many groundwater basins and the length of time they remain open in the subsurface (from decades to perpetuity unless the well is properly destroyed), contaminant migration similar to that described here is likely to occur at many locations and on a variety of spatial scales (well, wellfield and region). In many cases, the risks are not fully understood because important information regarding well operations and local hydrogeology is absent. The challenge of protecting groundwater quality under such circumstances is great and efforts to focus limited resources for investigation and corrective measures are needed. An available data set for a portion of northern California provides the opportunity to outline a possible approach.

Figure 12a summarizes a dataset recently constructed by the CADWR for geographic occurrence of water supply wells in the northern portion of California's Central Valley. There are 75,974 total wells of which 56,781 are domestic; 16,354 are for irrigation; 1,913 are for public supply and 926 are for industrial use. These wells vary in age from 1 year (2%) to more than 60 years (13%), with a median age of 36 years, and likely exist in a wide range of maintenance conditions. Depths of completion vary from 40 ft (12 m, 1%) to over 500 ft (152 m, 5%) with a median of 180 ft (55 m). The combination of hydrogeologic and well conditions (stratified alluvial sediments, pumping at depth, irrigation water applied at ground surface, poorer shallow groundwater quality in places and long well constructions) creates the potential for these wells to act as conduits for vertical migration of contaminants as described in the preceding. Figure 12b summarizes a worst-case calculation of the potential increase in effective vertical hydraulic conductivity for the groundwater system from water supply wells penetrating lower hydraulic conductivity strata. Details of the calculation are presented in the Appendix.

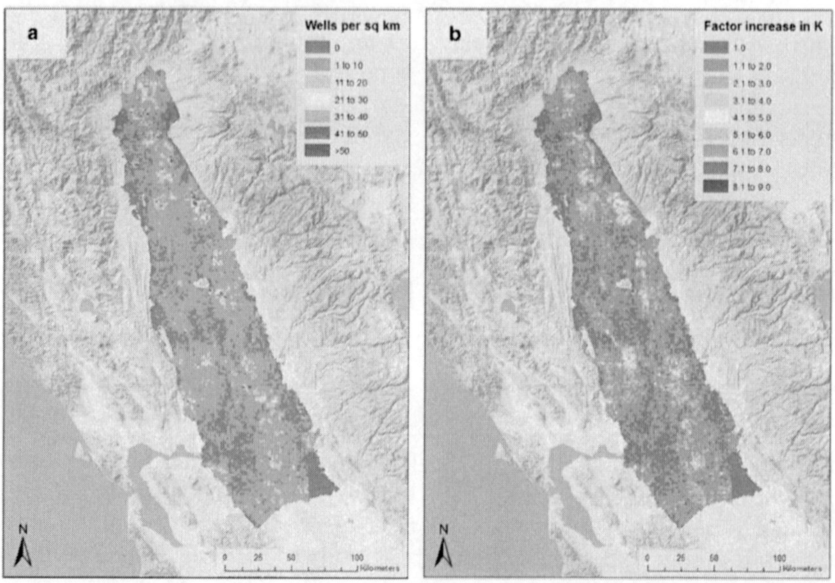

Figure. 12 Example screening approach for assessing potential impact of wells that act as conduits: **a** water well density in the northern Central Valley of California and **b** worst-case assessment of increase in vertical hydraulic conductivity of groundwater system resulting from water wells acting as vertical conduits for flow. A factor increase of 2 indicates a doubling of the vertical hydraulic conductivity

In areas of higher well density, the vertical hydraulic conductivity could increase by a factor of five to ten. Results similar to those presented on Fig. 12b could be combined with information on shallow water quality so areas that pose a higher risk to groundwater quality (high density of wells and poor shallow water quality) could be prioritized for investigation. In the high-priority areas, the effects of vertical flow through wells might be evaluated through routine water-quality monitoring upon renewed pumping after periods of inactivity. Further evaluation such as video logging and ambient flow measurement, could be performed as circumstances warrant. Responses to discovered problems, possibly repairs to casings or changes in pumping operations as discussed previously, could be implemented accordingly.

CONCLUSIONS

Table 3 summarizes the field examples previously presented. While most cases discussed here are in or near agricultural settings, the observations presented on the potential for wells to act as vertical conduits and degrade groundwater quality are also pertinent to urban or urbanizing (converting from agricultural to urban use) areas. Whenever the combination of hydrologic stresses and stratigraphy in a groundwater basin creates vertical hydraulic gradients and there is also vertical heterogeneity in concentrations of undesirable water quality constituents, the potential exists for contaminant migration through idle wells that are improperly constructed or maintained.

Table 3 Summary of case examples for wells acting as vertical conduits

Field example	No. 1	No. 2	No. 3	No. 4
Hydrostratigraphy	Alternating fine- and coarse-grained strata	Sand/aquitard	Alternating fine- and coarse-grained strata	Alternating fine- and coarse-grained strata
Hydrologic stress	Regional discharge to river	Deep pumping recharge operations	Deep pumping irrigation	Deep pumping irrigation
Vertical gradient[a]	Upward	−0.3 to −0.8	Downward	0 to −0.04
Screen length (m)	70 continuous	Upper: 8; lower: 15	18 total across 137	6 continuous
Gravel pack length (m)	106	Upper: 11; lower: 29	None	Confirmed: 15; likely: 46
Casing diameter (cm)	36	20	41	22
Flow rate (l/min)	132	300–1,360	Unknown	Unknown
Volume transferred (L)	Unknown	1.3×10^9	$>246 \times 10^6$ Impacted	Unknown but regional
Timeframe	May reverse seasonally	47 months	Unknown	Seasonal over decades
Contaminant present?	Yes	No	Yes	Yes

[a] Negative *vertical gradient* indicates downward flow

Wells that act as conduits may create water quality impacts for themselves as well as for the surrounding water-bearing zones and nearby wells. Therefore, special consideration should be given to the construction, maintenance and management of water supply wells when the conditions for intraborehole flow exist or are reasonably expected to exist. The information necessary for designing wells against vertical flow and water quality impacts has been available since at least the early 1960s (Johnson 1961; CADWR 1962); nevertheless, examples of the phenomenon appear to be fairly common. Continued and additional efforts are needed to protect groundwater quality from improperly constructed and maintained wells.

ACKNOWLEDGEMENTS

The author is grateful to the Orange County Water District for sharing data and results presented in field example No. 2 and to the California Department of Water Resources (Data Collection and Management Section, Northern Region Office) for sharing data regarding the geographic distribution of water supply well constructions in the northern Central Valley of California upon which the results in Fig. 12a, b are based. Graham Fogg, Thomas Harter and Jay Lund are also appreciatively acknowledged for discussions and editorial comments that significantly improved this manuscript. Christoph Weidner also provided valuable comments on the final manuscript. This work was supported by UC Water Security and Sustainability Research Initiative funded by the University of California Office of the President (UCOP; Grant No. 13941) and by the S.D. Bechtel Jr. Foundation through the UC Davis Center for Watershed Sciences.

APPENDIX

A screening approach was used to estimate the potential increase in effective vertical hydraulic conductivity discussed in section 'Regional approaches'. It is based upon summation of vertical flows in the region of consideration:

$$Q_{\text{regional}} = Q_{\text{strata}} + \sum_{j=1}^{n} (Q_{\text{well}})_j \qquad (1)$$

where:

$Q_{regional}$ is the total vertical flow through the groundwater system in the region

Q_{strata} is the vertical flow through strata in the region

Q_{well} is the vertical flow through a well acting as a conduit

n is the number of wells acting as conduits in the region

Substitution of Darcy's Law into Eq. (1) yields:

$$K_{\text{effective}}\, i_{\text{regional}}\, A_{\text{regional}} = K_{\text{strata}}\, i_{\text{regional}} \left(A_{\text{regional}} - \sum_{j=1}^{n} \left(A_{\text{well}}\right)_j \right) + \sum_{j=1}^{n} \left(A_{\text{well}}\, i_{\text{well}}\, A_{\text{well}}\right)_j \qquad (2)$$

where:

$K_{effective}$ is the effective vertical hydraulic conductivity for the groundwater system in the region

$i_{regional}$ is the average vertical hydraulic gradient for the region

$A_{regional}$ is the total area of the region

K_{strata} is the average vertical hydraulic conductivity of the strata in the region

A_{well} is the cross-sectional area of an individual well acting as a conduit

K_{well} is the effective vertical hydraulic conductivity of a well acting as a conduit

i_{well} is the vertical hydraulic gradient for a well acting as a conduit

The value of i_{well} is expected to be less than that of i_{regional} since there will be groundwater level drawdown where flow enters the well and mounding where flow exits the well.

Rearranging Eq. (2) provides an expression for the effective vertical hydraulic conductivity for the groundwater system:

$$K_{\text{effective}} = \left\{ K_{\text{strata}}\, i_{\text{regional}} \left[A_{\text{regional}} - \sum_{j=1}^{n} \left(A_{\text{well}}\right)_j \right] + \sum_{j=1}^{n} \left(K_{\text{well}}\, i_{\text{well}}\, A_{\text{well}}\right)_j \right\} \Big/ \left(i_{\text{regional}}\, A_{\text{regional}} \right) \qquad (3)$$

K_{well} in Eq. (3) is calculated using the Hagen-Poiseuille equation:

$$K_{\text{well}} = r^2 \rho g \big/ 8\mu \qquad (4)$$

where:

r is the inner radius of the well casing or screen

ρ is the groundwater density

g is the gravitational acceleration

μ is the groundwater viscosity

The values used for variables in the in the calculations were as follows:

$i_{regional}$ 0.4 as an example based upon author's experience

A_{total} 1.0 mile^2 (2.6 km^2) based upon data presented in Fig. 12a

$K_{regional}$ 10^{-5} cm/s as an example based upon author's experience

A_{well} varied based upon the database related to Fig. 12a

K_{well} 1.1×10^6 cm/s for a 20-cm diameter well and 2.5×10^6 cm/s for a 36-cm diameter well

i_{well} 7×10^{-6} as an example based upon author's experience (inferred from flows measured in wells)

n varied between 0 and 232 based upon the data presented in Fig. 12a

Importantly, the gradient in the regional strata (i_{regional}) results from the hydraulic stresses (pumping and recharge) and the resistance to flow caused by the layered sediments. The value used is consistent with observations in the field by the author. The gradient in the well (i_{well}) is much smaller and departs from that in the sediments as a result of head losses across the drawdown cone near the shallow screen, the mound up cone near the deep screen and the well screens themselves. These considerations reduce the hydraulic gradient that drives flow through the well.

A more-detailed approach, possibly similar to Silliman and Higgins (1990), might be applied if the necessary additional information is available and the hydrogeology can be represented as a simple layered system. However, data are often lacking and the goal of performing the calculations presented above is to identify areas where additional investigation and information collection would occur.

REFERENCES

1. Belitz K, Phillips SP, Gronberg JM (1993) Numerical simulation of ground-water flow in the central part of the western San Joaquin Valley, California. US Geol Surv Water Suppl Pap 2396
2. Bertoldi GL, Johnston RH, Evenson KD (1991) Ground water in the Central Valley, California: a summary report. US Geol Surv Prof Pap 1401-A
3. Bexfield LM, Jurgens BC (2014) Effects of seasonal operation on the quality of water produced by public supply wells. Ground Water 52(S1):10–24.

4. Bexfield LM, Jurgens BC, Crilley DM, Christenson SC (2012) Hydrogeology, water chemistry, and transport processes in the zone of contribution of a public-supply well in Albuquerque, New Mexico. US Geol Surv Sci Invest Rep 2011-5182
5. Burow KR, Dubrovsky NM, Shelton JL (2007) Temporal trends in concentrations of DBCP and nitrate in groundwater in the eastern San Joaquin Valley, California, USA. Hydrogeol J 15:991–1007.
6. CADWR (1962) Bulletin 74, recommended minimum water well construction and sealing standards for the protection of ground water quality. Preliminary edn. July 1962, State of California, Sacramento, CA
7. CADWR (1981) Bulletin 74–81, water well standards: State of California. December 1981. http://www.water.ca.gov/pubs/groundwater/water_well_standards__bulletin_74-81_/ca_well_standards_bulletin74-81_1981.pdf. Accessed 6 July 2016
8. CADWR (1991) Bulletin 74–90, supplement to bulleting 74–81, California well standards. June 1991. http://www.water.ca.gov/pubs/groundwater/ water_well_standards__bulletin_74-90_/ca_well_standards_bulletin74-90_1991.pdf. Accessed 6 July 2016
9. CADWR (2003) California laws for water wells, monitoring wells, cathodic protection wells, geothermal heat exchange wells. March 2003. http://www.water.ca.gov/pubs/groundwater/california_laws_for_water_wells__monitoring_wells__cathodic_protection_wells__geothermal_heat_exchange_wells_2003/ca_water_laws_2003.pdf. Accessed 6 July 2016
10. CADWR (2014) Bulletin 160–14, California water plan update 2013. October 2014. http://www.waterplan.water.ca.gov/cwpu2013/. Accessed 6 July 2016
11. Church PE, Granato GE (1996) Bias in ground-water data caused by well-bore flow in long-screen wells. Ground Water 34(2):262–273.
12. Clark BR, Landon MK, Kauffman LJ, Hornberger GZ (2008) Simulations of ground-water flow, transport, age, and particle tracking near York, Nebraska, for a study of transport of anthropogenic and natural contaminants (TANC) to public supply wells. US Geol Surv Sci Invest Rep 2007-5068
13. Corcho Alvarado JA, Barbecot F, Purtschert R (2009) Ambient vertical flow in long-screen wells: a case study in the Fontainebleau Sands Aquifer (France). Hydrogeol J 17:425–431. doi:10.1007/s10040-008-0383
14. Davis GH, Lofgren BE, Mack S (1964) Use of ground-water reservoirs for storage of surface water in the San Joaquin Valley California. US Geol Surv Water Suppl Pap 1618
15. Dragon K, Gorski J, Marciniak M, Kasztelan D (2009) Use of mathematical modeling to investigate inter-aquifer contamination by organic-rich water through an unplugged well (central Wielkopolska, Poland). Hydrogeol J 17:1257–1564. doi:10.1007/s10040-009-0432-4
16. Elci A, Molz FJ, Waldrop WR (2001) Implications of observed and simulated ambient flow in monitoring wells. Ground Water 39(6):853–862. doi:10.1111/j.1745-6584.2001.tb02473.x
17. Faunt CC (ed) (2009) Groundwater availability of the Central Valley Aquifer, California. US Geol Surv Prof Pap 1766
18. Fetter CW (1994) Applied hydrogeology. Prentice Hall, Englewood Cliffs, NJ
19. Fogg GE (1986) Groundwater flow and sand body interconnectedness in a thick multiple-aquifer system. Water Resour Res 22(5):1679–1694. doi:10.1029/WR022i005p00679
20. Freeze RA, Cherry JA (1979) Groundwater. Prentice Hall, Englewood Cliffs, NJ
21. Gass TE, Lehr JH, Heiss HW (1977) Impact of abandoned wells on ground water. US EPA Rep EPA/600/3-77-095, US Environmental Protection Agency, Washington, DC
22. Halford K (2009) AnalyzeHOLE: an integrated wellbore flow analysis tool. US Geol Surv Surv Tech Methods 4–F2
23. Halford KJ, Hanson RT (2002) Users guide for the drawdown-limited, multi-node well (MNW) package for the U.S. Geological Survey's modular three-dimensional

ground-water flow model, versions MODFLOW-96 and MODFLOW-2000. US Geol Surv Open File Rep 02-293
24. Halford K, Stamos C, Nishikawa T, Martin P (2010) Arsenic management through well modification and simulation. Ground Water 48(4):526–537. doi:10.1111/j.1745-6584.2009.00670.x
25. Hanson RT, Zhen L, Faunt CC (2004) Documentation of the Santa Clara Valley regional ground-water/surface-water flow model, Santa Clara County, California. US Geol Surv Sci Invest Rep 2004-5231
26. Harbaugh AW, Banta ER, Hill MC, McDonald MG (2000) MODFLOW-2000, the US Geological Survey modular ground-water model: user guide to modularization concepts and the ground-water flow process. US Geol Surv Open File Rep 00-92
27. Hart DJ, Bradbury KR, Feinstein DT (2006) The vertical hydraulic conductivity of an aquitard at two spatial scales. Ground Water 44(2):201–211. doi:10.1111/j.1745-6584.2005.00125.x
28. Horn JE, Harter T (2009) Domestic well capture zone and influence of the gravel pack length. Ground Water 47(2):277–286. doi:10.1111/j.1745-6584.2008.00521.x
29. Houben GJ, Hauschild S (2011) Numerical modeling of the near-field hydraulics of water wells. Ground Water 49(4):570–575. doi:10.1111/j.1745-6584.2010.00760.x
30. Houben GJ, Treskatis C (2007) Water well rehabilitation and reconstruction. McGraw Hill, New York
31. Izbicki JA, Christensen AH, Hanson RT, Martin P, Crawford SM, Smith GA (1999) US Geological Survey combined well-bore flow and depth-dependent water sampler. US Geol Surv Fact Sheet 196-99
32. Izbicki JA, Christensen AH, Newhouse MW, Smith GA, Hanson RT (2005a) Temporal changes in the vertical distribution of flow and chloride in deep wells. Ground Water 43(4):531–544. doi:10.1111/j.1745-6584.2005.0032.x
33. Izbicki JA, Christensen AH, Newhouse MW, Aiken GR (2005b) Inorganic, isotopic, and organic composition of high-chloride water from wells in a coastal southern California aquifer. Appl Geochem 20:1496–1517. doi:10.1016/j.apgeochem.2005.04.010
34. Izbicki JA, Metzger LF, McPherson KR, Everett RR, Bennett GL (2006) Sources of high-chloride water to wells, Eastern San Joaquin Ground-Water Subbasin, California. US Geol Surv Open File Rep 2006–1309
35. Izbicki JA, Stamos CL, Metzger LF, Halford KJ, Kulp TR, Bennett GL (2008) Source, distribution, and management of arsenic in water from wells, Eastern San Joaquin Ground-Water Subbasin, California. US Geol Surv Open File Rep 2008-1272
36. Izbicki JA, Wright MT, Seymour WA, McCleskey RB, Fram MS, Belitz K, Esser BK (2015) Cr(VI) occurrence and geochemistry in water from public-supply wells in California. Appl Geochem 63:203–217. doi:10.1016/j.apgeochem.2015.08.007
37. Jiménez-Martínez J, Arvenda R, Candela L (2011) The role of leaky boreholes in the contamination of a regional confined aquifer: a case study—the campo de Cartagena region, Spain. Water Air Soil Pollut 215:311–327. doi:10.1007/s11270-010-0480-3
38. Johnson (1961) Dual-aquifer wells pose problems. Johnson National Drillers Journal, January–February 1961
39. Johnson RL, Clark BR, Landon MK, Kauffman JL, Eberts SM (2011) Modeling the potential impact of seasonal and inactive multi-aquifer wells on contaminant movement to public water-supply wells. J Am Water Resour Assoc 47(3):588–596. doi:10.1111/j.1752-1688.2011.00526.x
40. Jurgens BC, Burow KR, Dalgish BA, Shelton JL (2008) Hydrogeology, water chemistry, and factors affecting the transport of contaminants in the zone of contribution of a public-supply well in Modesto, eastern San Joaquin Valley, California. US Geol Surv Sci Invest Rep 2008-5156
41. Jurgens BC, Fram MS, Belitz K, Burow KR, Landon MK (2010) Effects of groundwater development on uranium: Central Valley, California, USA. Ground Water 48(6):913–928. doi:10.1111/j.1745-6584.2009.00635.x

42. Jurgens BC, Bexfield LM, Eberts SM (2014) A ternary age-mixing model to explain contaminant occurrence in a deep supply well. Ground Water 52(S1):25–39. doi:10.1111/gwat.12170
43. Keys WS (1997) A practical guide to borehole geophysics in environmental investigations. CRC, Boca Raton, FL
44. Konikow LF, Hornberger GZ (2006) Modeling effects of multinode wells on solute transport. Ground Water 44(5):648–660. doi:10.1111/j.1745-6584.2006.00231.x
45. Konikow LF, Hornberger GZ, Halford KJ, Hanson RT (2009) Revised multi-node well (MNW2) package for MODFLOW ground-water flow model. US Geol Surv Tech Methods 6-A30
46. Lacombe S, Sudicky EA, Frape SK, Unger AJA (1995) Influence of leaky boreholes on cross-formational groundwater flow and contaminant transport. Water Resour Res 31(8):1871–1882. doi:10.1029/95WR00661
47. Landon MK, Clark BR, McMahon PB, McGuire VL, Turco MJ (2008) Hydrogeology, chemical characteristics, and transport processes in the zone of contribution of a public-supply well in York, Nebraska. US Geol Surv Sci Invest Rep 2008-5050
48. Landon MK, Jurgens BC, Katz BG, Eberts SM, Burow KR, Crandall CA (2009) Depth-dependent sampling to identify short-circuit pathways to public supply wells in multiple aquifer settings in the United States. Hydrogeol J 18(3):577–593. doi:10.1007/s10040-006-0148-7
49. Lu H, Liu T, Chen W, Peng T, Wang C, Tsai M, Liou T (2008) Use of geochemical modeling to evaluate the hydraulic connection of aquifers: a case study from Chianan plain, Taiwan. Hydrogeol J 16:139–154. doi:10.1007/s10040-007-0209-6
50. Ma R, Zheng C, Tonkin M, Zachara JM (2011) Importance of considering intraborehole flow in solute transport modeling under highly dynamic conditions. J Contam Hydrol 123:11–19. doi:10.1016/j.jconhyd.2010.12.001
51. Mansuy N (1999) Water well rehabilitation: a practical guide to understanding well problems and solutions. Lewis, New York
52. Mayo AL (2010) Ambient well-bore mixing, aquifer cross-contamination, pumping stress and water quality from long-screened wells: what is sampled and what is not? Hydrogeol J 18:823–837. doi:10.1007/s10040-009-0568-2
53. McCombs J, Fiedler AG (1927) Methods of exploring and repairing leaky artesian wells. US Geol Surv Water Suppl Pap 596-A
54. McMahon PB, Böhlke JK, Kauffman LJ, Kipp KL, Landon MK, Crandall CA, Burrow KR, Brown CJ (2008) Source and transport controls on the movement of nitrate to public supply wells in selected principal aquifers of the United States. Water Resour Res 44(4):W04401. doi:10.1029/2007WR006252
55. McMillan LA, Rivett MO, Tellam JH, Dumble P, Sharp H (2014) Influence of vertical flows in wells on groundwater sampling. J Contam Hydrol 169:51–60. doi:10.1016/j.jconhyd.2014.05.005
56. Mejia A, Cassiraga E, Sahuquillo A (2012) Influence of hydraulic conductivity and wellbore design in the fate and transport of nitrate in multi-aquifer systems. Math Geosci 44:227–238. doi:10.1007/s11004-012-9388-3
57. Phillips SP, Belitz K (1991) Calibration of a texture-based model of a ground-water flow system, western San Joaquin Valley, California. Ground Water 29(5):702–715. doi:10.1111/j.1745-6584.1991.tb00562.x
58. Phillips SP, Green CT, Burow KR, Shelton JL, Rewis DL (2007) Simulation of multiscale ground-water flow in part of northeastern San Joaquin Valley, California. US Geol Surv Sci Invest Rep 2007-5009
59. Reilly TE, Franke OL, Bennett GD (1989) Bias in groundwater samples caused by wellbore flow. J Hydraul Eng 115(2):270–276
60. Santi PM, McCray JE, Martens JL (2006) Investigating cross-contamination of aquifers. Hydrogeol J 14:51–68. doi:10.1007/s10040-00400403-8
61. Silliman S, Higgins D (1990) An analytical solution for steady-state flow between aquifers through an open well. Ground Water 28(2):184–190. doi:10.1111/j.1745-6584.1990.tb02245.x

62. Singleton MJ, Moran JE, Roberts SK, Hillegonds DJ, Esser BK (2010) California GAMA special study: an isotopic and dissolved gas investigation of nitrate source and transport to a public supply well in California's Central Valley, final report for the California State Water Resources Control Board, Rep LLNL-TR-427957. Lawrence Livermore National Laboratory, Livermore, CA
63. Smith SA, Comeskey AE (2010) Sustainable wells: maintenance, problem prevention, and rehabilitation. CRC, Boca Raton, FL
64. Sokol D (1963) Position and fluctuations of water level in wells perforated in more than one aquifer. J Geophys Res 68(4):1079–1080
65. Van Beek CGEM (2012) Cause and prevention of clogging of wells abstracting groundwater from unconsolidated aquifers. IWA, London
66. Vennard JK, Street RL (1982) Elementary fluid mechanics, 6th edn. Wiley, Chichester, UK
67. Viers JH, Liptzin D, Rosenstock TS, Jensen VB, Hollander AD, McNally A, King AM, Kourakos G, Lopez EM, De La Mora N, Fryjoff-Hyng A, Dzurella KN, Canada H, Laybourne S, McKenney C, Darby J, Quinn JF, Harter T (2012) Nitrogen sources and loading to groundwater. Technical report 2. In: Addressing nitrate in California's drinking water with a focus on Tulare Lake Basin and Salinas Valley groundwater, Report for the State Water Resources Control Board report to the legislature. Center for Watershed Sciences, University of California, Davis, CA
68. Vitale SA, Robbins GA (2016) Characterizing groundwater flow in monitoring wells by altering dissolved oxygen. Groundwater Monitor Remed 36(2):59–67. doi:10.1111/gwmr.12157
69. Wada Y, van Beek LPH, Bierkens MFP (2012) Nonsustainable groundwater sustaining irrigation: a global assessment. Water Resour Res 48(6):W00L06. doi:10.1029/2011WR010562
70. Welenco (1996) Water and environmental GeopEhysical well logs, volume 1: technical information and data, 8th edn. Welenco, Bakersfield, CA
71. Williamson AK, Purdic DE, Swain LA (1989) Ground-water flow in the Central Valley of California. US Geol Surv Prof Pap 1401-D
72. Yager RM, Heywood CE (2014) Simulation of the effects of seasonally varying pumping on intraborehole flow and the vulnerability of public-supply wells to contamination. Ground Water 52(S1):40–52. doi:10.1111/gwat.12150
73. Zheng C, Wang PP (1999) MT3DMS: a modular three-dimensional multispecies transport model for simulation of advection, dispersion, and chemical reactions of contaminants in groundwater systems: documentation and user's guide. US Army Corps Eng Rep SERDP-99-1, US Army Corp of Eng., Washington, DC
74. Zinn BA, Konikow LF (2007) Effects of intraborehole flow on groundwater age distribution. Hydrogeol J 15(4):633–643. doi:10.1007/s10040-006-0139-8

CHAPTER 3

SWAT and Wavelet Analysis for Understanding the Climate Change Impact on Hydrologic Response

Shesh Raj Koirala[1,2], Randall W. Gentry[1]

[1]Institute for a Secure and Sustainable Environment, University of Tennessee, Knoxville, USA; [2]Bechtel Corporation, Houston, USA.

ABSTRACT

Quantifying the hydrological response to an increased atmospheric carbon dioxide concentration and climate change is important in a watershed scale particularly from the application point of view. The specific objectives are to evaluate the climate change impact on the future water yield at the outlet of Clinch River Watershed upstream of Norris Lake in Tennessee, USA and see how the frequency of extreme water yield (e.g. flood) changes compared to present condition. The predicted future climate change by climate change scenarios A2 from community climate system model (CCSM) is applied. The model was calibrated using monthly average streamflow data from 1970 to 1989 and validated using similar data from 1990 to 2009 collected at a USGS gauging station 03528000. Changes in monthly average streamflow were estimated for long term (around 2099). Results were also interpreted in the time-frequency domain approach by showing how frequency of occurrence changes based on A2 scenario.

1. INTRODUCTION

Climate change will likely affect fundamental drivers of the hydrological cycle and has a wide range of consequences for human societies and ecosystems. According to Intergovernmental Panel on Climate Change (IPCC) fourth assessment report (AR4), observed global warming has been linked to changes in the large-scale hydrological cycles such as: changes in precipitation patterns, intensities, extremes, and changes in soil moisture and runoff. However, our ability to interpret changes in impact-relevant variables (e.g. changes in circulation, precipitation and extremes) remains limited [1]. More understanding of the processes driving the changes at different temporal and spatial scales could be beneficial. So IPCC AR4 recommended more research to improve the ability of models related to circulation and precipitation patterns, extremes, El-Nino and seasonal variability, hydrological cycle both in a short and long terms. It also recommended to increase the focus on regional and watershed scale climate study. Improved understanding of how climate change could influence the hydrological cycle and water resources in a regional and watershed scale is identified as a major research priority.

Since there is a regional variation of hydrologic conditions, the influence of climate change on hydrology in watershed scale may also vary between watersheds even under the same climate scenarios. Many recent studies indicate the importance of watershed scale climate variability due to global climate change particularly for long-term water resources planning and management [2, 3].

The hydrologic models should provide a link between climate changes and water yields through simulation of hydrologic processes within watersheds. However, most hydrologic models are unable to incorporate the climate change effect for simulation. The Soil and Water Assessment Tool (SWAT) [4] is one widely used model which has the capability of incorporating the climate change effect for simulation. In SWAT, it is possible to incorporate the general circulation model (GCM) projections of carbon dioxide concentration, precipitation and temperature changes. Hence SWAT was used in this study.

Most of the research related to the impact of climate change on water resources is focused on comparing the water yield in future, based on different climate change scenarios [5,6]. However, there is a limited research on how the frequency of occurrence of water yield changes compared to the present condition which in fact is particularly important from long term water resources planning and management point of view. In this study we attempt to explore this issue. The specific objectives are to evaluate the climate change impact on the future water yield at the outlet of

Clinch River Watershed upstream of Norris Lake in Tennessee and see how the frequency of extreme water yield (e.g. flood) changes compared to present condition.

The Intergovernmental Panel on Climate Change (IPCC) published a set of emission scenarios in the Special Report to serve as a basis for assessments of future climate change [1,2]. Among different scenarios, the A2 scenario was chosen for this study because it represents a plausible condition over the next century and this is one of the most widely simulated over all models [7]. The A2 scenario is characterized by a world of independently operating, self-reliant nations, continuously increasing population, regionally oriented economic development, and slow and more fragmented technological changes and improvements to per capita income [1]. It should be noted that not all modeling groups have runs for all emission scenarios.

The outline of the paper is as follows. Section 2 describes the research site and data used in the analysis. Section 3 presents the methodology with detailed description on statistical analysis and wavelet transform. Results are presented and discussed in Section 4 and Section 5 concludes the paper.

2. SITE AND DATA DESCRIPTION

The Clinch River is one of the tributaries of the Tennessee River. The Clinch River watershed in Tennessee, was selected for study because this watershed in the Tennessee River Basin was one of the basins which has not regulated and this may represent the natural watershed response. The watershed is a 3,818 square kilometer forested watershed (Figure 1).

Climate data required by the model are daily precipitation, minimum and maximum daily air temperature, solar radiation, wind speed and relative humidity. These daily climatic inputs can be entered from historical records, and/ or generated internally in the model using monthly climate statistics that are based on long-term weather records and are available internally in the SWAT database for US. In this study, daily total rainfall, daily maximum and minimum temperature are obtained from four weather stations within and near the watershed from 1970 to 2009. Monthly average discharge for calibration and validation was also used from 1970 to 2009 from USGS gauge station 03528000 at the outlet of the water shed upstream of the Norris Lake (Figure 2). Rest of the climate data required were generated internally in SWAT.

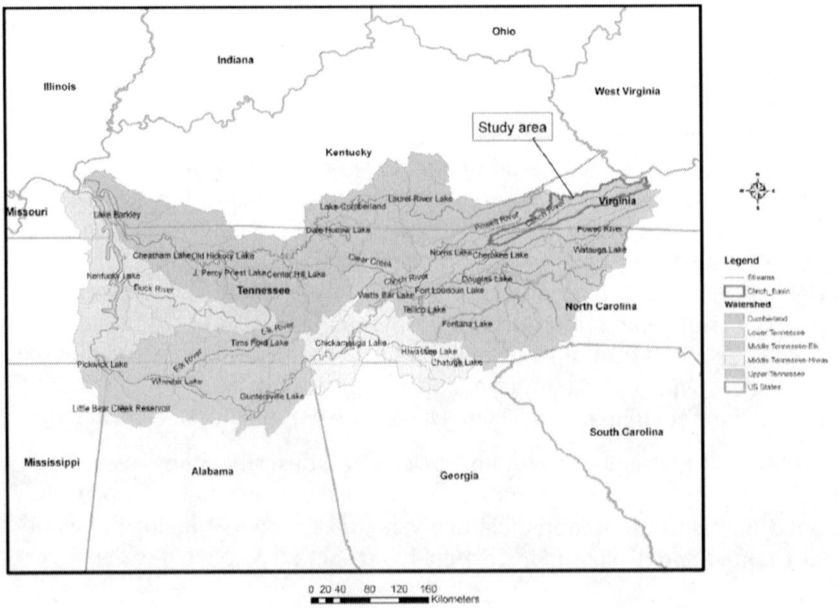

Figure 1. Study area location map.

Elevation data were obtained from USGS 30 m Digital Elevation Models (DEMs). Elevation ranges from 321 m to 1429 m from mean sea level (Figure 2). Landuse data were obtained from 1:24,000 NLCD database. Soil data were obtained from the State Soil Geographic (STASTGO) database.

3. METHODOLOGY

Before Delineation of watershed, subwatersheds and creation of stream networks were based on a USGS digital elevation model (DEM) with 30 m grid cell resolution by specifying the threshold drainage area and the watershed outlet. The threshold area is the minimum drainage area required to form the origin of the stream [6]. The watershed outlet was selected at the USGS gage station to compare the simulated and observed data. The threshold area of 8,000 ha created 19 subwatersheds as shown in Figure 2. Each of these subwatershed is further divided into Hydrological Response Units (HRUs) based on the defined threshold of landuse, soil and elevation characteristics. The defined threshold for landuse, soils and elevation were 10%, 10% and 5% respectively, which created 297 HRUs.

3.1. Statistical Analysis

The Coefficient of determination (R^2) is an indicator of the strength of the relationship between the observed and simulated values which ranges from 0 to 1. The Nash-Sutcliffe efficiency (E) indicates how the plot of the observed and simulated values fit the 1:1 line, which ranges from negative infinity to 1. The formula to calculate R^2 and E are given by [2,8,9],

$$R^2 = \left(\frac{\sum_{i=1}^{n} (O_i - \bar{O})(P_i - \bar{P})}{\sqrt{\sum_{i=1}^{n} (O_i - \bar{O})^2} \sqrt{\sum (P_i - \bar{P})^2}} \right)^2 \tag{1}$$

$$E = 1 - \frac{\sum_{i=1}^{n} (O_i - P_i)^2}{\sum_{i=1}^{n} (O_i - \bar{O})^2} \tag{2}$$

where O represents the observed value and P represents the predicted value. The over bar is the mean for the entire time period of evaluation, and i = 1, 2, 3···n, where n is the total number pairs of data.

3.2. Wavelet Analysis

The wavelet transform is a method of converting a function (or signal) into another form which either makes certain features of the original signal more amenable to study or enables the original data set to be described more succinctly [10]. In the wavelet transform, an energy spectrum describes in the frequency domain what the autocorrelation function expresses in the time domain. The amplitude of the spectrum, however, allows visualization of the frequency content of the time series and is more useful for comparing signals to each other and for isolating meaningful frequency peaks [11]. To perform the transform, a wavelet, which is a localized waveform and function that satisfies certain mathematical criteria, is needed. A Mexican-hat wavelet was chosen for the analysis because of our focus on the amplitude of the wavelet spectrum [12]. The Mexican-hat wavelet for time t is defined as:

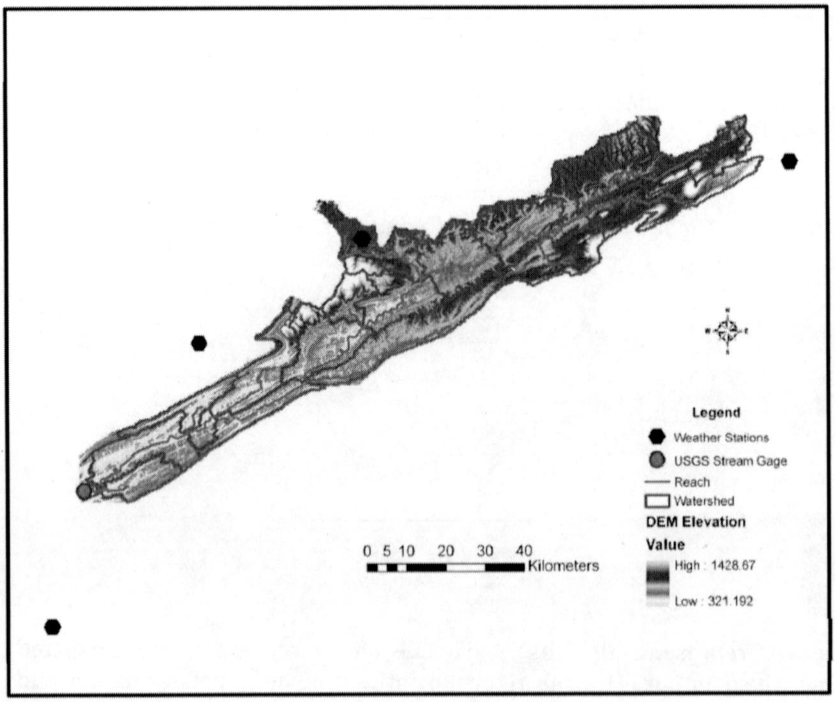

Figure 2. Digital elevation model of the study area.

$$\Psi(t) = \left(1 - t^2\right) e^{-\frac{t^2}{2}} \tag{3}$$

which is the second derivative of the Gaussian distribution function $e^{-\frac{t^2}{2}}$

Equation (3), is defined as the mother wavelet or analyzing wavelet. This is the basic form of the wavelet from which dilated and translated versions are derived and used in the wavelet transform as presented in Equation (4) The dilation and contraction are governed by the dilation parameter a, which is the distance between the center of the wavelet and its intersection with the time axis. The movement of the wavelet along the time axis is governed by the translation parameter b. The translated and dilated form of the altered mother wavelet is defined in equation (4).

$$\Psi\left(\frac{t-b}{a}\right) = \left[1 - \left(\frac{t-b}{a}\right)^2\right] e^{-1/2\left[\left(\frac{t-b}{a}\right)\right]^2} \tag{4}$$

From equation (4), a transformed signal, *x(t)*, can be defined using a range of a's and b's. The wavelet transform of a signal *x(t)* is thus defined as

$$T(a,b) = \frac{1}{\sqrt{a}} \int_{-\infty}^{\infty} x(t)\Psi\left(\frac{t-b}{a}\right)dt \tag{5}$$

The wavelet spectral power is estimated from this convolution integral using the following expression:

$$P(a,b) = \left|T(a,b)\right|^2 \tag{6}$$

A plot of the spectral power is obtained by varying a and b and is plotted against the time scale.

A MATLAB code developed by Torrence and Compo [12] was used for the analysis.

First the Coefficient of Determination (R^2) and The Nash-Sutcliffe efficiency (E) are determined from the calibrated and validated model input/output using the Equations (1) and (2). Monthly water yield and the projected monthly water yield time series from the model output are processed using the MATLB code for wavelet transform, which primarily uses the Equations (3)-(6).

4. RESULTS AND DISCUSSION

4.1. Sensitivity Analysis, Calibration and Validation

Potential evapotranspiration was modeled with the Penman-Monteith algorithm, while surface runoff was modeled with the Curve number and the variable storage channel routing approach. Initial model simulations were conducted using the default range of values for most model parameters using parameter solution (PARASOL) within SWAT-CUP. Table 1 shows the most sensitive parameters. Groundwater delay time (GW_DELAY), Manning's N value for the channel (CH_N2), baseflow alpha factor for bank storage (ALPHA_BNK), soil evaporation compensation factor (ESCO), effective hydraulic conductivity (CH_K2), initial SCS curve number (CN2), baseflow alpha factor (ALPHA_BF) and maximum canopy storage (CANMX) are most sensitive values which are statistically significant within 95% confidence level as shown in Table 1. These significant values are adjusted to find the best parameter values for calibration.

Following the sensitivity analysis, SWAT monthly calibration was carried out using the data from January 1970 to December 1989. The results were then assessed based on the visual agreement observed and simulated streamflow plots and the performance statistics generated i.e. R^2 and E which are 0.71 and 0.70 respectively (Figure 3).

Similar results were obtained for validation period which ranges from January 1990 through December 2009 with R^2 estimate of 0.66 and E estimate of 0.65 as shown in Figure 4.

Although the model overestimated the higher discharge values and slightly underestimated the baseflow, it is generally a good agreement between the observed and the simulated data as indicated by R^2 and E. The study area is in the karst zone which makes the simulation very complex. For comparison, in future we are also planning to perform more analysis using different approach such as wavelet analysis which may give better result in such complex karstic watershed.

4.2. Climate Change Scenario

For the A2 scenario test of climate change effects, the changes in temperature and precipitation were estimated using the CCSM downscaled data for the period of 2080 through 2099. Monthly changes in precipitation and temperature compared to present scenario are shown in **Table 2**. Summer and early fall (i.e. from May to September) months experience the highest increase in precipitation whereas there is a slight decrease in precipitation in December and January. There is always the increase in temperature ranging from 2.92 degree C in January to 5.22 degree C in June (Table 2). These changes were incorporated in the SWAT model and run the validated model to estimate the changes in water yield. Percentage change in average monthly water yield is shown in Figure 5.

Table 1. Sensitivity analysis of parameters.

Parameter	Description	T-stat	P-value
SOL_BD(1)	Moist bulk density	0.00	1.00
GWQMN	Threshold depth of water in shallow aquifer	−0.03	0.98
SFTMP	Snowfall temperature	0.36	0.72
SOL_K(1)	Sat. Hydraulic Conductivity	−1.50	0.13
SOL_AWC(1)	Available water capacity of soil layer	−1.84	0.07
GW_DELAY	Groundwater delay time	2.01	0.04
CH_N2	Manning's N value for the channel	3.65	0.00
ALPHA_BNK	Baseflow alpha factor for bank storage	−5.51	0.00
ESCO	Soil evaporation compensation factor	7.36	0.00
CH_K2	Effective Hydraulic conductivity in tributary	8.43	0.00
CN2	Initial SCS runoff curve number for moisture condition II	−16.54	0.00
ALPHA_BF	Baseflow alpha factor	−52.64	0.00

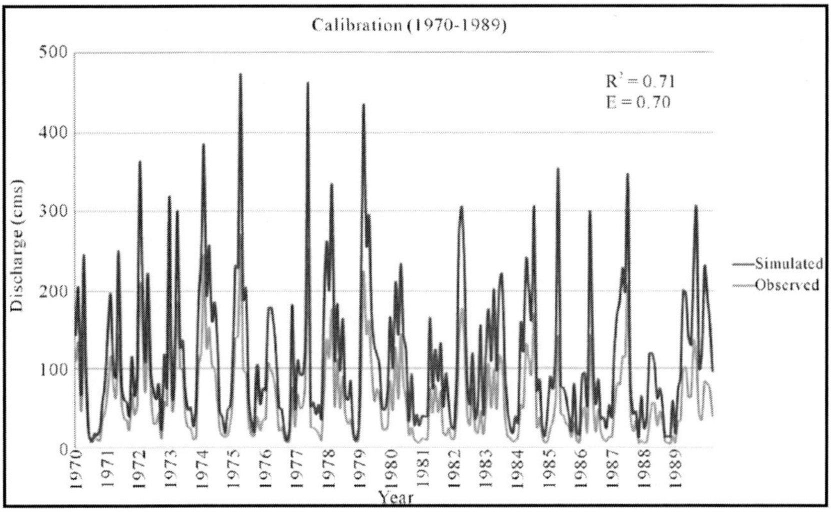

Figure 3. Calibration.

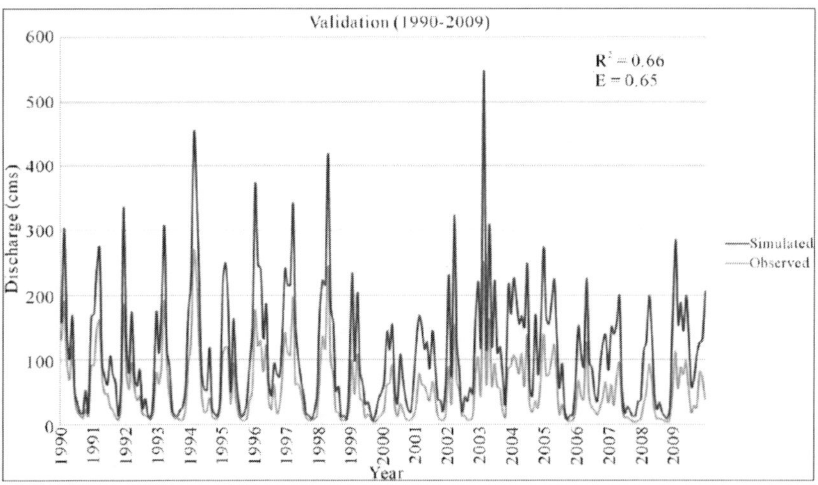

Figure 4. Validation.

Increase in monthly water yield is highest in September, which is about 200% more than current condition. This may have tremendous effect on downstream flooding.

Table 2. Projected change in precipitation and temperature in 2099 based on CCSM A2 scenario compared to current condition (1980-1999).

Month	% Change in precipitation	Change in temperature (degree C)
January	−1.53	2.92
February	1.73	3.74
March	4.80	3.89
April	6.58	3.67
May	16.13	4.83
June	23.10	5.22
July	34.02	4.34
August	31.32	3.85
September	43.92	3.97
October	0.82	4.84
November	1.86	4.19
December	−7.30	3.31

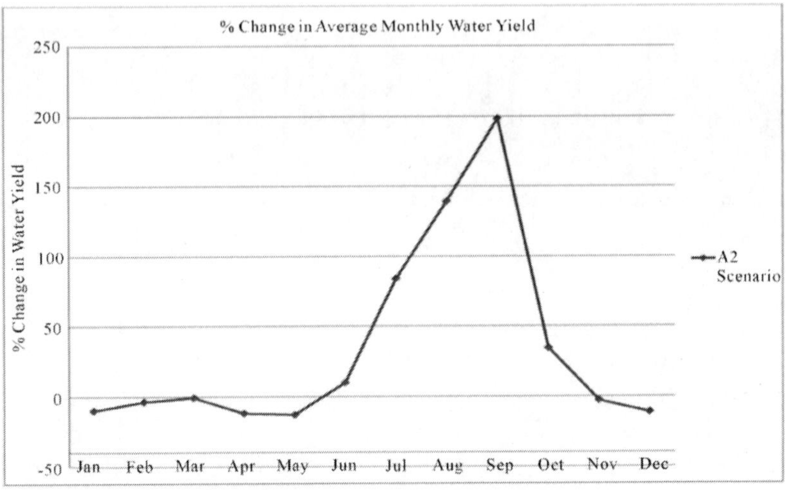

Figure 5. Percentage change in monthly water yield.

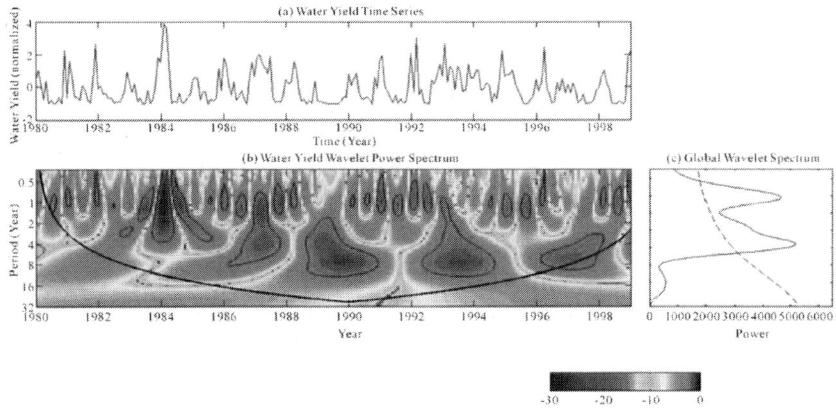

Figure 6. Monthly water yield (mm) from 1980 to 1999 (a) time series (b) corresponding wavelet spectrum (c) global wavelet spectrum.

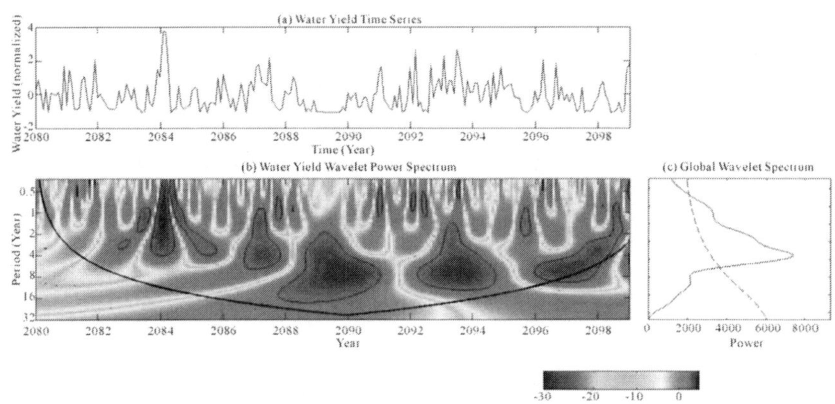

Figure 7. Projected monthly water yield (mm) from 2080 to 2099 based on A2 scenario (a) time series (b) corresponding wavelet spectrum (c) global wavelet spectrum.

4.3. Wavelet Analysis

Our results showed that the significant peaks for precipitation occurred at 12, 36 and 85 months for the baseline period (1980 to 1999) (Figure 6). However for future climate condition (i.e. A2 scenario, from 2080 to 2099), there is one additional significant periodicity at around 18 months (Figure 7). Similarly, there was only one significant periodicity of water yield in the current base line period but in the future climate change scenario, there is one additional significant periodicity at 18 months. These results indicate that there is of course a change in frequency of occurrence of extreme events (e.g. flood) in addition to the increase in rainfall and discharge events.

5. CONCLUSIONS

In this study, some assessments were performed to evaluate the climate change impact on the future water yield at the outlet of Clinch River Watershed upstream of Norris Lake in Tennessee and see how the frequency of extreme water yield (e.g. flood) changes compared to present condition. SWAT hydrologic model was used to simulate the rainfall runoff relation. The model was calibrated and then validated using the USGS stream flow data at the watershed outlet. The validated model was used to predict the future water yield based on community climate system model (CCSM) downscaled rainfall and temperature data for A2 climate change scenario. The results clearly indicate the increase in flooding events as well as its frequency of occurrence in the future. Therefore, if the climate change scenario adopted for this Shesh Raj Koirala, Randall W. Gentry study were to occur in the future, significant changes in the Tennessee and Cumberland Rivers could occur. There is a need to study implications of various mitigation strategies including the reviewing of downstream reservoir operation procedure to minimize the impact of climate change.

REFERENCES

1. S. Solomon, "Intergovernmental Panel on Climate Change, and Intergovernmental Panel on Climate Change," In: Working Group I, Climate change 2007: the physical science basis: contribution of Working Group I to the Fourth Assessment Report of the Intergovernmental Panel on Climate Change 2007, Cambridge, Cambridge University Press, New York, 996 p.
2. X. Zhang, R. Srinivasan and F. Hao, "Predicting Hydrologic Response to Climate Change in the Luohe River Basin Using the SWAT Model," Transactions of the ASABE, Vol. 50, No. 3, 2007, pp. 901-910.
3. L. D. Brekke, et al., "Climate Change Impacts Uncertainty for Water Resources in the San Joaquin River Basin, California," Journal of the American Water Resources Association, Vol. 40, No. 1, 2004, pp. 149-164.
4. J. G. Arnold, et al., "Large Area Hydrologic Modeling and Assessment Part 1: Model Development," Journal of the American Water Resources Association, Vol. 34, No. 1, 1998, pp. 73-89.
5. D. L. Ficklin, et al., "Climate Change Sensitivity Assessment of a Highly Agricultural Watershed Using SWAT," Journal of Hydrology, Vol. 374, No. 1-2, 2009, pp. 16-29.
6. M. Jha, et al., "Climate Change Sensitivity Assessment on Upper Mississippi River Basin Streamflows Using SWAT," Journal of the American Water Resources Association, Vol. 42, No. 4, 2006, pp. 997-1015.
7. N. S. Christensen and D. P. Lettenmaier, "A Multimodel Ensemble Approach to Assessment of Climate Change Impacts on the Hydrology and Water Resources of

the Colorado River Basin," Hydrology and Earth System Sciences, Vol. 11, No. 4, 2007, pp. 1417-1434.

8. D. R. Maidment, "Handbook of Hydrology," McGraw-Hill, New York, 1993.
9. D. R. Legates and G. J. McCabe, "Evaluating the Use of 'Goodness-of-Fit' Measures in Hydrologic and Hydroclimatic Model Validation," Water Resources Research, Vol. 35, No. 1, 1999, pp. 233-241.
10. P. S. Addison, "The Illustrated Wavelet Transform HandBook: Introductory Theory and Applications in Science, Engineering, Medicine and Finance 2002," Institute of Physics Publishing, Bristol, 353 p.
11. N. Massei, et al., "Investigating Transport Properties and Turbidity Dynamics of a Karst Aquifer Using Correlation, Spectral, and Wavelet Analyses," Journal of Hydrology, Vol. 329, No. 1-2, 2006, pp. 244-257.
12. C. Torrence and G. P. Compo, "A Practical Guide to Wavelet Analysis," Bulletin of the American Meteorological Society, Vol. 79, No. 1, 1998, pp. 61-78.

CHAPTER 4

A Compact Field Fluorometer and its Application to Dye Tracing in Karst Environments

Amaël Poulain[1] Gaëtan Rochez[1] Jean-Pierre Van Roy[2] Lorraine Dewaide[1] Vincent Hallet[1] Geert De Sadelaer[3]

[1]*Department of Geology, University of Namur, Rue de Bruxelles No. 61, B-5000 Namur, Belgium*

[2]*Department of Physics, University of Namur, Rue de Bruxelles No. 61, B-5000 Namur, Belgium*

[3]*RoyalMeteorological Institute of Belgium, Avenue Circulaire No. 3, B-1180 Bruxelles, Belgium*

ABSTRACT

Dye tracing is a classic technique in hydrogeology to investigate surface-water or groundwater flow characteristics, and it is useful for many applications including natural or industrial issues. The *Fluo-Green* field fluorometer has been successfully tested in a karst environment and is specifically suitable for in-cave karst water monitoring. Karst research often uses dyes to obtain information about groundwater flow in unexplored cave passages. The compact device, alternatively named *Fluo-G*, meets the requirements of cave media: small (10 × 16 × 21 cm), lightweight (0.75 kg without ballast) and simple in conception. It is easy for cavers to set up and handle compared to other sampling methods. The fluorometer records uranine, turbidity and temperature with a user-defined time-step

(1 min – 1 day). Very low energy consumption allows 9,000 measurements with six AA batteries. The device was calibrated and tested in the laboratory and in field conditions in Belgian karst systems. Results are in good fit with other sampling methods: in-situ fluorometers and automatic water sampling plus laboratory analysis. Recording high quality data (breakthrough curves) in karst with in-cave monitoring is valuable to improve knowledge of karst systems. Many hydrological and hydrogeological applications can benefit from such a low-cost and compact device, and finding the best compromise between resources and quality data is essential. Several improvements are possible but preliminary field tests are very promising.

Keywords: Tracer tests Field technique Fluorometer Groundwater monitoring Karst

INTRODUCTION

Many scientific fields use tracer tests to understand the flow organization and characteristics of water. This tool can answer multiple questions for both surface water and groundwater studies (Käss 1998). The fields of environmental management and engineering can also benefit from dye tracing to answer multiple questions about flow connections, catchment area, aquifer vulnerability, pollution and construction leakages. In the karst environment, they represent a powerful tool for the determination of an unexplored system configuration (Goldscheider et al. 2008). Hydraulic connections, catchment area or transit time can be easily obtained from dye tracing (Meiman et al. 2001; Lauber et al. 2014). In this context, fluorescent dyes tend to approach the "ideal tracer" as they are reasonably conservative, safe, inexpensive and highly detectable (Benischke et al. 2007). Uranine is for now the most commonly used tracer because of these characteristics.

The objectives and context of the dye tracing in karst areas are two key elements to determine the right sampling approach (Smart 2005). Qualitative and semi-quantitative methods like visual detection and activated charcoal answer geographical issues (connections, catchments, travel time) with limited material resources. Charcoal bags are broadly used because they are cheap and easy to implement but are subject to contamination. They are mainly valuable for preliminary tests and inaccessible sites (Goldscheider et al. 2008).

Quantitative methodologies give more valuable information because frequent water sampling allows the measurement of dye concentration in time. These data are instructive regarding the hydrogeological issue and for cave exploration. Automatic water samplers optimize the sampling task and improve the temporal resolution of data. Nevertheless, sampling, handling and analyzing water samples remains a significant task, even more with multiple sampling stations.

More recently, automatic field fluorometers have given the opportunity to make in-situ measurements of fluorescent dyes in water (Schnegg 2002). The temporal resolution, precision and accuracy are significantly enhanced. Furthermore, those devices reduce the need of frequent handling in the field due to automatic recording over an extended period of time. Another advantage of this in-situ measurement is the absence of transportation and analysis of bottled samples.

As discussed by Smart (2005), the dye tracing and specifically the sampling design has to be defined regarding the objectives and the scientific, socio-economic and logistic context. The first objective for cavers and karst researchers is geographical: flow routes, karst catchments, travel times. In addition, hydrogeological objectives can bring valuable data for cave research: velocity, dispersion, storage, conduits characteristics and possible retardation (Dewaide et al. 2016). In this context, cave researchers aim to run with limited technical, financial and personnel resources (Smart et al. 1998). Error tolerance and data resolution will depend upon the audience and the objectives but primarily it must be cost-effective. In-situ fluorometers tend to be the most valuable technique in terms of quality data relative to costs.

This paper presents the Fluo-Green, a recently developed fluorometer for hydrological investigations. The device can be used in variable environments but it has been specifically designed and tested for karst. In-cave monitoring can bring more detailed insights into the internal structure of karst aquifers (Goldscheider et al. 2008); nevertheless, cave access is often difficult and monitoring operations are few and under-exploited (Lauber et al. 2014).

The best compromise has to be found between quality data (precision and accuracy) and costs regarding the objectives and requirements of the research issue. The question of the benefit of increasing data quality compared to the customer needs has to be asked. For "simple" geographical and hydrogeological issues, experimenter's expectations can be lowered with simplified material.

The Fluo-Green device, or Fluo-G, was designed by G. De Sadelaer in 2015. Laboratory calibrations and field tests were successfully conducted in 2015 and 2016 at the University of Namur in cave systems of Belgium. Comparisons with other sampling methods in variable sampling environments allowed the evaluation of the field performances of this new device.

Description of the Fluo-G

The Fluo-G fluorometer is a two-in-one device with in-situ fluorescence measurement and data logger in the same box (Fig. 1). The device is compact (10 × 16 × 21 cm) and extra light (0.75 kg without ballast). The measurement is made through a transparent casing with an excitation-detection unit. The excitation unit comprises a set of two LEDs, and wavelengths are selected for uranine excitation (470 nanometers [nm], 6,300 millicandela [mcd]) and turbidity measurement (625 nm, 6,500 mcd). The Fluo-G is designed to measure uranine only, since it is the most common fluorescent dye in karst hydrogeology and cave exploration. Other fluorescent dyes could easily be measured by choosing another LED wavelength for the excitation unit and applying the adequate calibration procedure. The detection unit is right next to the excitation unit and has a RGB (red-green-blue) sensor perpendicular to the light source. Detection and correction of turbidity and a water temperature probe are also integrated.

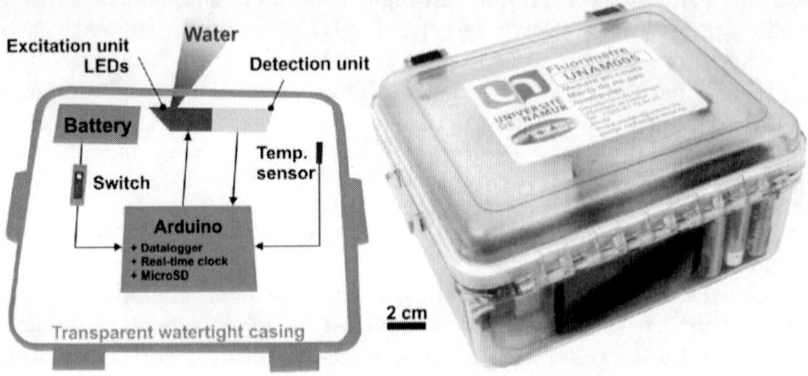

Figure. 1 Photograph and schematic view of the Fluo-Green fluorometer

Excitation and detection sequences are controlled by an Arduino coupled to a data logger shield with a microSD memory card. A real-time clock is used to wake up the fluorometer and run the measurement at a defined time-step. A minimum time step of 1 min allows to record high quality data for further interpretation. The energy consumption is 0.2 mAh per measurement and the device is turned off between measurements. Six AA batteries (1,800 mAh) give 30 days of lifetime with a measurement time-step of 5 min. Before each sequence of excitation, the detection unit makes one dark measurement to allow a possible daylight correction. Main characteristics of the Fluo-G are given in the Table 1. A comparison is made with the GGUN FL30 of Albillia Switzerland since it is a standard commercial instrument for in-situ dye-tracing monitoring.

Table 1 Properties of the Fluo-Green and comparison with the GGUN FL30 (Schnegg 2002)

Properties	Fluo-Green	GGUN FL30 (Schnegg 2002)
Size	10 × 16 × 21 cm	Probe : 16 × 17 cm
		Logger : 29 × 25 × 19 cm
Weight	0.75 kg	Probe : 7.3 kg
		Logger : 7.2 kg
		Cable : 1 kg/10 m
Internal ballasts	2 × 1.0 kg	–
Channels	2 (470 and 625 nm)	4 (370, 470, 525 and 660 nm)
	Uranine and turbidity	3 tracers + turbidity
Turbidity measurement	0.08–200 NTU	0.02–400 NTU
Time resolution	1 min – 1 day	2 s (1 channel) – 15 min
Duration of measurement sequence	Total measurement = 9.4 s	2 s/channel
Consumption	77 mA (measurement)	50 mA (measurement)
	0 mA (stand by)	1.5 mA (stand by)
Detection resolution (uranine)	0.06–0.09 ppb	0.02 ppb
Saturation threshold	3,000 ppb (theoretical)	2,500 mV
Contamination	Impossible	Impossible
Temperature probe resolution (field measurement)	0.06 °C	0.01 °C
Max depth of use	Tested at −4 m	−50 m

As mentioned in the Table 1, the main advantages of Fluo-G are the compact size and the low energy consumption allowing extended monitoring. Disadvantages are a small number of detectable dyes (GGUN FL30 has more channels) and a smaller maximal depth of use. The latter is limited by the tightness of the transparent casing, which is one of the main possible deficiencies of the Fluo-G and has to be improved in the future. GGUN-FL30 also has detection threshold and resolution that are somewhat better than the Fluo-G. Advantages and disadvantages of the Fluo-G compared with other monitoring methods will be discussed later.

Data acquisition is automatically started at connection of the battery. No external connections are needed during normal use of the fluorometer. Raw measurement data are stored on the internal microSD memory card as a text file. The data can be read into spreadsheet software to apply calibration parameters and get the results in ppb or µg/liter.

Laboratory calibration

Standard calibration procedure

Calibration of the Fluo-G is recommended before every experiment to guarantee data quality. The calibration procedure evaluates the response of the sensor with respect to the water of the monitored site. The background fluorescence can also be removed to avoid misinterpretation and get the real concentration of fluorescent dye.

This calibration is applicable when using a single color dye tracer. Although it is possible to use other tracers, the device was calibrated for uranine. Uranine has a very low minimal detectable signal, so the sensitivity is high compared to other tracers (Smart et al. 1998). Due to cost compromises, no extra filters were installed in the optical detection path, which will lead to a supplementary offset signal (see 'Correction for stray light' section).

Figure 2 shows the relationship between the concentration of the tracer (u) and the G channel output signal of the detector that has been fitted with a second order polynomial. The R^2 is 0.999 so the model fits the experimental data quite perfectly. The general equation for the calibration is:

$$u = C1 \times G^2 + C2 \times G + C3 \tag{1}$$

where the following terms and definitions apply:

u (ppb) Concentration of the uranine dye

G (counts) The corrected ADU value detected by the green channel (see 'Correction for stray light' section; ADU: analogue digital units, the result of an A/D conversion, also called arbitrary digital units)

C1, C2, C3 Constants derived after calibration curve fitting. Here the minimum resolution of the sensor is 0.09 ppb/count. A routine calibration can be limited to a three-point calibration at 1, 10 and 100 ppb and an offset measurement at 0 ppb.

Correction for stray light

Stray light from surrounding light and the excitation LED enters the detector directly, via reflections or by water turbidity. This interference will produce a variable (high) offset signal at the output of the detector. It is necessary to correct this interference in order to obtain reliable data.

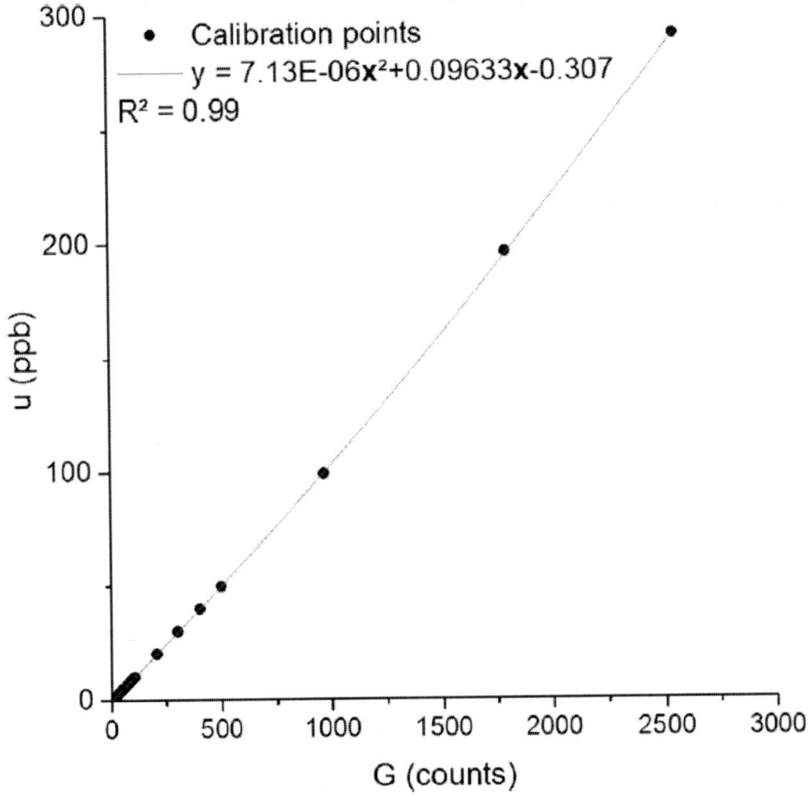

Figure. 2 Calibration curve for the Fluo-Green showing the relationship between uranine concentration (u) and ADU counts of the detection unit (G)

A classical filter fluorometer uses filters with a high optical density to avoid any offset signal. The Fluo-G setup has a limited optical density, and a reference channel is used to correct the green detector output signal (ADU). Due to the quasi-identical optical path, stray light is detected identically by all three detector channels (red, green and blue) with exception of a gain difference and a very small zero offset. This property makes a good offset correction possible. The compensation will lead to a corrected G channel with a zero error of ±1 count. Correction of surrounding light (daylight or artificial lights) is done by a dark background measurement with all LED off (no excitation). The green detector output (ADU) measured with no excitation is subtracted from the signal measured when the excitation is on. The correction amplitude of artificial lighting is visible on Fig. 5.

Turbidity

Turbidity will influence the measured tracer emission signal in two ways:

- Light from the excitation LED will be scattered to the emission detector (offset). This offset is already corrected by the method already described.
- Excitation and emission light will be absorbed (gain). The influence of turbidity on the measured uranine concentration has to be determined during the turbidity calibration, and a correction is applicable.

Turbidity is measured in a wavelength that is not influenced by tracer excitation/emission spectra; here a red LED at 625 nm is used.

Field tests

Rather than laboratory tests and calibrations, field tests were needed to validate the fluorometer. Precision and accuracy are important parameters to be tested in the cave environment to evaluate their consistency with the objectives and context of the tracing (Smart 2005). The Fluo-G has been tested in the framework of hydrogeological studies in 2015 and 2016 at the University of Namur (Belgium). Four experiments were conducted into different karst systems of southern Belgium (Fig. 3). Different sampling environments were selected:

- A surface karst resurgence with daylight interference
- An underground river with a pebble riverbed, high water current and artificial lighting
- An underground river with a mud riverbed and very low water current
- Drip-water from a stalactite

The Fluo-G was coupled to additional sampling methods in order to compare monitoring results. An automatic water sampler and laboratory spectrofluorometer was used for test No. 1. A GGUN FL30 fluorometer (Schnegg 2002) was used for test No. 2. All fluorometers and the spectrofluorometer were calibrated using the site blank water and the same uranine.

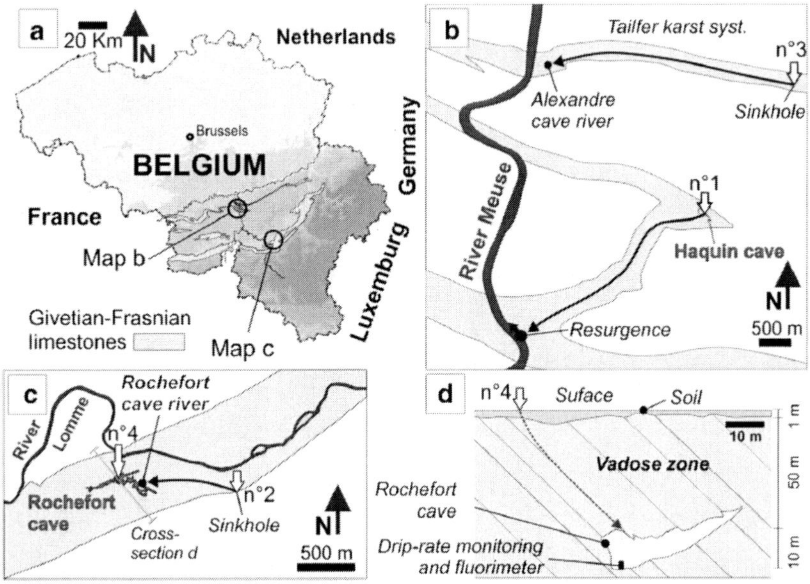

Figure. 3 Maps of the study sites. **a** Location of the test sites within the southern Belgium limestones. **b** Location of test Nos. *1* and *3* in the Frasnian limestones of the Meuse valley. **c** Location of test Nos. *2* and *4* in the Givetian limstones of the Lomme Valley. **d** Cross-section in the *Rochefort cave* showing the surface-to-cave dye tracing (No. *4*) in the Givetian limestones

Test No. 1: surface karst resurgence with daylight interference

The Haquin karst system (Lustin) was investigated during the first test (Fig. 3b). It is a 3-km-long sinkhole-resurgence system in Frasnian limestones of southern Belgium. The resurgence was equipped with an automated water sampler with a 1-h time step during 48 h. Water samples were analyzed with an Agilent spectrofluorometer at the University of Namur. Three Fluo-G were installed in the resurgence with a 5-min time-step configuration.

The three Fluo-G breakthrough curves show similar results after correction with calibration formulas (Fig. 4). In-situ measurements are also in good fit with results from the laboratory spectrofluorometer. The first arrival-of-tracer events are similar and the maximum concentration difference is 0.2–0.8 ppb. Those results show the ability of the Fluo-G to give reliable data in terms of precision (the three devices give the same results) and accuracy (as the spectrofluorometer value is expected to be the most accurate measurement).

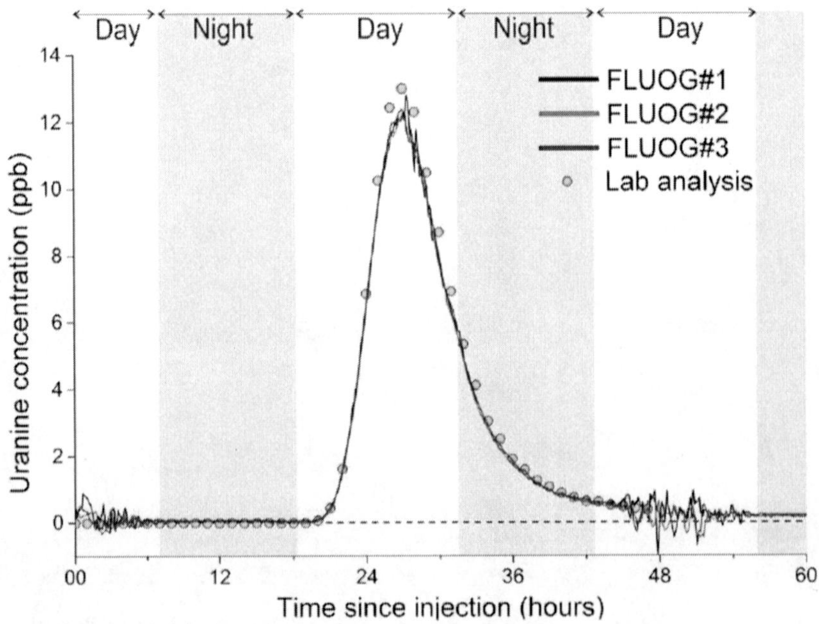

Figure. 4 Breakthrough curves for test No. 1. Comparison between spectrofluorometer laboratory analysis on water samples (*green dots*) and three Fluo-G in-situ fluorometers (*FLUOG#1, 2, 3*). The influence of daylight on the Fluo-G measurement is visible

An important daylight noise is observable in all the breakthrough curves. The dark measurement allows a partial correction and the BTC is easily distinguishable from the noise; nevertheless, a daylight coverage should be implemented for surface measurements. Additional smoothing of the data can easily remove residual daylight noise.

Test No. 2: underground river with pebble riverbed, high water current and artificial lighting

The second field test was conducted in the Givetian limestones of the Lomme karst system (Fig. 3c). The Fluo-G and a GGUN-FL30 field fluorometers were installed in the Rochefort cave underground river (50 L/s). This river has high current and a pebble riverbed. Uranine was injected in a small sinkhole 1 km upstream and the GGUN FL30 fluorometer (Schnegg 2002) was used for comparison as it is a reference in modern submersible fluorometers (Goldscheider et al. 2008).

Figure 5 shows the strong similitude between the GGUN FL30 and the Fluo-G BTCs. Basic features of the restitution are the same: time of first arrival (24 h), peak time (33 h) and maximum concentration (difference of 0.7 ppb). Total restitution difference is 3.3% between the first arrival and $t = 100$ h. Stray light induced by artificial lighting in the cave is visible on the uncorrected Fluo-G breakthrough curve (Fig. 5a). The dark measurement allows for the removal of this interference as illustrated in Fig. 5b. A signal noise of ± 0.05 ppb is visible on the Fluo-G BTC. This noise is less visible on the GGUN curve; however, this variability does not affect the interpretation of the results.

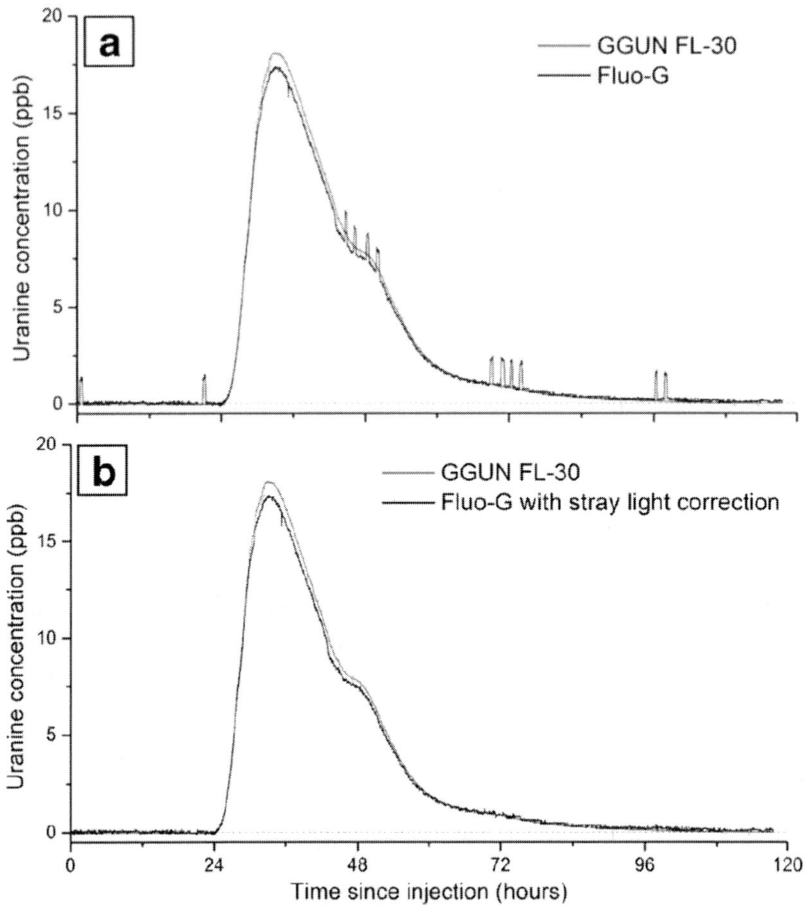

Figure. 5 Breakthrough curves for test No. 2. Comparison between in-situ fluorometers *Fluo-G* and *GGUN FL30*. **a** Fluo-G data without stray light correction. **b** Fluo-G data with stray light correction (via dark measurement)

Test No. 3: underground river with mud riverbed and low water current

The third test was made in the Tailfer karst system (Fig. 3b), which is similar to the Haquin system (test No. 1) in terms of geologic context and size. The Fluo-G has been installed in the small underground river of the Alexandre cave. This river has a muddy riverbed and almost no visible water current. High water turbidity was observed during the measurement. The BTC of this experiment is displayed in Fig. 6.

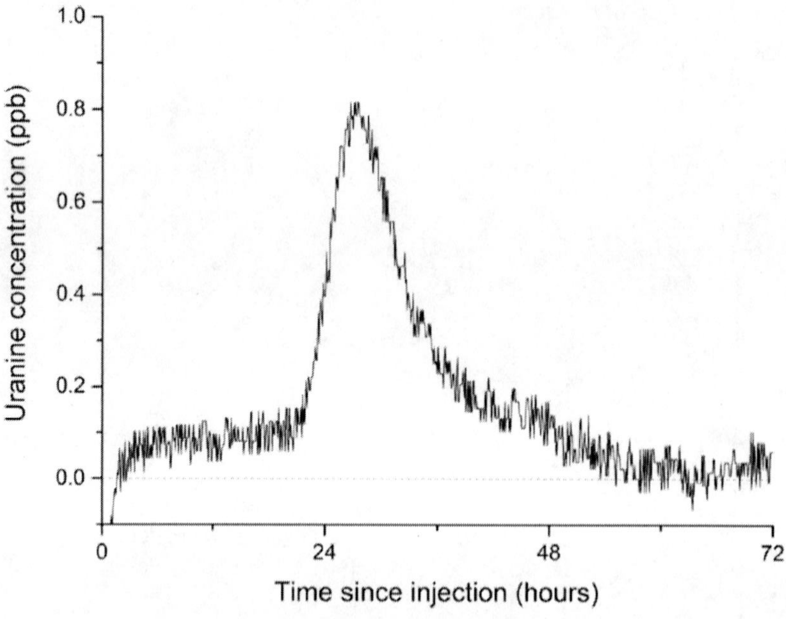

Figure. 6 Breakthrough curve (BTC) for test No. 3. Result shows a typical BTC of an advective karst system despite a tracer concentration lower than 1 ppb

The shape of the BTC is characteristic of a karst stream dominated by advection and dispersion. The dye restitution is clearly visible despite the very low tracer concentration. The maximum concentration is 0.8 ppb with a background fluorescence signal ranging from 0 to 0.1 ppb. The result of this test shows the ability of the Fluo-G to detect small concentrations of uranine despite the basic design of the excitation/detection unit. This low restitution threshold allows one to use less tracer which is both a financial advantage and also avoids visual contamination. Figure 6 also shows the measurement error with the Fluo-G, ranging from 0.06 to 0.09 ppb. Although this could become an issue to detect small changes in fluorescence signal, it can be easily overcome by using more dye during injection. For this case, the error/peak concentration ratio is 1/8 and allows a precise determination of BTC characteristics.

Test No. 4: drip-water monitoring for vadose zone tracing

A vadose zone tracing test was performed in the Rochefort cave of the Lomme karst system (Fig. 3c,d). Uranine was injected at the surface and the monitoring (drip-rate, uranine) was made under a perennial stalactite (5–25 L/h), 50 m below. The Fluo-G was submerged into a small bath (5 L) collecting drip-water.

Figure 7 shows the parameters recorded with the Fluo-G in the stalactite drip-water: battery voltage, water temperature and uranine concentration. Additional parameters such as surface rainfall, surface temperature and stalactite drip-rate are also displayed in order to properly interpret the Fluo-G results.

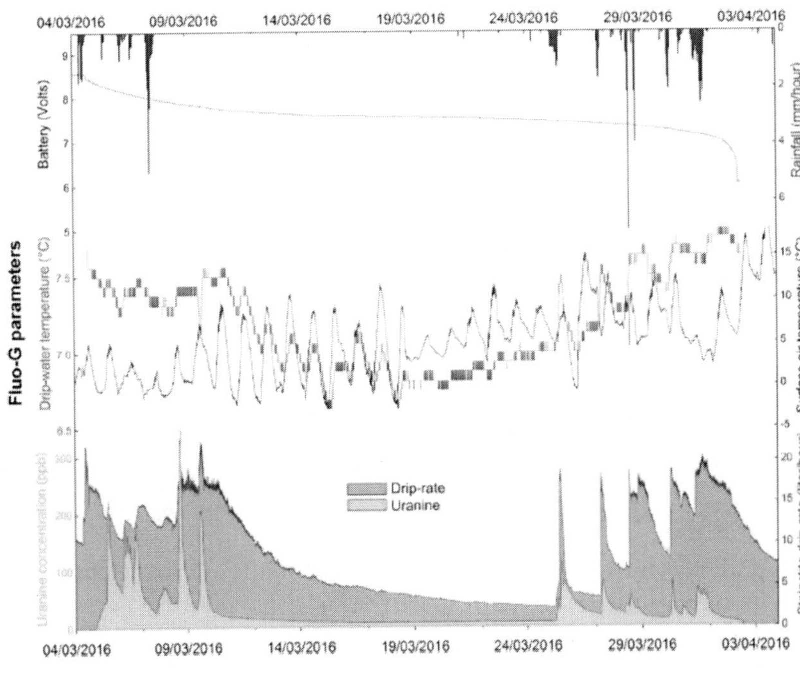

Figure. 7 Surface and cave parameters recorded during test No. 4. *Fluo-G parameters* are displayed on the *left axis*: *battery* voltage (Volts), *drip-water temperature* (°C) and *uranine concentration* (ppb). Other parameters are displayed on the *right axis*: *surface rainfall* (mm/h), *surface air temperature* (°C) and *stalactite drip-rate* (L/h)

The Fluo-G successfully records uranine concentration and water temperature for 30 days; an additional battery can be used, allowing 2 months of measurement with a 5-min time-step. Extended lifetime is due to the very low energy consumption of Fluo-G and is crucial for long-term experiments or remote study sites (Poulain et al. 2015). The small size of the device allows one to make the measurements in a small bath, which tends to avoid concentration buffering for drip-water monitoring. The temperature signal recorded by the fluorometer shows a resolution of 0.06 °C, which is precise enough to allow a good correlation with surface air temperature in this case. Field temperature measurement with the GGUN shows a resolution of 0.01 °C, which can be useful in case of low variability signals.

Advantages and disadvantages of the Fluo-G

This new device has the same kinds of advantages as any other field fluorometer: extended autonomy, data resolution, no water bottle to handle, no contamination or tracer degradation, no freezing sensitivity and reduced dye tracer mass to inject. Basically, the direct measurement of the tracer concentration simplifies the handling on the field as described by Schnegg and Doerfliger (1997) for the GGUN FL-30. For both kinds of devices, the acquisition, calibration and display of data are easy.

As mentioned previously, the Fluo-G has been especially designed regarding the specific objectives and context of karst tracing with geographical/hydrogeological issues. Some advantages arise given its conception. The size and portability make it easy to handle by only one person even in difficult caves. It can be turned-on at the surface and carried in a small bag. The other kinds of devices are often heavy and/or oversized for most of the caves and are better used in karst springs or easily accessible karst systems. The simplified design guarantees a low energy consumption and allows an extended autonomy with small batteries, which can be useful for long-term tracing in remote areas. Finally, the production costs tend to be minimal which should represent a major advantage for potential users.

The Fluo-G is under constant development and improvements are necessary despite the first good results. The capabilities of this version are intentionally limited compared to currently available commercial solutions. It was designed for uranine only since it is the most used dye by speleologists and karst hydrogeologists. It does not have the capabilities to measure simultaneous dyes like the GGUN FL30. Another disadvantage of the Fluo-G is the data collection. Since it is a 2-in-1 system, the device must be out of the water for data reading; however, a simple handling allows a quick reading of the data and the rebooting of the system with minimum data losses. Another possible issue is the sealing of the box in case of extreme flood events.

CONCLUSIONS

The Fluo-G field fluorometer is a new kind of compact field fluorometer for hydrogeological use. The intent was to create a simple, compact and easy-to-use device. Standard components were required to build a low-cost and simple product. The fluorometer has been tested in a karst environment but can also be used in other hydrological environments that require dye-tracing methodology.

Karst systems and caves offer great opportunities for scientific research and exploration to understand both the functioning and organization of groundwater. In-cave dye tracing can give valuable additional information about the internal structure and hydraulic functioning of karst aquifers (Lauber et al 2014); however, specific conditions inside caves require particular dispositions, which is especially the case for tracer test techniques.

Besides the classic advantages of automatic field fluorometers compared to charcoal bags and water sampling, this device is small, extra-light and energy efficient. The handling is very simple for calibration, data acquisition and results display. The Fluo-G does not have all the capabilities of similar devices used for fluorescence measurement. The purpose was to design a simple and low-cost product for cavers and karst researchers. Uranine, turbidity and temperature are the three parameters that can be measured.

The fluorometer has been successfully tested in the laboratory and in the field and the results are in good fit with other methods in terms of dye concentration, data resolution, precision and accuracy. While the main goal was to get quality data in line with the specific issues in karst research, the next step is to ensure the durability of the device regarding the aggressive karst environment. The Fluo-G has so far been tested in different case studies in Belgium. Applications to a larger number of environments, also with different field settings and specific issues will help to highlight capabilities and limitations of the device and will suggest further improvements.

ACKNOWLEDGMENTS

The development of the Fluo-Green was made possible through the financial support of the Belgium National Fund for Scientific Research (FRS-FNRS) in the framework of the KARAG Project (www.karag.be). The authors would also like to acknowledge N. Goldscheider and an anonymous reviewer for their valuable comments, which helped to improve this manuscript.

REFERENCES

1. Benischke R, Goldscheider N, Smart C (2007) Tracer techniques. In: Goldscheider N, Drew D (eds) Methods in karst hydrogeology. International Contributions to Hydrogeology. Taylor and Francis, London, pp 147–170
2. Dewaide L, Bonniver I, Rochez G, Hallet V (2016) Solute transport in heterogeneous karst systems: dimensioning and estimation of the transport parameters via multi-sampling tracer-tests modelling using the OTIS (One-dimensional Transport with Inflow and Storage) program. J Hydrol 534:567–578.
3. Goldscheider N, Meiman J, Pronk M, Smart C (2008) Tracer tests in karst hydrogeology and speleology. Int J Speleol 37(1):27–40
4. Käss W (1998) Tracing technique in geohydrology. Balkema, Rotterdam, The Netherlands, 600 pp
5. Lauber U, Ufrecht W, Goldscheider N (2014) Spatially resolved information on karst conduit flow from in-cave dye tracing. Hydrol Earth Syst Sci 18:435–445
6. Meiman J, Groves C, Herstein S (2001) In-cave dye tracing and drainage basin divides in the Mammoth Cave Karst aquifer, Kentucky. In: Kuniansky E (ed) U.S. Geological Survey karst interest group proceedings. US Geol Surv Water Resour Invest Rep 01-4011, pp 179–185
7. Poulain A, Rochez G, Bonniver I, Hallet V (2015) Stalactite drip-water monitoring and tracer tests approach to assess hydrogeologic behavior of karst vadose zone: case study of Han-sur-Lesse (Belgium). Environ Earth Sci 74:7685–7697. doi:10.1007/s12665-015-4696-9
8. Schnegg PA (2002) An inexpensive field fluorometer for hydrogeological tracer tests with three tracers and turbidity measurement. In: Bocanegra E, Martinez D, Massone H (eds) XXXII IAH and ALHSUD Congress on Groundwater and Human Development. Mar del Plata, Argentina, October 2002
9. Schnegg PA, Doerfliger N (1997) An inexpensive flow-through field fluorometer. In: Proceedings of the 6th Conference on Limestone Hydrology and Fissured Media. La Chaux-de-Fonds, Switzerland, August 1997
10. Smart C (2005) Error and technique in fluorescent dye tracing. In: Sinkholes and the engineering and environmental impacts of karst. ASCE, Reston, VA, pp 500–512
11. Smart C, Zabo L, Calvin Alexander E Jr, Worthington SRH (1998) Some advances in fluorometric techniques for water tracing. Environ Monit Assess 53:305–320

CHAPTER 5

The Source of Arsenic and Nitrate in Borrego Valley Groundwater Aquifer

Mohammad Hassan Rezaie-Boroon*, Jessica Chaney, Bradley Bowers

California State University, Los Angeles, Geosciences and Environment Department, Los Angeles, USA

ABSTRACT

Groundwater in California is very precious, yet what we can withdraw is often contaminated with natural and anthropogenic pollution sources. We have examined the Borrego Valley (BV) ground- water (N = 6 wells) in southern California to understand the source of arsenic and nitrate in some of its groundwater production wells. The results show that the arsenic values range from <2 ppb to 12.2 ppb and the nitrate values from <1 ppm to 10.2 ppm for different wells respectively. The results showed that the arsenic concentration increased 270% for the well # ID1-10 since 2004 and showed an increase of 63% since 2013 respectively. For other wells the results showed an increase of 147% and 72% since 2001. The nitrate concentration has jumped 42% in concentration since last year in one of the wells. The objective of this study is to understand the nature and source of arsenic and nitrate in BV groundwater aquifer as to how this change in arsenic and nitrate concentration occurs through the time. The arsenic retention in the sediments is highly variable and controlled by local processes as a result of natural weathering process of metamorphic bedrock. The second results from the development of strongly reducing conditions at near-neutral pH values, leading to the desorption of arsenic from mineral oxides and to the reductive dissolution of Fe and Mn oxides, also leading to arsenic release. The high arsenic concentrations in some groundwater wells in Borrego Valley CA require the need for reconnaissance surveys in mineralized areas of fractured crystalline basement. Net groundwater extraction values are based on an irrigation efficiency of 78 percent with 14 to 22 percent irrigation return. We believe that the return flow from irrigational activity could be one of the major sources of nitrate transferring the agricultural contaminants such as nitrate to Borrego Valley aquifer.

Keywords: Hydrogeology, Hydrogeochemistry, Groundwater, Borrego Valley, Return Flow

1. INTRODUCTION

Groundwater is a major source of public drinking-water supply for many of the cities located in the Colorado Desert hydrogeologic province of California [1]. While drought and water shortage have become an increasingly serious problem in California especially in arid semi arid regions, over pumping of the aquifer has resulted in depletion of the volume of groundwater in storage within the aquifer system and a decline of groundwater levels in Borrego Valley (BV) California [2]. Equally important is the water quality to human health, which gains a great deal of interest with increasing demand for drinking water. Past research studies show higher concentrations of arsenic and nitrate is found in the ground water aquifer within fractured crystalline bedrock in BV [3]. Additional studies show that health risk based on the presence of arsenic in the groundwater does not exclusively occur in southern California but in the western United States [4]. In terms of nitrate concentration, the nitrate values in the BV study area groundwater wells were less than the health-based benchmark values (10 mg/l), but the results indicate the fluctuations of nitrate values possibly depending on the oxidation-reduction state of the groundwater [5].

High concentrations of nitrate can adversely affect human health, particularly the health of infants. Thus, the understanding of groundwater chemistry, occurrence and distribution of chemical constituents in groundwater are valuable for identifying the natural and human factors affecting water quality [6]. The controlled processes on chemical composition of groundwater include evaporation, dissolution/precipitation, weathering of silicates, oxidation-reduction, sorption and exchange reactions, transformation of organic matter, and mixing processes [7].

Similarly, research studies indicate that the mobilization of arsenic and nitrate in sedimentary aquifers of these regions may be, in part, a result of changes in the geochemical environment due to agricultural irrigation [8]. Prior to the current study, the groundwater quality in the BV region was investigated by the US Geological Survey (USGS) from December 2008 to March 2010 and by Borrego Valley Water District (BVD) in the last 10 years.

Generally, all studies dealing with arsenic are in good agreement that the arsenic originates either from natural sources (geogenic) or industrial activities and/or fertilizer runoff (anthropogenic). Nevertheless, the processes leading to the release of arsenic in groundwater aquifer are not yet fully understood. The literature synthesis indicates various hypothesis concerning the release of arsenic 1) by oxidation of arsenic-bearing pyrite [9] -[11] , 2) by competitive exchange of phosphate from fertilizers [12] and 3) by reduction of iron hydroxides under anoxic conditions [13] [14]. Our study, however, provides a baseline of groundwater quality in BV using the current ongoing research results and literature synthesis addressing the source and origin, trend, and pattern of contamination in BV aquifer. Furthermore, it is important to identify the enduring processes in the environment for the nitrate and arsenic concentration values in BV groundwater and to recognize the cause for fluctuation of this value through the time. Therefore, the main objective of this study is to find the exact source and the pattern of arsenic and nitrate in BV aquifer. Furthermore, the purpose of this research study is to examine the ground-water chemistry in a groundwater basin, which historically contains high arsenic concentrations. And also, it is important to identify geochemical patterns that would help explain the occurrence of high arsenic concentrations. We believe that the distribution of arsenic in the BV aquifer depends on the partitioning between the aqueous and solid phase. Therefore processes causing changes in the aquifer geochemistry need to be considered since they directly affect the mobility of arsenic.

2. MATERIALS AND METHODS

2.1. Study Area

The Borrego Valley (BV) is a desert area with surface area of approximately 616 km^2 (~238 mi^2). BV is located in southern California (in the northeastern portion of San Diego County), within the Colorado Desert geomorphic province, (in the westernmost part of the Sonoran Desert), approximately 137 km (85 miles) northeast of San Diego (**Figure 1**). BV has an arid climate with an average annual precipitation ranging from 76 to 152 mm (3 to 6 inches). Most of the precipitation occurs during the winter months, between November and April [2].

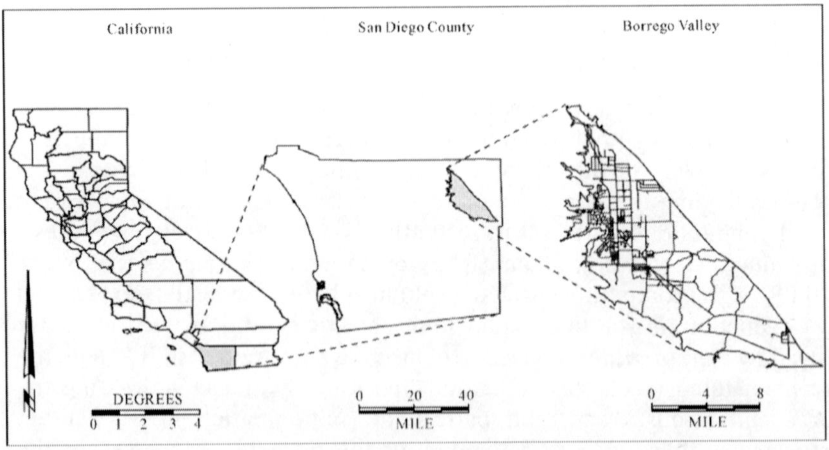

Figure 1. Regional location map of Borrego Valley, CA (modified after [2]).

2.2. Geological and Hydrogeological Setting

2.2.1. Topography

BV is bounded on the north and northeast by Coyote Mountain, on the west and southwest by the San Ysidro Mountains and Pinyon Ridge, and on the south by the Vallecitos Mountains (**Figure 2**). The east side of BV is bordered by the Coyote Creek fault, with the Borrego Badlands beyond, and Borrego Mountain. The primary sources of recharge to the BV groundwater basin are the multiple creeks and intermittent streams that drain to BV from the surrounding mountainous areas, comprising more than 1036 km^2 (400 square miles) of watershed [2]).

The ~285 km^2 (110 square mile) valley floor ranges in elevation from 365.76 m (1200 feet) at the northern end of the valley, in the vicinity of Ocotillo Flat near Coyote Canyon, to approximately 140 m (460 feet) in the vicinity of Borrego Sink. The elevation of the valley floor generally ranges from about 243.80 to 304.8 m (800 to 1000 feet) around the margins of the basin, and declines towards the center of the valley to Borrego Sink. The surrounding mountains are characterized by steep slopes with elevations ranging from 243.80 to 304.8 m (800 to 1000 feet), where the mountain slopes reach the valley floor, to mountain peaks ranging in elevation from approximately 518.16 m (1700 feet) at Yaqui Ridge (southeast extension of

Pinyon Ridge) to over 6500 feet at Hot Springs Mountain (in the San Ysidro Mountains). Figure 2 characterized region with surrounding mountains and groundwater level elevation.

2.2.2. Hydrology, Geology, and Hydrogeologic Setting

The geology, hydrogeology and water resources of BV have been studied and reported by various authors since the early 1900's. In regards to water resource issues in the BV, there are three features to be considered 1) the watershed or drainage basin that includes the surrounding mountains from which runoff from rainfall in the mountains is drained into the valley and recharges the aquifer via canyons and other creeks. Coyote Canyon, at the northwest end of the valley, is the most significant drainage feature, 2) the valley floor, which is bounded and defined by mountains and canyons on the north, west and by the Borrego Badlands to the east and 3) the groundwater basin, which underlies the valley floor [15].

The BV located within the BV groundwater basin (12,000 AFY), encompasses a large section of the Anza- Borrego Desert State Park The groundwater basin is identified by DWR as groundwater basin number 7 - 24 with a surface area of about 150,000 acres and includes both the main Borrego and Lower BV area [16].

The BV Basin is filled with up to 2400 feet of poorly consolidated to unconsolidated sediments resting on the basement granite. The USGS Report 82 - 855 identified an upper, middle and lower aquifer [17]. The alluvial sediments filling the basin originated from the weathering action of the rocks in the surrounding mountains. The base of the groundwater basin is a complex of the oldest geologic units in the vicinity of BV and is comprised of the Cretaceous granitic and the Triassic or older meta-sedimentary rocks of the Southern California batholith.

The basement complex is the ultimate base of the aquifer system and crops out on the north, west and south of the basin as well as at Borrego Mountain [2]. The upper aquifer is comprised of Holocene to Pleistocene age alluvial fan, playa and eolian deposits. The unit is about 1000 feet in thickness at the north end of the basin. Specific yields for these deposits range from 15 to 25 percent [16]. This aquifer is the principal source of groundwater in BV and well yields are about 2000 gal/min [18]. The middle aquifer is of Pleistocene age and composed of continental deposits, including moderately consolidated sand, gravel, and boulders. Thickness ranges to 700 feet in the middle aquifer and is thickest near the center of the valley, but thins towards the southeast [18]. Specific yield ranges from 5 to 10 percent in the middle aquifer [16]. Groundwater is easily extracted from this aquifer. The lower aquifer is also of Pleistocene age and consists of continental deposits, including moderately consolidated sand, gravel, and boulders. According to DWR (1984) the lower aquifer reaches 1800 feet in thick- ness in the south-central part of the valley. Specific yield ranges from 1 to 5 percent in the lower aquifer [16].

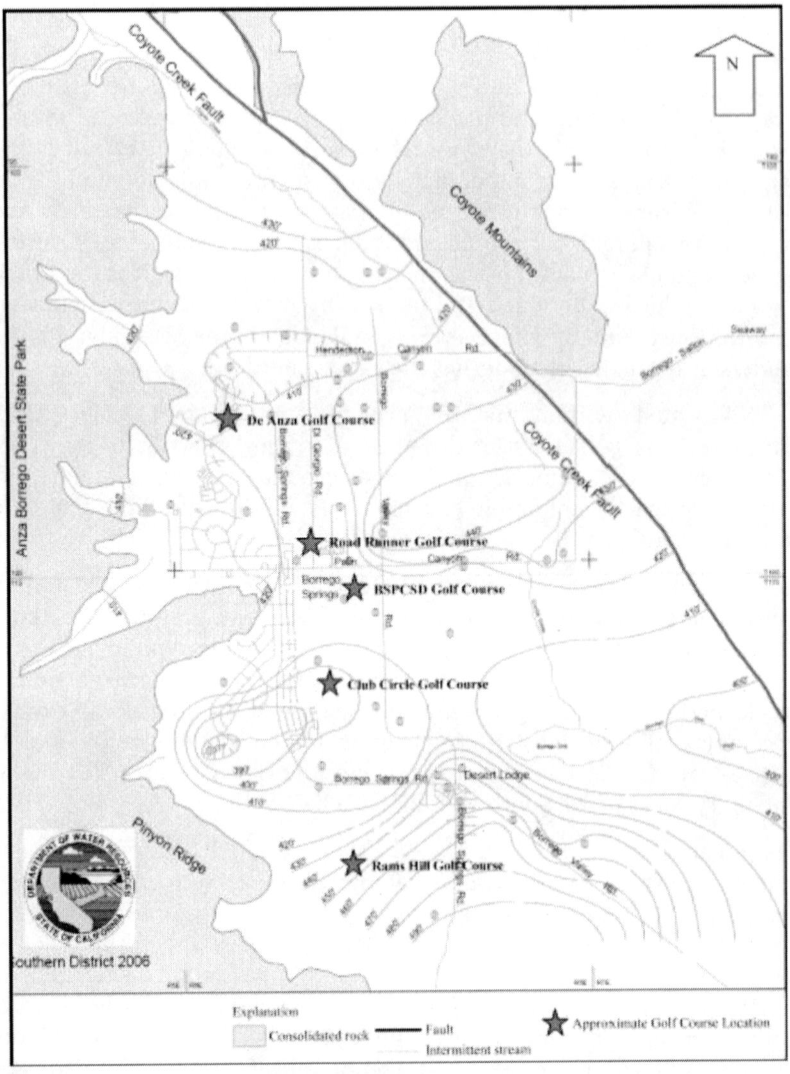

Figure 2. Base map modified from USGS 7.5 minute topographic maps. Groundwater contours area generalized representation of the regional static water level interpreted from 35 wells measured in Spring 2005. Water levels are interpreted to represent unconfined conditions [2] [16] (Courtesy of DWR, 2005; Netto, 2001).

2.3. Sampling Preparation and Sampling Well Locations

A total of six groundwater production well were sampled in the study area. Groundwater sampling was conducted according to standard procedures [19] [20] in June 2014 in BV. The production wells were chosen specifically based on the history of contamination records for arsenic and nitrate for each well. The sampling wells include ID1-10, ID1-12, ID4-10, ID4-11, ID4-18, and Wilcox, which were located in the western edge of BV (**Table 1** and **Figure 3**).

In the field, before sampling groundwater wells, each well was pumped continuously to purge at least three casing-volumes of water from the well and also until pH and temperature readings stabilized [21]. A total of 12 groundwater samples including blank and replicates samples were collected and sent to the lab for arsenic and nitrate analysis using brand new sample containers (Teflon® bottles). The containers were clearly labeled with well identification number, date of collection, type of parameter to be analyzed, the preservative type used, whether they were field filtered using 0.45 μm cellulose filters or field unfiltered. The water samples were acidified with 0.7 N HNO_3 (pH = 2) for the arsenic analysis. For the quality assurance (QA) and quality control (QC) purposes, we used duplicates, blank samples, and standard reference materials according to EPA standards. Field blank and duplicate samples were collected on site to evaluate positive bias as a result of contamination during sample handling or analysis as well as for the inconsistency in results. Physiochemical analysis such as salinity, pH, Eh, and temperature were determined on site using portable pH-meter (Accumet-AP71) and YSI (550A).

All instruments were calibrated (2-point calibration) just before sampling trip using calibration standard solution (Fisher Scientific) three times. Arsenic was measured with inductively coupled plasma and mass spectrometry (ICP-MS/Perkin Elmer; NexION®350) and nitrate was analyzed using ion chromatography (Dionex Ion Chromatography ICS-3000) at Associates Labs in Orange County, California using EPA standard methods (for arsenic EPA 200.8 & EPA 3010A, and for nitrate EPA 300.0). We noted well yield, duration of pumping prior to sampling, depth, pump data, water use, and possible contaminants near the well on water sampling protocol sheets.

In addition, we obtained from Borrego Water District (BWD) office all GPS coordinates for the precise well locations, renewed well completion data, and history.

Upon completion of the sampling trip, the recorded data was updated in the BV groundwater database. The **Table 1** shows all well GPS coordinates and their related information such as elevation, depth, Eh, pH values.

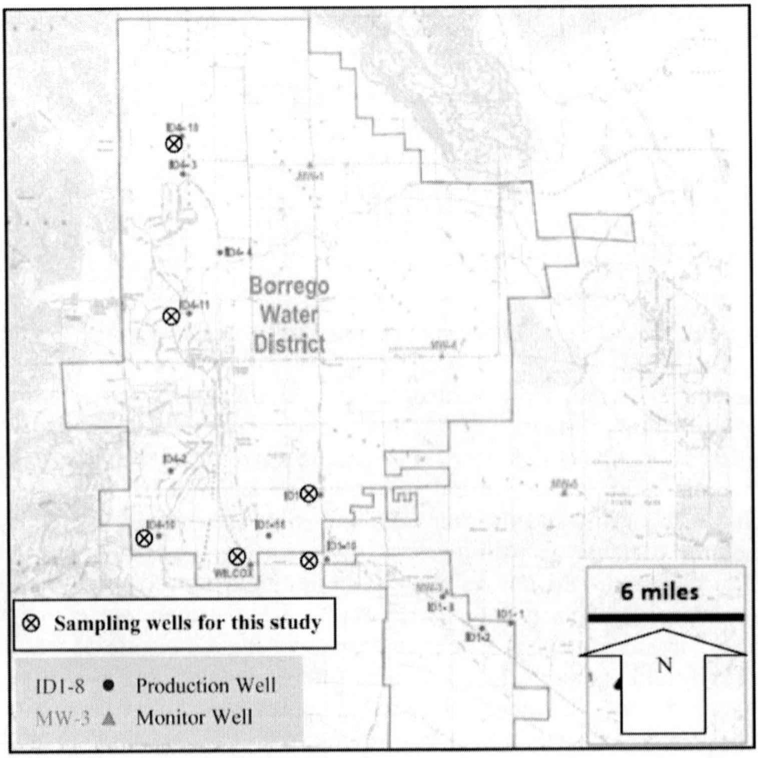

Figure 3. Sampling locations (modified after [15]).

Table 1. List of sampled production wells in BV with well characteristics. N/A means here as the data "not available". The supply wells data is the courtesy of BWD (Source: BWD, 2014). N/A: not available.

Well ID	Latitude	Longitude	Land surface elevation (ft)	Well data								
				Total depth (ft)	Eh (mV)**	pH**	Completed depth* (ft)	Water level* (ft)	Diameter (in)	Well yield GPM	Well yield GPM**	Total perforated (ft)
ID4-18	33.306749390	−116.384712840	691.06	N/A	−25.27	6.97	570	301.2	12	250	170	N/A
ID4-11	33.267497550	−116.383355020	614.06	800	23.5	6.91	770	216.6	14	900	900	310
ID1-12	33.226028270	−116.348315130	532.24	768	−45.77	7.37	580	143.2	14	900	950	320
ID4-10	33.218318120	−116.392224090	830.29	630	35.3	7.17	630	451	8	100	100	210
ID1-10	33.211789010	−116.346811970	594.74	816	−71.8	7.65	392	236.1	12	350	350	210
Wilcox	N/A	N/A	N/A	N/A	−61.7	7.49	N/A	304.2	N/A	150	170	N/A

*Data was collected on Oct. 2013 by BWD. **Observed on the sampling day.

2.4. Quality Assurance and Quality Control Methods

The purpose of QA is to identify which data best represent environmental conditions and which may have been affected by contamination or bias during sample collection, processing, storage, transportation, and (or) laboratory analysis. 1) Field blank samples were collected to assess contamination during sample handling or analysis, 2) replicate samples were collected to assess variability, 3) matrix-spike tests were done in lab to assess to assess bias of laboratory analytical methods.

The primary purpose of collecting blanks is to evaluate the magnitude of potential contamination of samples with compounds of interest during sample collection, processing, transport, and (or) analysis.

Replicate samples were collected to assess the precision of the water-quality data. Estimates of data precision are needed to assess whether differences between concentrations in samples are because of differences in groundwater quality or because of variability that may result from collecting, processing, and analyzing the samples.

Addition of a known concentration of a constituent (spike) to a replicate sample enables the analyzing laboratory to determine the effect of the matrix, in this case groundwater, on the analytical technique used to measure the constituent. The known compounds added in matrix spikes are the same as those being analyzed in the method. This enables an analysis of matrix interferences on a compound-by-compound basis. For this study, matrix spikes were added by the laboratory performing the analysis.

3. RESULTS

The results in this paper are based on our current ongoing investigation of BV aquifers and from the existing records. We have examined the BV groundwater (N = 6 wells) in southern California to understand the source of arsenic and nitrate in some of its aquifers. The results show the arsenic values range from <2 ppb to 12.2 ppb and the nitrate values from <1 ppm to 10.2 ppm for different wells respectively. The preliminary investigation shows that the arsenic and nitrate concentration has been fluctuating through the time however, the numbers for some wells are much higher than the existing record was published in the last 10 years. The results showed the BV supply well such as ID1-10 had 270% increase in arsenic concentration since 2004 and 63% since 2013.

Moreover, some showed 147% and 72% increase in arsenic since 2001. According to the BWD (Borrego Water District) the results for this well and

Wilcox show a steadily increase in concentration since 2004 (**Figure 4**). The nitrate concentration has jumped 42% in concentration since last year in one of the wells. The concentration of nitrate and arsenic varies from well to well; however for some slightly above EPA State MCL (10 ppb for arsenic and 10 ppm for nitrate) (**Figure 4**). The nitrate values for all sampling wells were lower than MCL of drinking water and human health risk standard/and public health regulation (45 mg/l according to 22 CCR §63341). Total dissolved solids (TDS) and arsenic samples were taken every three years as per Department of Public Health Regulations and nitrates samples were taken annually. The arsenic results for 2014 (dry season) showed the well number ID4-10 with lowest concentration (<2 ppb) and well number ID1-10 with highest concentration (12.2 ppb) respectively. The concentration ranking (ppb) was ID1-10 > Wilcox > ID1-12 > ID4-11 > ID4-10. The nitrate results showed that the well number ID4-10 has the highest concentration (10.2 ppm) and ID4-18 the lowest concentration (2.47 ppm). The concentration ranking according to our study include ID4-10 > ID1-11 > ID4-18. The ranking of nitrate mean (n = 6) concentration (ppm) values ± SD for all previous annual records for nitrate from 2006 and 2014 showed the following ranking: 2012 (5.5 ± 3) > 2013 (5.3 ± 3.2) > 2014 (5.1 ± 4.4) > 2007 (4.8 ± 3.3), 2006 (4.7 ± 3.7) > 2010 (4.0 ± 3.4) and >2011 (4.1 ± 2.8). This record obviously shows a fluctuation pattern within 8 years of monitoring. For arsenic the records showed different pattern, a constantly and continuously increasing values in the last 10 years. The ranking of arsenic mean (n = 3) concentration (ppb) values ± SD for all previous records every three year for a 10-year period from 2004 to 2014 showed the following ranking: 2014 (6.3 ± 3.9) > 2013 (4.9 ± 2.3) > 2010 (4.4 ± 3.8) > 2007 (3.4 ± 3.1) > 2004 (3.0 ± 0.7) (**Table 2**).

The results for all other wells form GAMA website was also investigated to compare them with our results and to understand the trend. GeoTracker GAMA integrates and geographically displays water quality data from multiple sources. Through the GeoTracker website all arsenic, nitrate, and TDS values for the BV groundwater basin 7 - 24 were evaluated and analyzed statistically. The mean concentration for all of the wells in BV water using supply and DWR wells was considered [16].

The GAMA values for arsenic and nitrate were all lower than health-based benchmark (10 ppb for arsenic and 10 ppm for nitrate respectively). The GAMA arsenic concentration (ppb) data showed mean values for arsenic 3.93 for 2013 (n = 8), 3.45 for 2010 (n = 10) and 2.34 for 2007 (n = 14). The trend for these values were positive showing an increase in nitrate values through the time despite the fact that these values were all below MCL. The nitrate values showed that the annual rate of concentration from 2006 to 2014 changing through the time.

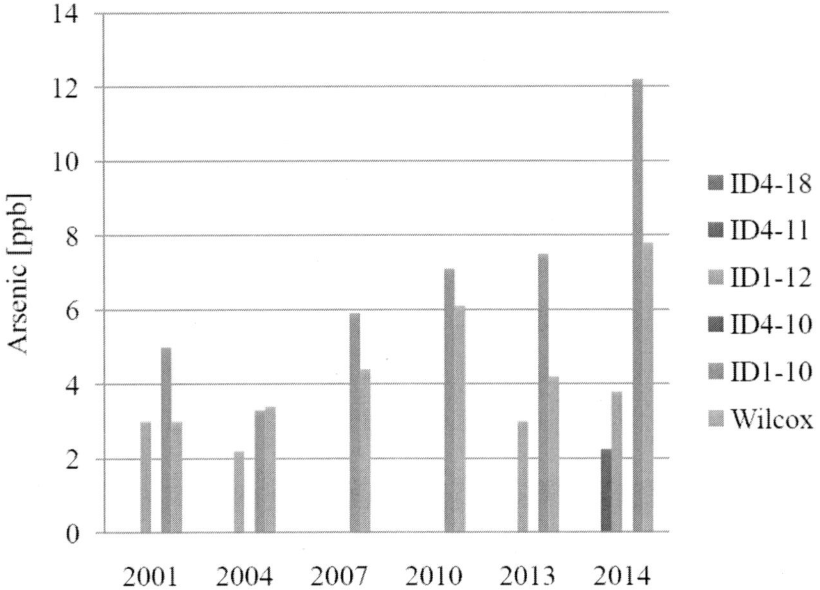

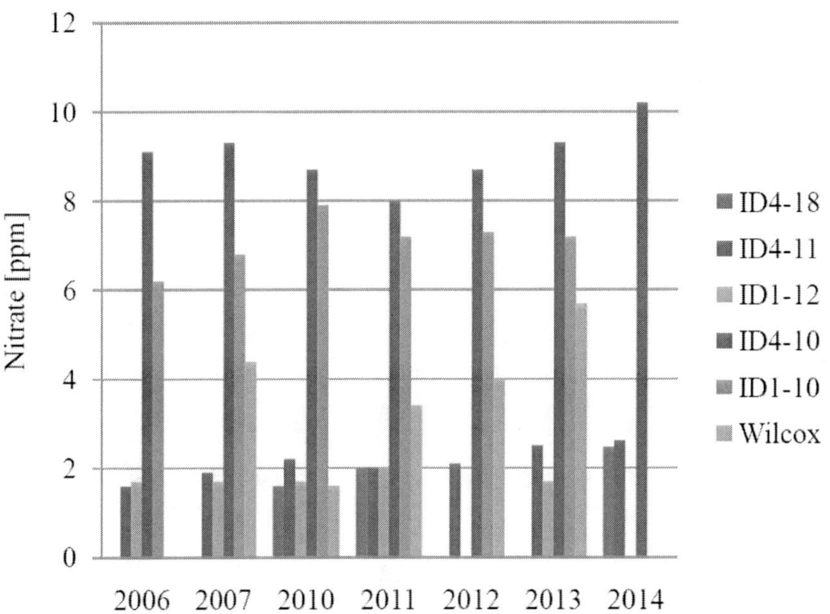

Figure 4. The concentration for arsenic (left) and nitrate (right) for different wells samples in June 2014. The values for 2006 to 2013 is the courtesy of BWD.

Table 2. The arsenic, nitrate, and TDS concentration for different years and from different wells based on authors results (2014) and data from BWD. N/A = not available.

	ID4-18	ID4-11	ID1-12	ID4-10	ID1-10	Wilcox	Mean	SD	Min	Max
Nitrate [ppm]										
2006	<1.0	1.6	1.7	9.1	6.2	<1.0	4.7	3.7	1.6	9.1
2007	N/A	1.9	1.7	9.3	6.8	4.4	4.8	3.3	1.7	9.3
2010	1.6	2.2	1.7	8.7	7.9	1.6	4.0	3.4	1.6	8.7
2011	2	2	2	8	7.2	3.4	4.1	2.8	2.0	8.0
2012	<2	2.1	<1.7	8.7	7.3	4	5.5	3.0	2.1	8.7
2013	<2	2.5	1.7	9.3	7.2	5.7	5.3	3.2	1.7	9.3
2014	2.47	2.61	N/A	10.2	NA	NA	5.1	4.4	2.5	10.2
Arsenic [ppb]										
2001	N/A	N/A	3	N/A	5	3	3.7	1.2	3.0	5.0
2004	N/A	N/A	2.2	N/A	3.3	3.4	3.0	0.7	2.2	3.4
2007	N/A	N/A	0	N/A	5.9	4.4	3.4	3.1	0.0	5.9
2010	N/A	N/A	0	N/A	7.1	6.1	4.4	3.8	0.0	7.1
2013	N/A	N/A	3	N/A	7.5	4.2	4.9	2.3	3.0	7.5
2014	N/A	2.23	3.79	N/A	12.2	7.8	6.3	3.9	2.2	12.2
Total dissolved solid (TDS) [ppm]										
2001	620	390	290	530	460	250	423.3	141.7	250.0	620.0
2004	558	318	246	459	274	200	342.5	137.8	200.0	558.0
2007	590	390	260	490	250	210	365.0	151.8	210.0	590.0
2010	620	370	240	510	240	220	366.7	166.3	220.0	620.0
2013	620	340	270	500	280	230	373.3	153.6	230.0	620.0

The concentration values for nitrate were all below health-based benchmark except few that were close to 10 ppm. The GAMA nitrate concentration (ppm) data showed mean values for nitrate 9.48 for 2012 (n = 10) 4.44 for 2013 (n = 11) 4.23 for 2010 (n = 12), 4.03 for 2008 and 2009 (n = 15 and 13), and 1.57 for 2014 (n = 4).

The Total Dissolved Solids (TDS) values were evaluated for ID4-18, ID4-11, ID4-10, ID1-10, and Wilcox by BWD. We also used the GAMA data conducting statistical review using GAMA model for all other wells in groundwater basin 4-27. While all GAMA values are below the state MCL level (1000 ppm) the studied TDS levels has increased since 2010 for ID1-10, ID 1-12, and ID1-18, and Wilcox. The ID4-10 and ID4-11, and GAMA values showed a drop in TDS values.

The BV groundwater basin was studied extensively in the past years yet none revealed the sources of arsenic and nitrate. However, in our study, we investigate the source of arsenic and nitrate contamination in water through geochemical evolution of groundwater in the BV groundwater basin using hydrogeochemical data, conceptual hydrochemical models and facies. While the physical and chemical parameters of groundwater are important in classifying and assessing water quality [22] , the lithology, solution kinetics, and groundwater flow patterns of the aquifer [23] control the hydrochemical facies. To that end, we discuss the hydro-geochemistry facies of the BV groundwater in study area, which we believe is mainly controlled by water-rock interaction.

The primary aquifer in the BV study area is the upper aquifer of a three-tiered, alluvium-filled aquifer system that is underlain by Cretaceous granitic and the triassic or older metasedimentary rocks [24]. The metasedimentary rocks are generally described as biotite schist, gneiss, quartzite, with sparse limestone and dolomite. Furthermore, the upper aquifer is composed of alluvial, playa, eolian, and fan deposits of Pleistocene to Holocene age and the middle and lower aquifers consist of continental deposits of moderately consolidated gravel, sand, and boulders of Pleistocene age [16]. According to [25] the BV Groundwater aquifer is considered mostly as calcium-sodium-bicarbonate or sodium-mixed anion type waters. Their findings show that bicarbonate accounted for 20% to 80% of the total anions, chloride, and sulfate each accounted for 10% to 60%. In addition, sodium and potassium accounted for 40% to 100% of the cations [25]. The Ca^{2+}, Mg^{2+}, SO_4^{2-}, and HCO_3^- may be derived mainly from weathering of carbonate rocks. The latter represent a part of BV lacustrine unit of Palm Spring Formation in the central portion of the valley, where the lacustrine unit was encountered in well logs. The well logs were obtained from the BWD and from published literature after [26].

3.1. Arsenic

The most common forms of arsenic in groundwater are their oxy-anions, arsenite (As^{+3}) and arsenate (As^{+5}). Both cations are capable of adsorbing to various subsurface materials, such as iron oxides and clay particles. Iron oxides are particularly important to arsenate fate and transport as iron oxides, which are abundant in the subsurface and arsenate strongly adsorbs to these surfaces in acidic to neutral waters. Furthermore also suggests arsenic is more soluble in an alkaline and anoxic environment for instance in ID1-10 (pH = 7.65) and for Wilcox (pH = 7.49) (**Figure 5(e)**). Alkaline condition may produced by the weathering of silicate minerals, which also are to reduce the acidity [27] in BV aquifer. This increase in the pH to an alkaline condition will cause both arsenite (As^{+3} the predominant species under moderately reducing conditions) and arsenate (As^{+5} the predominant species in oxygenated water) to desorb. Hence, they can be expected to be very mobile in an alkaline environment [28]. The pH-Eh condition may also control dissolved arsenic in aqueous environment [29] , where under oxidizing conditions or high Eh conditions (Eh > +100 mV), iron oxide coatings on soil/sediment particles adsorb and so the concentrations of arsenic in solution usually drops (low). When Eh is low (Eh < −250 mV), sulfate in groundwater is reduced; iron, arsenic, and other trace metals may precipitate as sulfide minerals, and the associated aqueous concentrations again are low [29]. According to [29] in the "danger zone" between these extremes (~0 < Eh < −200 mV), the iron oxides and sulfide phases dissolve; any arsenic associated with them is released into the groundwater. The BV groundwater aquifer showed the Eh for the supply well ID1-10 and Wilcox was −71.8 and −61.7 respectively (**Figure 5(f)**). The positive correlation between arsenic and sulfate suggests that arsenic maybe released in groundwater as the result of the weathering of sulfide minerals (**Figure 5(d)**). We observed a positive correlation between arsenic and pH and arsenic with sulfate. There is no correlation between dissolved arsenic and both iron and manganese in the groundwater using GAMA data (**Figure 5(g), Figure 5(h)**). Note that the 3 points on graph are the mean values of GAMA data for 2007 (Fe [n = 14], As [n = 14]), for 2010 (Fe [n = 12], for As [n = 10]); and for 2013 (Fe [n = 15] and As [n = 8]).

Remediation strategies for dissolved arsenic often drive significant shifts in oxidation-reduction potential or Eh. Therefore, arsenic may be either dissolved or precipitated in conjunction with certain treatment technologies. As the release of arsenic requires reducing conditions, a logical remediation suggest that aeration of the arsenic contaminated water might reverse the effect, forcing iron oxide ($Fe(OH)_3$) to precipitate, but remove some of dissolved arsenic as it adsorbs to the precipitating oxide surface [30].

3.2. Nitrate

The quality of groundwater in the primary aquifer system may be different from that in the shallower or deeper water-bearing zones; shallow groundwater may be more vulnerable to surficial contamination [25]. In terms of nitrate, the results of our study and the synthesis of GAMA data for nitrate from 1980 to 2000 from similar wells throughout the BV groundwater basin as long-term data screening have demonstrated that the higher nitrate concentration in the water of shallow wells occurs in dry years (and dry months, and the lower nitrate concentration in wet years (and wet months September through February). Much work indicates that nitrate concentration values in wells are closely related to the amount of precipitation and its flow in the ground [31] [32]. According to the graph (**Figure 6**) depicting nitrate data from GAMA and precipitation records

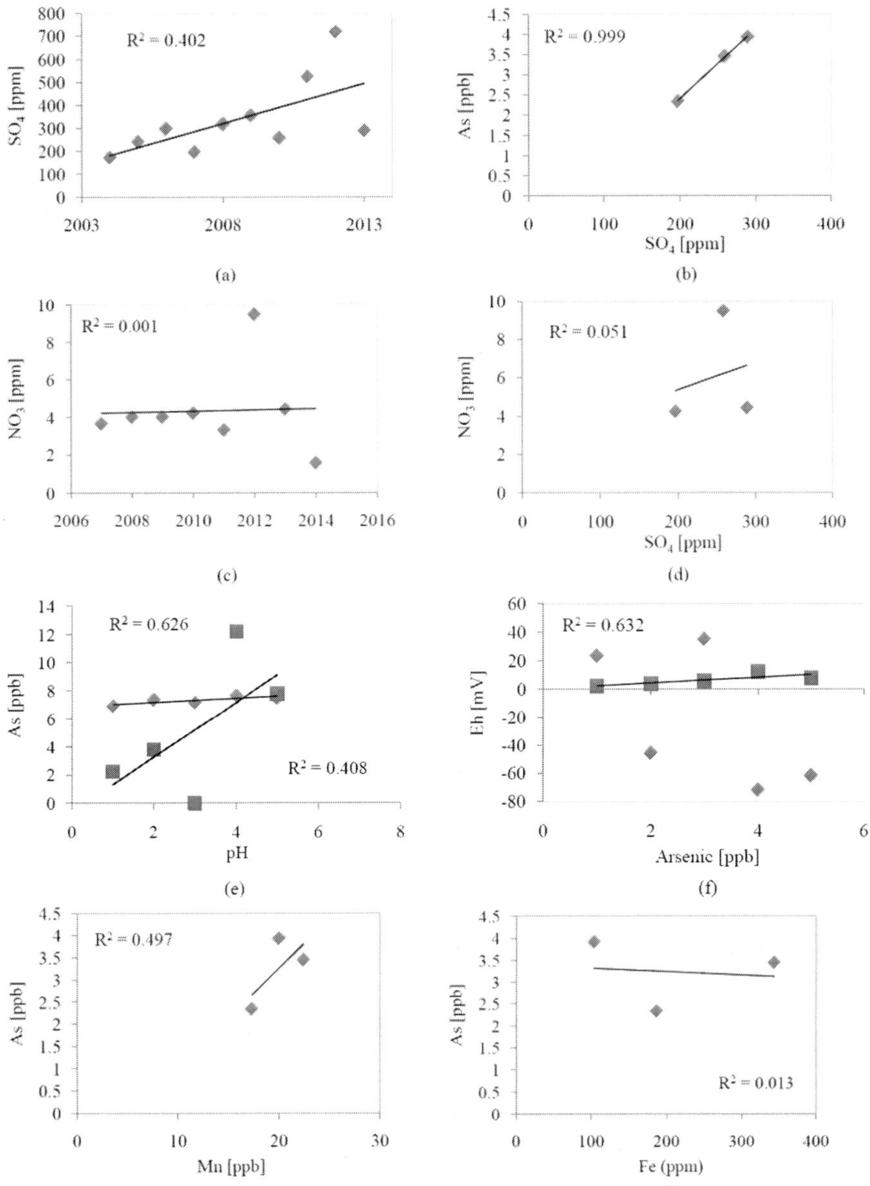

Figure 5. Clockwise from top-left: (a) Historical record for sulfate mean concentration (n = 152) from 2004 to 2014 based on GAMA ($R^2 = 0.40$); (b) Correlation of arsenic and sulfate concentration values from mean data record of 2007 (SO_4 [n = 25], for As [n = 14]), 2010 (SO_4 [n = 16], As [n = 10]); and for 2013 (for SO_4 [n = 20] and for As [n = 8]); (c) Historical record for nitrate concentration from 2007 to 2014 based on mean GAMA data (n =

98). The graph shows fluctuation pattern for nitrate; (d) There is a weak correlation between nitrate and sulfate. The mean values of NO_3 and SO_4 were used for 3 different years (SO_4 [n = 25], NO_3 [n = 18]), for 2010 (SO_4 [n = 16], NO_3 [n = 12]); and for 2013 (for SO_4 [n = 20] and for NO_3 [n = 11]); (e) Correlation of As with pH (n = 6); (f) Correlation of As with Eh (n = 6); (g) Correlation of Mn versus As using mean GAMA data for 2007 (Mn [n = 14], As [n = 14]), 2010 (Mn [n = 10], As [n = 10]); and for 2013 (for Mn [n = 9] and for As [n = 8]); (h) Correlation between Fe and As using mean GAMA data for 2007 (Fe [n = 14], As [n = 14]), for 2010 (Fe [n = 12], for As [n = 10]); and for 2013 (Fe [n = 15] and As [n = 8]).

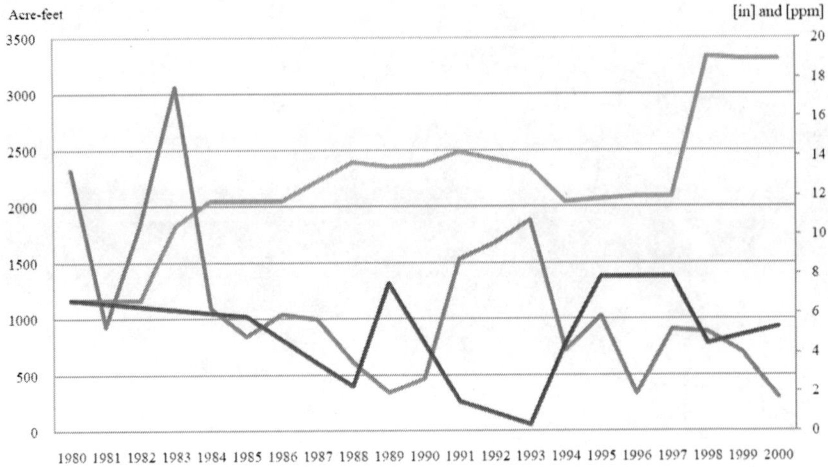

Figure 6. Comparing the precipitation, nitrate concentration from different wells and return flow values from 1980 to 2000. The return flow is the portion of the water pumped from the BV aquifer for irrigation of citrus groves and for golf course grass, which returns to the aquifer via seepage through the vadose zone. There was no data recorded for nitrate from 1981 to 1985.

from 1980-2000, groundwater pollution in fields with little fertilization and high amount of precipitation is lower due to dilution.

Past research study also suggest nitrate may be introduced to BV aquifer by 4 major recharge systems in BV 1) by stream recharge or the infiltration of runoff (example San Felipe Creek) containing fertilizer and other nutrients runoff from the several creeks and intermittent streams that drain to the valley from the mountains of the surrounding watershed in wet season. The stream runoff was estimated by Netto, 2001, approximately 600 acre-feet to approximately 26,000 acre-feet annually and average 3860 acre-feet per year during the period 1945 through 2000 [2] ; 2) by bedrock recharge, which occurs as subsurface seepage into the aquifer from the fractured crystalline basement rocks surrounding the basin, this may accelerate the nitrate migration downward due to gravity flow; 3) the underflow from adjacent basins, where groundwater flow through saturated alluvial sediments containing high level of nitrate from adjacent basins or through alluvial filled channels entering BV aquifer. This may cause mixing of an aquifer with low nitrate concentration with adjacent aquifer with the high nitrate concentration. According to [17] the underflow along San Felipe Creek was estimated to be approximately 32 acre-feet per year (af/yr) [17] ; 4) irrigation return flow or artificial recharge; it is that portion of the water pumped from the BV aquifer for irrigation of citrus groves and for golf course grass, which returns to the aquifer via seepage through the vadose zone. An estimated irrigation return flow, using the chloride mass balance technique was conducted for 2 Mojave Desert sites [33] , was calculated for the citrus grove of 22 percent of the applied irrigation rate or approximately 888 m^3/y (0.72 af/yr/ acre) and for the golf course, the return rate on the golf course of 14 percent of the applied irrigation rate, or approximately 1036 m^3/y (0.84 af/yr/ acre).

All other irrigation in BV was assumed to generate 14 percent return flow [2]. We anticipate that groundwater flow conditions can vary from year to year, depending on factors such as the amount of precipitation, the amount of regional pumping from the aquifers, seasonality changes in hydraulic gradient, and changes in recharge and discharge and irrigation pattern [34]. Therefore, it is not surprising that nitrate levels may be fluctuating by local and regional dynamics of groundwater flow affecting variation in nitrate concentration throughout the year or from year to year (**Figure 6**). The graph shows higher precipitation record results in lower nitrate concentration between1980 and 2000 while the return flow of irrigation values gradually increasing through the same time period.

Beside natural sources, human activities on land surface and in the subsurface can affect groundwater quality. Land-use data commonly are used to identify potential effects of such activity because each type of land use typically is associated with specific activities and certain changes in water quality. Land use in the study unit is 92 percent natural, 2.7 percent agricultural, and 5.7 percent urban, according to classifications from USGS National Land Cover Data [35]. Recreation has become the second most intensive use of groundwater in BV. The primary use of water for recreational activities in BV is for irrigation of the several golf courses in BV. The gross irrigation applied across all golf courses is estimated to be approximately 6 feet per year [2]. Wells ID4-2 and ID4-3 have been taken out of production by BWD and no longer used due to high nitrate concentration recorded above MCL. According to topographic map, these wells are located approximately on the east side of De Anza, Club Circle, and

BSPCSD Golf Course (**Figure 2**). Moreover, treatment as well as reducing the return flow of Golf Course irrigation and fertilizer application may also contribute to lower contamination of nitrate in BV groundwater aquifer.

4. CONCLUSION

The groundwater basin system of BV is comprised of three hydraulically distinct aquifers that will likely provide sufficient supply if properly managed. However, nitrate and arsenic contamination in groundwater may be a concern for the study area. The results of this preliminarily study indicate that nitrate concentration is not exceeding regulatory limits (MCL = 45 ppm) but shows some sign of increase of arsenic in only one well (around and/or slightly above 10 ppb). This may not be ruled out as an imminent threat but we suggest that further monitoring of the nitrate and arsenic levels (e.g. wet season water sampling) is necessary to evaluate the temporal evolution and distribution of nitrate and arsenic concentrations within the aquifers underlying the region. Under these circumstances, future studies using nitrogen, oxygen, and carbon isotopes are also required to better assess the migration pattern of the specific sources and contributions of nitrate to the BV groundwater aquifer.

ACKNOWLEDGEMENTS

Mohammad Hassan Rezaie-Boroon, Jessica Chaney, Bradley Bowers The authors thank Dr. Tim Ross at Department of Water Resources. This research study would not have been possible without his assistance. We thank also Mr. Jerry Rolwing, General Manager at Borrego Water District for allowing the groundwater well sampling and access to wells, and also working with us tirelessly on sampling day, providing us valuable information for this study regarding groundwater use and information about wells in Borrego Valley area.

REFERENCES

1. Water Education Foundation (2006) Where Does My Water Come from? Sacramento, Calif., Water Education Foundation. http://www.watered.org/watersources/default.asp
2. Netto, S.P. (2001) Water Resources of Borrego Valley San Diego County, California. Thesis, Master of Science in Geological Sciences, 159 p.

3. Mathany, T.M., Wright, M.T., Beuttel, B.S. and Belitz, K. (2012) Groundwater-Quality Data in the Borrego Valley, Central Desert, and Low-Use Basins of the Mojave and Sonoran Deserts Study Unit, 2008-2010—Results from the California GAMA Program. U.S. Geological Survey Data Series 659, 100 p.
4. Welch, A.H., Lico, M.L.S. and Hughes, J.L. (1988) Arsenic in Ground Water of the Western United States. Groundwater, 26, 333-347. http://dx.doi.org/10.1111/j.1745-6584.1988.tb00397.x
5. Fetter, C.W. (2008) Contaminant Hydrogeology. 2nd Edition, Waveland Press Inc., Long Grove, 500 p.
6. Azaza, F.H., Ketata, M., Bouhlila, R., Gueddari, M. and Riberio, L. (2010) Hydrogeochemical Characteristics and Assessment of Drinking Water Quality in Zeuss-Koutine Aquifer, Southeastern Tunisia. Environmental Monitoring and Assessment, 174, 283-298. http://dx.doi.org/10.1007/s10661-010-1457-9
7. Carrillo-Rivera, J.J., Varsányi, I., Kovács, L.Ó. and Cardona, A. (2007) Tracing Groundwater Flow Systems with Hydrogeochemistry in Contrasting Geological Environments. Water, Air, & Soil Pollution, 184, 77-103. http://dx.doi.org/10.1007/s11270-007-9400-6
8. Spalding, R.F. and Exner, M.E. (1993) Occurrence of Nitrate in Groundwater—A Review. Journal of Environmental Quality, 22, 392-402. http://dx.doi.org/10.2134/jeq1993.00472425002200030002x
9. Mandal, B.K., Chowdhury, T.R., Samanta, G., Mukherjee, D.P., Chanda, C.R., Saha, K.C. and Chakraborti, D. (1996) Impact of Safe Water for Drinking and Cooking on Five Arsenic-Affected Families for 2 Years in West Bengal, India. Science of the Total Environment, 218, 185-201. http://dx.doi.org/10.1016/S0048-9697(98)00220-4
10. Ravenscroft, P., McArthur, J.M. and Hoque, V. (2001) Geochemical and Palaeohydrological Controls on Pollution of Groundwater by Arsenic. In: Chappell, W.R., Abernathy, C.O. and Calderon, R., Eds., Arsenic Exposure and Health Effects IV, Elsevier Science Ltd., Oxford, 53-77.
11. Schreiber, M.E., Gotkowitz, M.B., Simo, J.A. and Freiberg, O.G. (2003) Mechanisms of Arsenic Release to Ground Water from Naturally Occurring Sources, Eastern Wisconsin. In: Welch, A.H. and Stollenwerk, K.G., Eds., Arsenic in Ground Water, Kluwer Academic Publishers, Dordrecht, 259-280. http://dx.doi.org/10.1007/0-306-47956-7_9
12. Acharyya, S.K., Lahiri, S., Raymahashay, B.C. and Bhowmik, A. (2000) Arsenic Toxicity of Groundwater in Parts of the Bengal Basin, in India and Bangladesh: Role of Quaternary Stratigraphy and Holocene Sea-Level Fluctuation. Environmental Geology, 39, 1127-1137. http://dx.doi.org/10.1007/s002540000107
13. McArthur, J.M., Banerjee, D.M., Hudson-Edwards, K.A., Mishra, R., Purohit, R., Ravenscroft, P., Cronin, A., Howarth, R.J., Chatterjee, A., Talukder, T., Lowry, D., Houghton, S. and Chadha, D.K. (2004) Natural Organic Matter in Sedimentary Basins and Its Relation to Arsenic in Anoxic Ground Water: The Example of West Bengal and Its Worldwide Implications. Applied Geochemistry, 19, 1255-1293. http://dx.doi. org/10.1016/j.apgeochem.2004.02.001
14. Rezaie-Boroon, M.H., Gnandi, K. and Folly, K.T. (2011) Presence and Distribution of Toxic Trace Elements in Water and Sediments of the Southern Togo Rivers Watershed, West Africa. PSP Environmental Bulitin, 20, 1853-1865.
15. Mills, W.R. (2009) Integrated Water Resources Management Plan. Final Report, 128 p.
16. California Department of Water Resources DWR (2004) California's Groundwater Individual Basin Descriptions. California Department of Water Resources Bulletin 118, Lucerne Valley. http://www.water.ca.gov/pubs/groundwater /bulletin_118 /basindescriptions/7-19.pdf
17. Moyle Jr., W.R. (1982) Water Resources of Borrego Valley and Vicinity, California; Phase 1-Definition of Geologic and Hydrologic Characteristics of Basin. U.S. Geological Survey Open-File Report 82-855, 39 p.

18. Mitten, H.T., Lines, G.C., Berenbrock, C. and Durbin, T.J. (1988) Water Resources of Borrego Valley and Vicinity, San Diego County, California: Phase 2-Development of a Ground-Water Flow Model. U.S. Geological Survey Water- Resources Investigations Report 87-4199, 27 p.
19. Smedley, P.L., Nicolli, H.B., Macdonald, D.M.J., Barros, A.J. and Tullio, J.O. (2002) Hydrogeochemistry of Arsenic and Other Inorganic Constituents in Groundwater from La Pampa, Argentina. Applied Geochemistry, 17, 259-284. http://dx.doi.org/10.1016/S0883-2927(01)00082-8
20. Wilde, F.D. (2004) Cleaning of Equipment for Water Sampling (Ver. 2.0). U.S. Geological Survey Techniques of Water-Resources Investigations, Book 9, Chap. A3. http://pubs.water.usgs.gov/twri9A3/
21. Wilde, F.D., Radtke, D.B., Gibs, J. and Iwatsubo, R.T. (1999) Collection of Water Samples (Ver. 2.0). U.S. Geological Survey Techniques of Water-Resources Investigations, Book 9, Chap A4. http://pubs.water.usgs.gov/twri9A4/
22. Jahin, H.S. and Gaber, S.E. (2011) Study of Groundwater Quality in El-Kharga Oasis, Western Desert, Egypt. Asian Journal of Water, Environment and Pollution, 8, 1-7.
23. Raju, N.J., Ram, P. and Dey, S. (2009) Groundwater Quality in the Lower Varuna River Basin, Varanasi District, Uttar Pradesh. Journal of the Geological Society of India, 73, 178-192. http://dx.doi.org/10.1007/s12594-009-0074-0
24. Dibblee Jr., T.W. (1984) Stratigraphy and Tectonics of the San Felipe Hills, Borrego Badlands, Superstition Hills, and vicinity. In: Rigsby, C.A., Ed., The Imperial Basin-Tectonics, Sedimentation, and Thermal Aspects, Pacific Section SEPM, Los Angeles, 31-34.
25. Parsons, M.C., Hancock, T.C., Kulongoski, J.T. and Belitz, K. (2014) Status of Groundwater Quality in the Borrego Valley, Central Desert, and Low-Use Basins of the Mojave and Sonoran Deserts Study Unit, 2008-2010: California GAMA Priority Basin Project. USGS Scientific Investigations Report: 2014-5001, 88 p.
26. Threet, R.L. (1972) Hydrogeology of Southern Borrego Valley. Prepared 13 April 1972 for: George J. Kuhrts III, Borrego Springs, California.
27. Pazand, K. and Fereidoni Sarvestani, J. (2013) Hydrogeochemical Investigation in an Arid Region of Iran (Tabas, Central Iran). Environmental Earth Sciences, 70, 743-752. http://dx.doi.org/10.1007/s12665-012-2162-5
28. Henkel, S. and Polette, D. (1999) Arsenic in Ground Water of the Willamette Basin, Oregon. U.S. Geological Survey Water-Resources Investigations Report 98-4205, 27 p.
29. Stein, C.L., Brandon, W.C. and McTigue, D.F. (2005) Arsenic Behavior under Sulfate-Reducing Conditions: Beware of the "Danger Zone". EPA Science Forum 2005: Collaborative Science for Environmental Solutions, 16-18 May 2005, Washington DC.
30. Andrews, J.E., Brimblrcombe, P., Jickells, T.D., Liss, P.S. and Reid, B.J. (2003) An Introduction to Environmental Chemistry. 2nd Edition, Wiley-Blackwell, Hoboken, 296 p.
31. Gopal, B.K. (1990) Investigation of Nitrate Contamination in Shallow Ground Waters near Wood Ward, Oklahoma. Ground Water Quality and Agricultural Practices. Lewis Publishers, Boca Raton, 247-264.
32. Rutkoviene, V., Kusta, A. and Èesoniene, L. (2005) Environmental Impact on Nitrate Levels in the Water of Shallow Wells. Polish Journal of Environmental Studies, 14, 631-637.
33. Prudic, D.E. (1994) Estimates of Percolation Rates and Ages of Water in Unsaturated Sediments at Two Mojave Desert sites, California-Nevada. U.S. Geological Survey Water Resources Investigations Report 94-4160, 19 p.
34. Murgulet, D. and Tick, G.R. (2009) Assessing the Extent and Sources of Nitrate Contamination in the Aquifer System of Southern Baldwin County, Alabama. Environmental Geology, 58, 1051-1065. http://dx.doi.org/10.1007/s00254-008-1585-5

35. Vogelmann, J.E., Howard, S.M., Yang, L., Larson, C.R., Wylie, B.K. and Van Driel, N. (2001) Completion of the 1990s National Land Cover Data Set for the Coterminous United States from Landsat Thematic Mapper Data and Ancillary Data Sources. Photogrammetric Engineering & Remote Sensing, 61, 650-662.

CHAPTER 6

Identifying Groundwater Recharge Connections in The Moscow (USA) Sub-Basin using Isotopic Tracers and a Soil Moisture Routing Model

Jasper Candel[1,2], Erin Brooks[3], Ricardo Sánchez-Murillo[4], George Grader[5], Roel Dijksma[1]

[1] *Hydrology and QuantitativeWater Management Group,WageningenUniversity, Wageningen, The Netherlands*

[2] *Soil Geography and Landscape Group, Wageningen University,Wageningen, The Netherlands*

[3] *Biological & Agricultural Engineering, University of Idaho, Moscow, ID, USA*

[4] *National University of Costa Rica, Heredia, Costa Rica*

[5] *PRISEM Geoconsulting, 823 W 2nd Ave, Spokane, WA 99203, USA*

ABSTRACT

Globally, aquifers are suffering from large abstractions resulting in groundwater level declines. These declines can be caused by excessive abstraction for drinking water, irrigation purposes or industrial use. Basaltic aquifers also face these conflicts. A large flood basalt area (1.1×10^5 km^2) can be found in the Northwest of the USA. This Columbia River Basalt Group (CRBG) consists of a thick series of basalt flows of Miocene age. The two major hydrogeological units (Wanapum and Grand

Ronde formations) are widely used for water abstraction. The mean decline over recent decades has been 0.6 m year^{-1}. At present day, abstraction wells are drying up, and base flow of rivers is reduced. At the eastern part of CRBG, the Moscow sub-basin on the Idaho/Washington State border can be found. Although a thick poorly permeable clay layer exists on top of the basalt aquifer, groundwater level dynamics suggest that groundwater recharge occurs at certain locations. A set of wells and springs has been monitored bi-weekly for 9 months for $\delta^{18}O$ and $\delta^{2}H$. Large isotopic fluctuations and *d*-excess values close to the meteoric water line in some wells are indicating that recharge occurs at the granite/basalt interface through lateral flow paths in and below the clay. A soil moisture routing (SMR) model showed that most recharge occurs on the granitic mountains. The basaltic aquifer receives recharge from these sedimentary zones around the granite/basalt interface. The identification of these types of areas is of major importance for future managed-aquifer recharge solutions to solve problems of groundwater depletion.

Keywords: Stable isotopes USA Basalt aquifers Groundwater recharge/ water budget Numerical modelling

INTRODUCTION

Declining groundwater levels are a global concern, as a large part of the human population depends on fresh water from aquifers (UNESCO 2009). Van Loon et al. (2016) showed that humans are an important factor in the causes of droughts and the depletion of reservoirs. The depletion of reservoirs is mainly a result of groundwater pumping for drinking-water supply (Willis and Garrod 1998), irrigation (Amelung et al. 1999; Foster et al. 2004; Konikow and Kendy 2005; Hoque et al. 2007; Qureshi et al. 2010; Wada et al. 2012) and industrial use (Hayashi et al. 2009). Globally, groundwater withdrawal amounts to 750–800 km^3 year^{-1} and exceeds annual recharge in many places of the world (Shah et al. 2000).

Urgent action is needed to slow down the depletion of groundwater reservoirs. Currently, managed aquifer recharge (MAR) methods are available to recover depleted aquifers (Dillon 2005; Maliva et al. 2014). Examples are known from many aquifers in the world—e.g. Burdekin Delta and Angas-Bremer area in Australia (Gerges et al. 2002; Dillon 2009); Satlasana and Kodangipalayam in India (Gale et al. 2006); and the Columbia River Basalt Group aquifer in USA (Eaton et al. 2009). However, in order to improve the success of MAR projects (Gale et al. 2006), advanced aquifer characterization is needed to identify suitable locations for MAR practices (Dillon 2005; Rahman et al. 2012; Maliva et al. 2014). Basaltic aquifers which are widely used for water abstraction face water conflicts between irrigation demand, drinking-water supply and sustainable aquifer management (Macdonald et al. 1995; El-Naqa et al. 2007; Pavelic et al. 2012)—for example on the basaltic Deccan Traps in India, irrigation by groundwater has developed from shallow dug wells to mechanical abstraction since the 1950s. In the Maharashtra state in India alone, the total abstraction volume from more than 1.5 million wells has increased by over 700 % since then (Macdonald et al. 1995).

One of the areas where severe groundwater depletion exists is the Moscow-Pullman region on the border of Idaho and Washington State in the USA (Fig. 1). Here, groundwater decline started with exploitation of upper aquifers in the early 20th century (Laney et al. 1923). The Moscow sub-basin is part of the Palouse basin, which is located on the eastern part of the Columbia River Basalt Group (CRBG; Hooper 1982). The CRBG aquifer is a main source of fresh water in the area (Tolan et al. 1989). Despite curtailed groundwater pumping since 1992, aquifer levels have continued to decline at 40–60 cm year^{-1} (Robischon 2006, 2007). The Moscow sub-basin boundary is characterized by eolian loess deposits overlying eastward-dipping basalts to the west with granitic highlands and Miocene/Pliocene sediments to the east. These unique geologic zones interfinger in the subsurface (Fig. 2). Groundwater levels in the shallow basaltic aquifer (Wanapum Formation) recover in response to reduced pumping, indicating that recharge occurs to the shallow aquifer (Leek 2006). The presence of tritium in water samples recently taken from the Wanapum Formation and in some wells in the upper portion of the deeper basaltic aquifer (Grande Ronde), suggest that both these aquifers received recharge to some extent over the last 50 years (Carey 2011).

Identifying groundwater-recharge-source areas is critical to local managers and municipalities for targeting management which preserves and potentially enhances aquifer recharge. Previous work suggested that no vertical recharge occurs in the Moscow sub-basin (arrow A in Fig. 2). This was concluded based on the presence of thick impermeable clay deposits (up to 30 m) that were identified through deep well logs (Lum et al. 1990; Bush 2005; Fairley et al. 2006). Rather than vertical percolation, lateral saturated flow occurs on top of these impermeable clays (Brooks et al. 2004). Using a spatially distributed hydrologic analysis approach, Dijksma et al. (2011) showed that it is possible that recharge occurs only in certain regions. They, and Bush (2005), suggested that the greatest potential for recharge might be via spatially discrete areas of high permeability (e.g. paleo-channels) draining from forested regions on the Moscow Mountain range front, recharging upper and lower aquifers at the granite/basalt interface (arrow B and C in Fig. 2). In addition, Dijksma et al.

(2011) and Fairley et al. (2006) found evidence in boreholes of the existence of thick layers of coarse material, of which some end up at the surface as perennial springs (De Graaf 2011). Using a numeric groundwater model, De Graaf (2011) confirmed that aquifer recharge through paleo-channels could provide a relatively large part of the aquifer recharge that occurs in the Paradise Creek watershed. It is very likely that paleo-channels are present and provide lateral conduits of groundwater recharge to the deeper aquifers (Brooks and Grader 2011). One of the remaining challenges is the identification of these groundwater recharge source areas in order to use these lateral conduits in MAR applications.

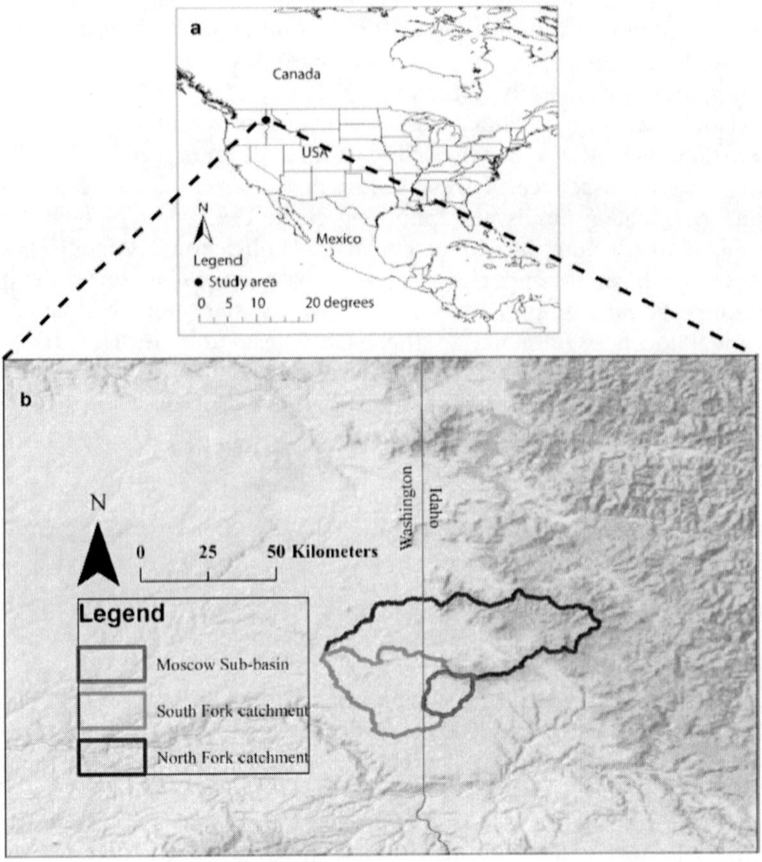

Figure. 1 A The Moscow sub-basin (study area), located in the Palouse Basin, on the border of Idaho/Washington state, USA. B The Palouse Basin, subdivided in the North Fork and South Fork of the Palouse River. The *red line* shows the Moscow sub-basin Figures 1 and 3 contains poor quality of text. Otherwise, please provide replacement figure file.correct

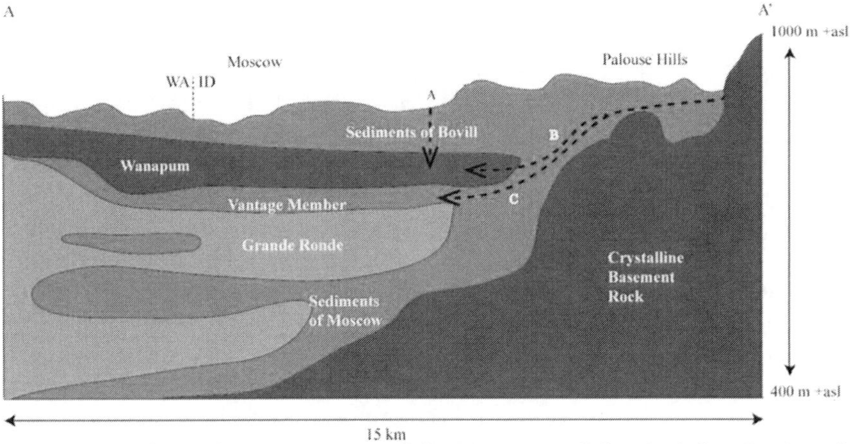

Figure. 2Geological cross-section of the Moscow sub-basin (after Bush and Garwood 2005). Letters *A–C* and *dashed arrows* correspond to the recharge pathways mentioned in the text. Cross-section is based on approximate W–E *dashed line* A–A′ in Fig. 4

The aim of this study is to identify these groundwater recharge pathways in the Moscow sub-basin. A soil moisture routing (SMR) model (Dijksma et al. 2011) was used first, to provide a spatial representation of percolation below the root zone across the entire basin. Secondly, stable isotopes $\delta^{18}O$ and δ^2H as well as deuterium excess (*d*-excess) were used to identify linkages between the surface water and groundwater systems. $\delta^{18}O$ and δ^2H have widely been recognized as useful tracers in providing insights into water movements in watersheds (McDonnell and Kendall 1992; Kendall and MacDonnell 1998). Seasonal and event-based variability in $\delta^{18}O$ and δ^2H in rain and snowmelt provide a unique composition which can be traced in streamflow and observed in groundwater wells. $\delta^{18}O$ and δ^2H tracers have been used in the Moscow sub-basin to describe variability in groundwater sources; however, since only one sample was taken from each well at different moments in time (Larson et al. 2000; Carey 2011; Moxley 2012) these studies were not able to investigate temporal variability in $\delta^{18}O$ and δ^2H. If a hydrologic connection exists between two different sources of water (e.g. groundwater, precipitation or surface water), then it is expected that the down-gradient water source would have a similar but damped $\delta^{18}O$ and δ^2H that is lagged in time relative to up-gradient sources (Changnon 1987; Soulsby and Tetzlaff 2008; Katsuyama et al. 2010; Speed et al. 2010; Wassenaar et al. 2011). If a hydrologic connection between different sources of water exists, temporal fluctuations in $\delta^{18}O$ and δ^2H will be hydrologically similar (Scanlon et al. 2002).

In addition to $\delta^{18}O$ fluctuations, the $\delta^{18}O$ can be compared to the Global Meteoric Water Line (GMWL: $\delta^2H = 8 \times \delta^{18}O + 10$), that serves as a reference to determine deviations of the relation between $\delta^{18}O$ and δ^2H (Craig 1961). The offset of 10 is determined by kinetic isotope fractionation that occurs during non-equilibrium processes such as evaporation (Craig 1961; Cappa et al. 2003). This offset can be calculated by the *d*-excess ($d = \delta^2H - 8 \times \delta^{18}O$), and is a measure for the relative proportions of δ^2H and $\delta^{18}O$ that are affected by evaporation (Dansgaard 1964). The correlation with precipitation is higher when the *d*-excess of a water source is close to 10 ‰. A *d*-excess close to 10 ‰ also indicates that precipitation has infiltrated and percolated below the root zone quickly, and has not been subject to evaporation (Froehlich et al. 2001). Sánchez-Murillo et al. (2015) measured $\delta^{18}O$ and δ^2H in precipitation (*N* samples = 203) and streamflow (*N* = 244 and *N* = 195, in Crumarine Creek and the South Fork of the Palouse River, respectively) from June 2011 to January 2014 (Fig. 3). Their record provides the isotopic baseline necessary to investigate hydrologic connections to groundwater.

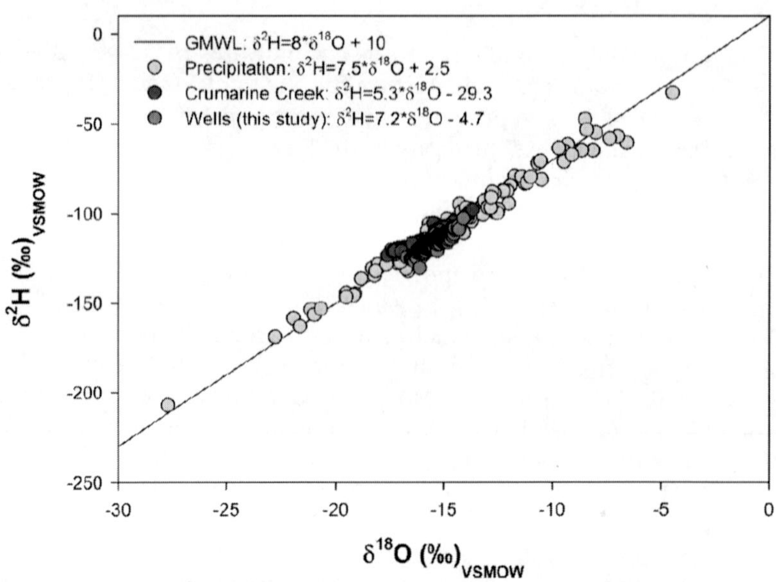

Figure. 3Local meteoric water lines for various water sources collected in this study in comparison to the global meteoric water line (GMWL). Crumarine Creek and precipitation is according to Sánchez-Murillo et al. 2015

AREA DESCRIPTION

Figure 4a shows the digital elevation model for the Moscow sub-basin, including the drainages from Moscow Mountain (part of the east–west trending Palouse Range). These drainages include Paradise Creek and the South Fork of the Palouse River. Discharge measurements by the US Geological Survey (USGS) of Paradise Creek near the University of Idaho Campus from 1979 to 2012 indicate the mean annual discharge of 0.12 mm/year. The South Fork of the Palouse River at Pullman (Washington) has a mean annual discharge of 0.40 mm/year, which includes the Paradise Creek discharge (US Geological Survey 2012). Non-irrigated agriculture is the dominant land use in the Moscow sub-basin (71 %); forest (15 %), grass fields (6 %), and urban areas (8 %) are the other land use types present (Fig. 4b; Bara and Shaw 1995).

Pullman-Moscow groundwater systems are fairly complex with multiple basalt flows underlain by and laterally resting against protruding basement metamorphic/granitic ridges. Basin margin paleo-valleys occur at different scales, and are stacked and altered through geologic time by successive basalt dam effects and are interwoven with clay and sand/gravel lacustrine and stream/ swamp inter-beds (Grader 2011). High-porosity interbeds (associated with subordinate paleo-channels) are known to occur at lower stratigraphic levels (Grader 2011).

The extent of the upper CRBG basalt flows (Wanapum Basalt) is indicated in Fig. 4. Underlying older, thicker Grande Ronde flows occur westward of this transitional line (shown in cross-section, Fig. 2). Associated aquifers use the same names "upper Wanapum" or "lower Grande Ronde" formations, but involve heterogeneous geology. The CRBG basalt flows are inter-fingered with Miocene and younger sediments (Latah Formation including Sediments of Moscow, Vantage Member, and Sediments of Bovill), and are a result of alluvial deposition when basalt flows blocked former drainage patterns (Fig. 2). A detailed Quaternary surficial geological map is dominated by eolian deposited loess silts (Palouse Formation) as shown in Fig. 5. Arable stoneless soils show that no ice age tillites were deposited at this latitude; however, major volcanic airfall ashes derived from the Cascade Mountains were admixed into both Miocene and Quaternary sediments. Lower through upper Miocene sediments are a mixture of all textures from clay to sand and gravel (Lum et al. 1990; Bush 2005; Fairley et al. 2006). This research focuses on recharge in this part of the basin, where paleo-channels of different orders of size should be contained by and will laterally migrate within the larger-scaled subsurface paleo-valleys.

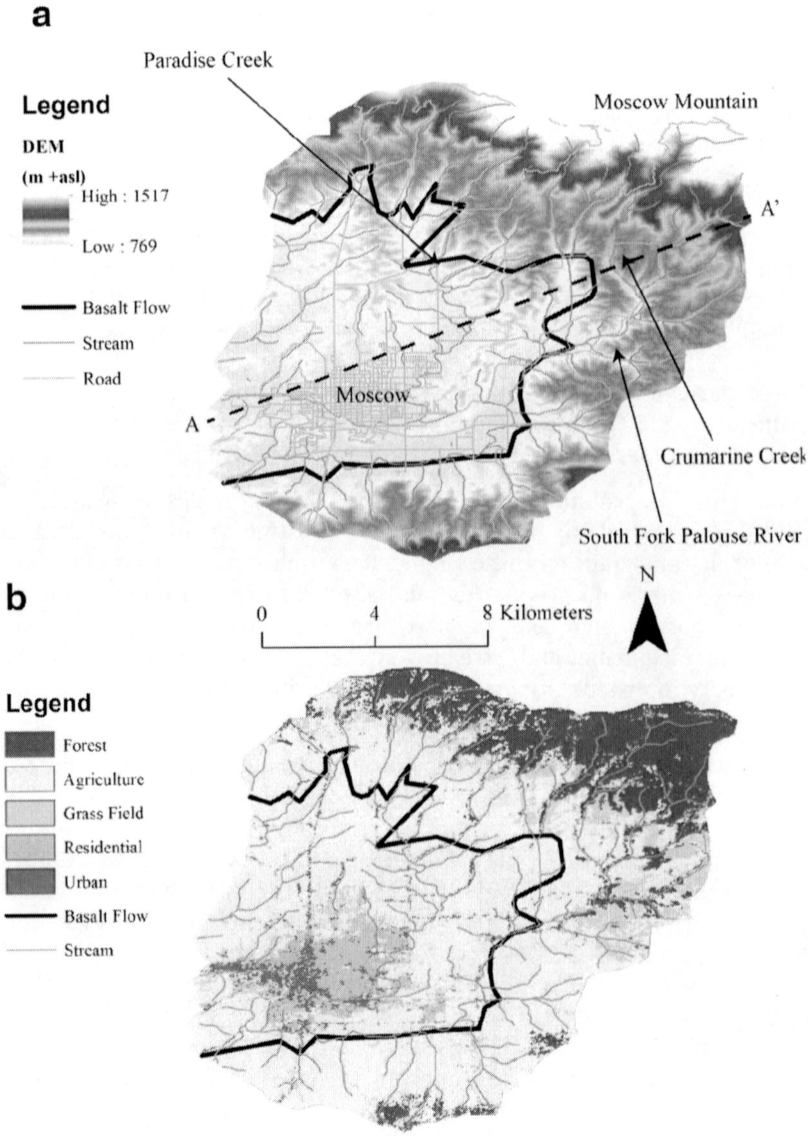

Figure. 4 A Digital elevation model (DEM) of the Moscow sub-basin. *Black line* indicates the extent of the CRBG basalt flows, defined by Grader (2011). The *dashed line* A–A′ represents the geological cross-section as shown in Fig. 2, B Land use map of the Moscow sub-basin

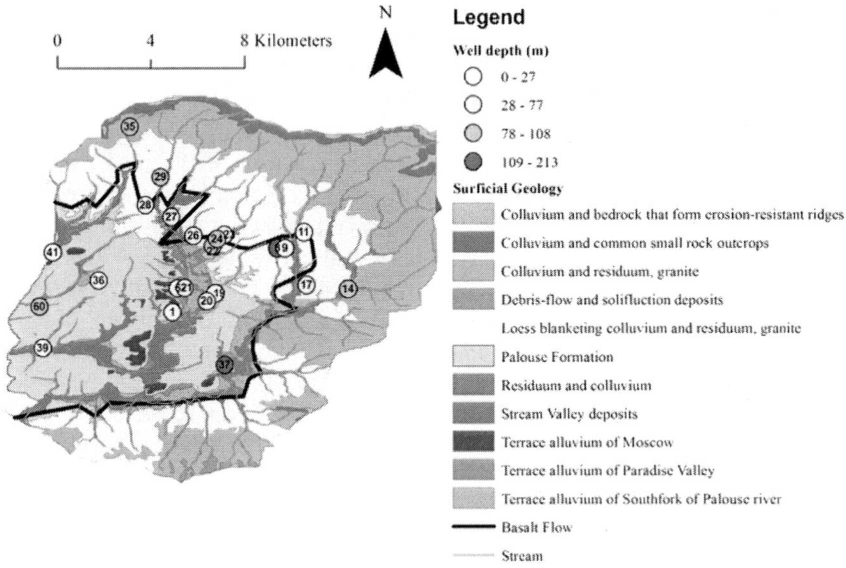

Figure. 5 Surficial geological map of the Moscow sub-basin including the well locations, well sample IDs (*1–60*) and well depth (Othberg et al. 2001)

The Wanapum and Grande Ronde aquifers are deformed, and tectonic features or feeder dikes may be present resulting in isolated groundwater reservoirs with low connectivity (McVay 2007). Locally open horizontal and vertical fractures are present and are, at present, clay- or even sand-filled. These fractures have not been extensively studied and are likely to vary spatially (Fairley et al. 2006). Unpredictable rock/sediment relationships are the result of invasive basalts that flowed over wet sediments as observed in ancient lakes or fluvial streams along paleo-valley boundaries. Previous attempts to predict groundwater flow in the Wanapum and Grande Ronde aquifers by numerical modelling (Barker 1979; Lum et al. 1990; Hansen et al. 1994; Whiteman et al. 1994; Vaccaro 1999; Reidel et al. 2003) were often inconsistent due to the combination of both aquifer's hydrogeological heterogeneity and the complex groundwater extractions from different locations and intervals (Leek 2006; McVay 2007; Bennett 2009).

MATERIAL AND METHODS

In this study, the hydrologic connections between precipitation, streamflow, and groundwater are investigated by testing groundwater in 22 wells and 2 springs throughout the Moscow sub-basin (Fig. 5; Table 1). These locations have been selected based on a preliminary sampling of 49 wells and 3 springs (Candel 2014). Well information was derived from the online database of the Idaho Department of Water Resources (IDWR; Idaho Department of Water Resources 2013). When well information was not present in the online database, it was derived from the well owners. Each of the wells and springs were sampled once every 2 weeks. In total, 352 samples have been analysed for $\delta^{18}O$ and $\delta^{2}H$. Wells were selected based on geology, well depth (Fig. 4b), and proximity to streams, with the intention to have distributed samples throughout the Moscow sub-basin (Idaho Department of Water Resources 2013). Water samples are taken directly from outdoor spigots after flushing the system of old water. Water lines were flushed until the water temperature was constant. Water samples were collected in 30-ml glass E-C borosilicate bottles with TFE-lined caps (Wheaton Science Products, USA) and stored upside down at 5 °C. Bottles were filled with no head space and covered with parafilm (Thermo Scientific, USA) to avoid exchange with atmospheric moisture.

Table 1 Characteristics of all wells and springs that were sampled

Well sample ID	Well ID	Bedrock	Depth		Elevation			Long	Lat
			Well (m)	Water (m)	Well top (m + asl)	Well bottom (m + asl)	Water level (m + asl)		
1	279844	SOB	24	20	812	788	792	−116.975467	46.745203
5	279433	B	72	53	818	746	765	−116.973363	46.754049
6	Spring	–	0	–	818	818	–	−116.973363	46.754049
8	411379	G	213	184	842	628	658	−116.935307	46.768930
9	279954	SOB	77	74	832	755	758	−116.932524	46.768905
11	279918	G	27	24	849	821	825	−116.925279	46.774895

14	–	–	102	–	823	721	–	−116.908495	46.753806
17	280306	G	43	37	829	786	792	−116.924215	46.755273
19	Spring	–	0	–	809	809	–	−116.958980	46.752131
20	279727	B	55	44	800	745	756	−116.962160	46.749262
21	279905	SOB	46	22	811	765	789	−116.970663	46.754251
22	280057 and 280008	SOB	105	42	817	712	775	−116.959883	46.769872
23	280053	SOB	62	13	827	765	814	−116.955420	46.773923
24	338130	SOB	93	84	832	739	748	−116.958287	46.772420
26	279678	B	47	42	817	771	775	−116.967314	46.773636
27	389781	G	67	53	824	757	772	−116.975827	46.780564
28	279492	SOB	73	71	834	762	763	−116.985709	46.785009
29	279708	G	93	83	839	746	756	−116.979895	46.795657
35	–	G	94	5	950	857	945	−116.991942	46.814497
36	280524	SOB	19	14	815	796	801	−117.003918	46.756973
37	–	B	150	32	797	647	765	−116.955450	46.725566
39	–	B	21	20	774	753	755	−117.025350	46.731821
41	–	B	27	27	798	770	771	−117.021437	46.767460
60	–	V	108	107	800	692	693	−117.026589	46.747202

The bedrock codes are defined as: *SOB* Sediments of Bovill, *B* basalt, *G* granite, *V* Vantage Member

Stable isotope analyses were conducted at the Idaho Stable Isotope Laboratory, University of Idaho (Idaho, USA) using a cavity ring-down spectroscopy (CRDS) water isotope analyser L1120-i (Picarro, USA). Laboratory standards, previously calibrated to the reference waters Vienna Standard Mean Ocean Water and Standard Light Antarctic Precipitation (VSMOW2-SLAP2), were glacier water ($\delta^2H = -255.0$ ‰, $\delta^{18}O = -30.8$ ‰), commercial bottled water ($\delta^2H = -64.2$ ‰, $\delta^{18}O = -8.3$ ‰), and Moscow tap water ($\delta^2H = -125.5$ ‰, $\delta^{18}O = -15.4$ ‰). Glacier water and commercial bottled water were used to normalize the results to the VSMOW2-SLAP2 scale, while the Moscow tap water was a quality control standard. The laboratory instrument precision on average was ±0.5 ‰ (1σ) for δ^2H and ±0.1 ‰ (1σ) for $\delta^{18}O$. The estimated *d*-excess analytical uncertainty was ±0.6 ‰ (1σ).

A check to the suitability of using outdoor spigots was completed by quantifying changes in $\delta^{18}O$ with time at five spigots. The temporal variation in $\delta^{18}O$ after 2, 5 and 10 min of flushing was always smaller than the standard error of the Picarro Instrument, indicating that differences found in isotopic composition cannot be explained by different water being sampled within the pipelines of the wells.

Soil-moisture routing model

In this study, a SMR model was used to provide spatial predictions of the mean annual percolation rates below the root zone throughout the Moscow sub-basin. The SMR model used is a spatially distributed, grid-based hydrologic model which operates within a geographic information system (GIS) environment, originally developed at Cornell University (Frankenberger et al. 1999). The model simulated interception, evapotranspiration, subsurface lateral flow, deep vertical percolation and saturation-excess surface runoff through a multi-layer soil profile. Snow accumulation and melt is simulated using an energy balance approach (Brooks and Boll 2005; Brooks et al. 2007). The SMR model has been developed as a spatially explicit management tool and therefore has been developed to rely primarily on publicly available data with minimal calibration (Frankenberger et al. 1999; Brooks et al. 2007). It uses spatial explicit topographic, land cover, and soil maps to represent the hydrologic mass balance within the root zone of a watershed. It is particularly well suited in landscapes having restrictive soil horizons (e.g. argillic and fragipan) where saturation excess runoff is generated by the subsurface lateral redistribution of water following variable source area hydrology concepts (Brooks et al. 2004, 2007; McDaniel et al. 2008). Despite the modest input requirements, modelling results are at least as good as more complex hydrology models (Johnson et al. 2003; Mehta et al. 2004;

Dijksma et al. 2011). Brooks et al. (2007) demonstrated the ability of the model to accurately represent the development of shallow, dynamic perched water tables over restrictive fragipan soil horizons, including snow accumulation and melt, and surface runoff, from a small grassland catchment on the eastern edge of the Moscow sub-basin.

Dijksma et al. (2011) applied and assessed the ability of the SMR model to represent the hydrology of the Paradise Creek watershed. The Paradise Creek watershed is also located in the Moscow sub-basin (Fig. 2), and covers approximately one third of the total (150 km^2) area. The Nash-Sutcliffe efficiency (NSE; Nash and Suthcliffe 1970) between observed and simulated streamflow from Paradise creek watershed was 0.57, which is a "good" result according to the qualitative assessment of NSE by Foglia et al. (2009). More details can be found in Dijksma et al. (2011).

Since the focus of this study was on the entire Moscow sub-basin, the spatial extent of the Dijksma et al. (2011) model extended to cover the entire region. The model was extended using the same topographic, land cover, soil maps and parameters as had been used as by Dijksma et al. (2011). The daily average percolation was calculated by SMR for each 30 × 30 m cell in the Moscow sub-basin, over the period 2001–2008.

RESULTS

Model

As seen in Fig. 6, the SMR model indicates that percolation rates vary spatially with the greatest average percolation rate of 2 mm d^{-1}, whereas there are other areas that have negligible percolation. The overall average daily percolation over the Moscow sub-basin was 0.50 mm d^{-1}.

Isotopes

Table 2 shows the results of the $\delta^{18}O$ analysis for all wells and springs. The $\delta^{18}O$ values for the wells varied from −16.7 ‰ in well 23 to −14.1 ‰ in well 39. The temporal fluctuations in $\delta^{18}O$ varied in magnitude at each of the sampled wells and springs. The variability in $\delta^{18}O$ was significant and consistent across many of the locations. Figure 7a shows highly dynamic $\delta^{18}O$ fluctuations at four wells that had the highest standard deviation in $\delta^{18}O$ (wells 9, 14, 24 and 27). In contrast, $\delta^{18}O$ in some other wells is relatively stable. Figure 7b shows relatively stable $\delta^{18}O$ fluctuations at four wells that had the lowest standard deviation in $\delta^{18}O$ (wells 11, 35, 36, 37). The standard deviation correlates with the maximum range in $\delta^{18}O$ values

($R^2 = 0.91$). Figure 8b shows the spatial distribution of the maximum range in $\delta^{18}O$ in all wells and springs. Similarly, Fig. 8c shows the spatial distribution of the average *d*-excess in all wells and springs. Interestingly the fluctuations in $\delta^{18}O$ indicated by Fig. 8b, c do not necessarily correlate with well depth (Fig. 8a). For example, deep wells (e.g. wells 35, 29, 9, 14) can have high fluctuations in $\delta^{18}O$ and a high *d*-excess and shallow wells can have stable fluctuations and a low *d*-excess (e.g. wells 1, 41, 36).

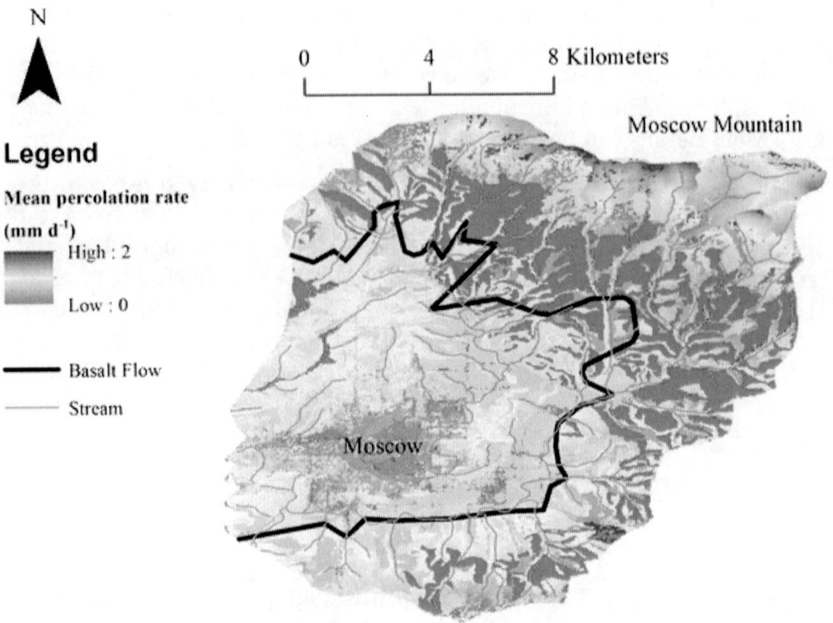

Figure. 6Daily average percolation in the Moscow sub-basin in m d^{-1}. A result from the SMR model

Table 2 Results of water samples from all wells and springs, from 17-04-2013 until 02-01-2014. *SD* standard deviation

Well sample ID	∂¹⁸O ratio			*d*-excess	Mean temp.	Mean EC	*n*
	Mean	SD	Range				
	(‰)	(‰)	(‰)	(‰)	(°C)	(μS cm^{-1})	
1	−15.24	0.13	0.47	7.0	11.4	227	18
5	−16.05	0.11	0.87	7.4	11.6	282	19
6	−14.53	0.10	1.64	8.0	12.7	287	14

8	−15.89	0.20	0.67	7.8	12.3	249	19
9	−15.76	0.30	1.20	7.8	13.9	227	19
11	−15.37	0.10	0.54	8.2	11.0	172	17
14	−15.40	0.28	1.05	8.5	12.8	457	18
17	−15.39	0.10	0.37	8.4	15.1	126	14
19	−15.10	0.18	0.66	7.6	12.8	211	20
20	−15.64	0.10	0.43	8.0	13.4	278	17
21	−15.01	0.16	0.63	6.9	11.7	389	18
22	−15.80	0.13	0.50	7.5	14.2	241	18
23	−16.46	0.12	0.46	7.0	16.1	326	13
24	−15.34	0.25	0.80	7.4	14.0	220	18
26	−15.45	0.19	0.76	7.9	13.6	215	17
27	−15.10	0.22	0.56	7.4	11.8	222	18
28	−15.43	0.10	0.42	8.3	12.1	201	18
29	−15.55	0.22	0.63	8.1	11.9	377	18
35	−15.20	0.10	0.55	9.1	11.9	262	17
36	−14.90	0.10	0.41	7.4	13.6	430	18
37	−16.14	0.07	0.44	8.8	14.1	257	19
39	−14.89	0.26	0.96	6.4	13.7	607	17
41	−15.15	0.12	0.44	7.7	11.7	256	17
60	−14.15	1.12	0.34	−6.9	10.9	268	5

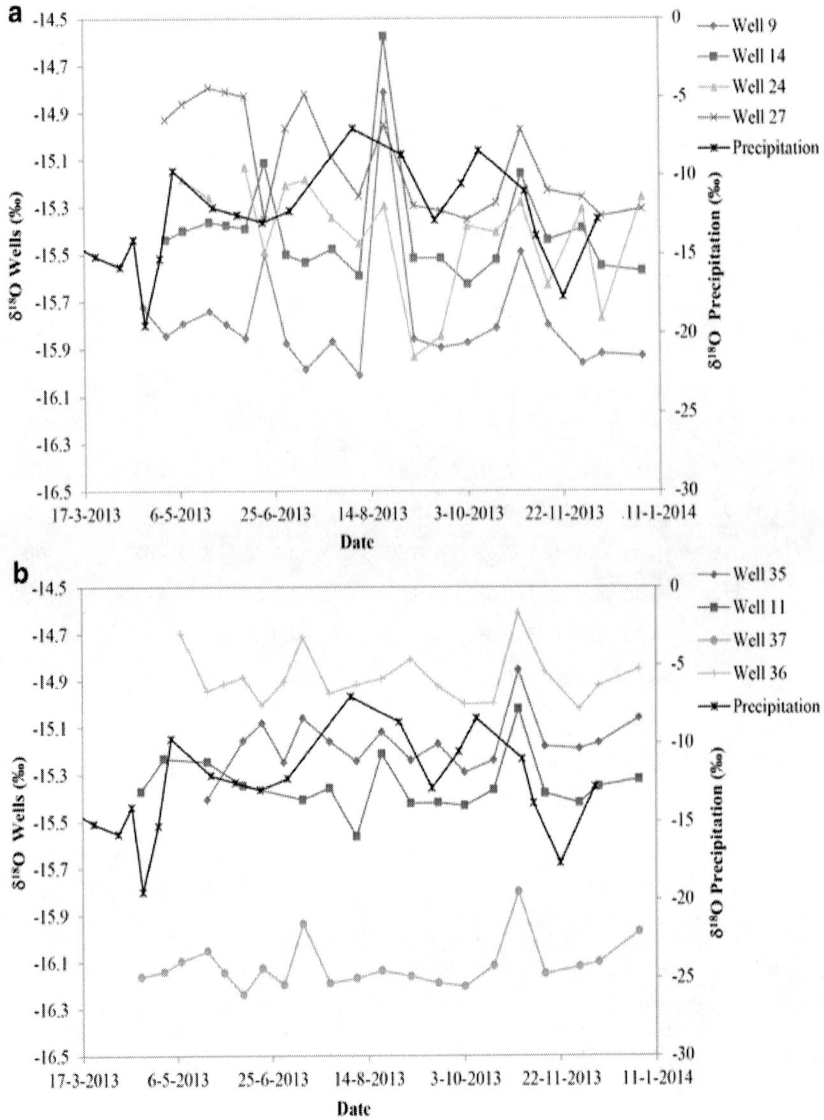

Figure. 7 A $\delta^{18}O$ readings for the four most dynamic wells in the study, including the precipitation $\delta^{18}O$, B $\delta^{18}O$ readings for the four most stable wells in the study, including the precipitation $\delta^{18}O$

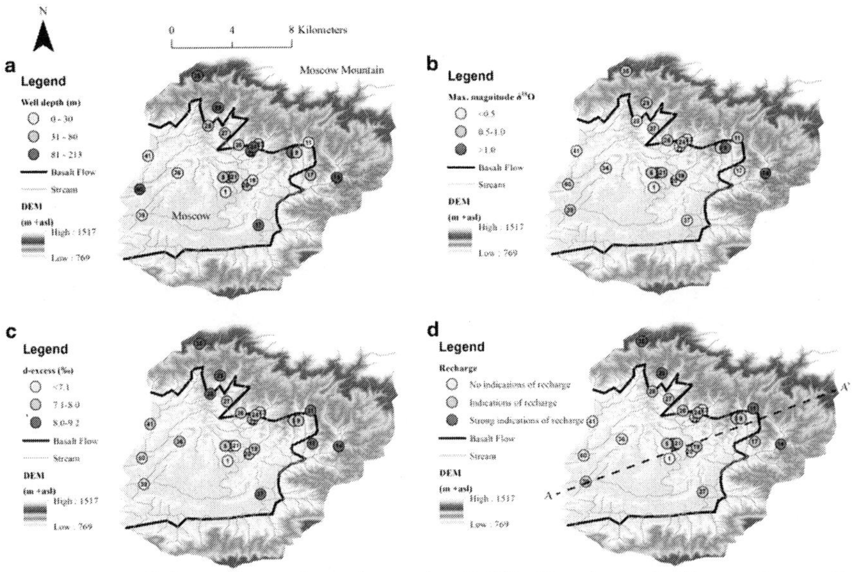

Figure. 8Well locations plotted on the DEM. Numbers indicate the well sample IDs. A the depth of the wells, B the maximum magnitude of the $\delta^{18}O$ composition, C the average *d*-excess for all samples, D indications of recharge

DISCUSSION

Recharge in the Moscow sub-basin

According to the SMR model, most of the percolation occurs at higher elevations near Moscow Mountain. The relatively large percolation at the higher elevations can be partially attributed to the fact that Moscow Mountain receives nearly twice the precipitation (1,270 mm) as the city of Moscow located at relatively low elevation in the study area (WRCC 2013). The model predicts zero percolation in areas having strong hydraulically restrictive argillic soil characteristics. Instead of water moving vertically through these horizons, perched water tables develop in these soils and the water then will run laterally downslope, producing runoff at toe-slope positions. Soils having low percolation exist throughout a large part of the eastern Palouse region between Moscow and the Moscow Mountain, where thick clay layers are present (Lum et al. 1990; Fairley et al. 2006; Dijksma et al. 2011). Percolation is also limited within the city of Moscow due to impervious surfaces which route water through storm drain networks.

The $\delta^{18}O$ data indicate that connections exist between precipitation, surface water and groundwater. It is suggested here that there are two characteristics in the $\delta^{18}O$ data that indicate recharge. Firstly, the magnitude of $\delta^{18}O$ fluctuations, whereby if fluctuations are large, then the groundwater is hydrologically connected to a source of water causing these variations in $\delta^{18}O$. This different source can either be groundwater recharge from precipitation or surface water. As seen in Fig. 7, there is considerable overlap in the range of $\delta^{18}O$ values in wells exhibiting both relatively large and small fluctuations, which suggests that no distinction in recharge can be made based on the average annual $\delta^{18}O$; hence, the time record is needed for this distinction. Interestingly, wells that have small $\delta^{18}O$ fluctuations still have a dampened $\delta^{18}O$, suggesting a slower connection or smaller amount of recharge may be occurring in these wells. The second characteristic that suggests that recharge is occurring is if the *d*-excess is close to 10 ‰. If water percolation occurs rapidly without excessive evaporation, then the groundwater will have a *d*-excess close to 10 ‰.

In order to examine the spatial patterns in recharge, three ranges were defined for each of these recharge indicators varying from no indications of recharge to strong indications of recharge. These categories are shown in Fig. 8b, c with different colors. Yellow, orange and red represent that the indications for recharge are absent, present and strong, respectively. Each of these categories was given a value of 0 to 2, respectively. Figure 8d shows the sum of these values for both recharge indicators in similar categories ranging from yellow (0–1), orange (2–3) and red (4). This method, which assumes that both recharge indicators are equally important, provides a clear overview of potential groundwater recharge source areas in the Moscow sub-basin.

In order to provide a perspective of the importance on well depth and location relative to specific geologic features and groundwater recharge, a geological cross-section is provided of the well locations across the Moscow sub-basin (see Fig. 9). The well locations are simplified and schematically presented in the 2D cross-section. The color of each well represents the same recharge potential as used in Fig. 8d. In order to properly interpret these data, it is important to recognize that well logs indicate most of these wells pull groundwater from coarse interbeds below poorly permeable (e.g. thick clay) layers (Idaho Department of Water Resources 2013). The wells showing the strongest indications of recharge (35, 11, 29 and 14) are all located along the granite/basalt interface near Moscow Mountain and vary largely in depth (Figs. 8d and 9). The location of these wells agrees well with the simulated percolation maps from the SMR model. Further west towards Moscow, the indications of recharge decline, although both shallow springs exhibit strong indications of recharge (6, 19). These springs are likely supplied by upstream recharge. Some wells show no indications of recharge (1, 22, 23, 36). These wells are all relatively shallow and get water from the Sediments of Bovill, which lay above the Wanapum. As most of the water is recharged at the granite/basalt interface (arrow B and C in Fig. 2), a relatively small amount of water recharges the top of the Wanapum where these wells are located. The lack of recharge indications in these wells, combined with the SMR output, also underlines that little to no vertical percolation and recharge pathways occur in this region (arrow A in Fig. 2). Moreover, wells located relatively far from the recharging area do not exhibit any indications of recharge (see Wanapum and Vantage Member wells 20, 41 and 60); however, some Wanapum wells (39, 5) have indications of recharge, explained by the heterogeneity of the Wanapum aquifer (Leek 2006; McVay 2007; Bennett 2009). Overall, the isotope and modelling data both suggest that groundwater recharge is occurring at the basalt/granite interface near the forested uplands in the eastern fringe of the Moscow sub-basin. The well data suggest that this groundwater recharge flows laterally below the poorly permeable argillic horizons and thick clay deposits typically found in the lower agricultural production regions in the Moscow Sub-basin. This lateral flow likely occurs through complex coarse interbed layers or conduits at the granite/basalt interface where these geological units interfinger. Some of these lateral flow paths end up as surficial springs (Dijksma et al. 2011), whereas others likely recharge both aquifers. Other than the forested uplands, the modelling and isotope data indicate that little vertical percolation occurs over much of the Moscow sub-basin. This agrees well with the Fairley et al. (2006) groundwater characterization study which concluded little vertical recharge occurs over most of the Moscow sub-basin due to the presence of widespread, thick clay deposits; however, this study provides added insight that the groundwater recharge is likely occurring as lateral flow beneath these clay deposits along the granite-basalt interface.

Temporally the $\delta^{18}O$ peaks in groundwater are lagged and damped relative to $\delta^{18}O$ peaks in precipitation. Although the dataset is relatively short, Fig. 7a indicates there is a 2–5-week lag between a $\delta^{18}O$ peak in precipitation and a $\delta^{18}O$ peak in groundwater. This lagged $\delta^{18}O$ suggests that the recharge response time may be as short as 2–5 weeks. A longer more complete dataset would be required to determine a characteristic recharge response time for specific regions in the basin.

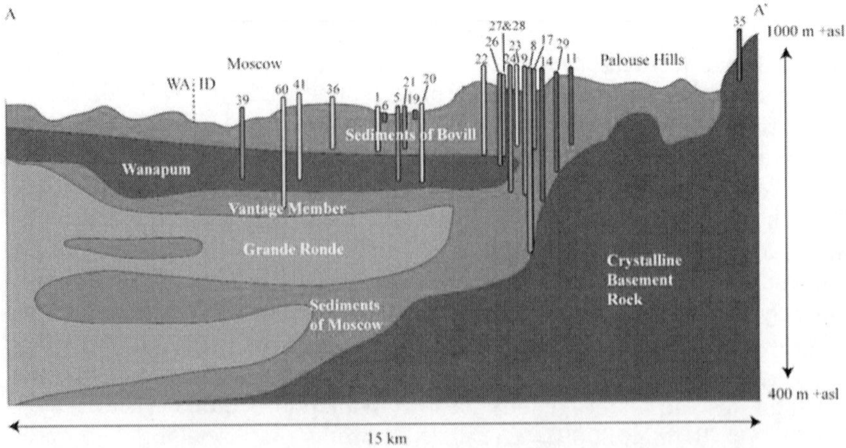

Figure. 9Geological cross-section of the Moscow sub-basin (after Bush and Garwood 2005). All wells and springs have been included. The well locations are simplified and schematic. The *numbers* corresponds to the well sample ID. The *well colours* represent the same as in Fig. 8d. *Red* is strong indications of recharge, *orange* is indications of recharge, *yellow* is no indications of recharge. Cross-section is based on approximate W–E *dashed line* A–A' in Fig. 4

Wider implications

This study shows that permeable linear elements on the interface between impermeable (granitic) bedrock and permeable (basalt) formations can act as recharge pathways. Inducing recharge by artificial infiltration ponds may help to reduce the water level decline or even cause groundwater level rise in such basaltic aquifers. This principle also can be applied to other basaltic regions such as Deccan Traps. Isotope fractioning can help finding suitable locations for infiltration ponds.

CONCLUSION

Biweekly isotope measurements of 22 wells and two springs combined with a soil moisture routing (SMR) model showed strong indications of recharge in the proximity of the Moscow Mountain. The SMR model results suggest that precipitation percolates mainly at the Moscow Mountain, where maximum daily average percolation rates are reached (up to 2.0 mm d^{-1}). The Moscow Mountain provides the source of aquifer recharge, from where water flows through lateral conduits below the poorly permeable layers recharging the basalt aquifers at the granite/basalt interface where these geological units interfinger. This study shows the importance of granite/basalt interface areas for recharge to the basalt aquifers. These locations may be the main sources from where these types of aquifers receive recharge. As the local economy and many individuals depend on groundwater from basalt aquifers, these interface areas should be further explored towards the possibilities of using managed aquifer recharge and aquifer storage methods.

ACKNOWLEDGEMENTS

This research is supported by the Palouse Basin Aquifer Committee (PBAC). Many thanks go to Steve Robischon for arranging the support by PBAC for this research. Furthermore, we thank Craig Woodruff and Todd Anderson for their support during the isotope sampling, as well as all well owners for letting us sampling their wells. Finally, thanks to the anonymous reviewers who gave great input and helped improve this article.

REFERENCES

1. Amelung F, Galloway DL, Bell JW, Zebker HA, Laczniak RJ (1999) Sensing the ups and downs of Las Vegas: InSAR reveals structural control of land subsidence and aquifer-system deformation. Geology 27:483–486.
2. Bara TJ, Shaw D (1995) Multi-resolution land characteristics consortium documentation notebook, ManTech Environmental Technology, Fairfax, VA.
3. Barker RA (1979) Computer simulation and geohydrology of a basalt aquifer system in the Pullman-Moscow basin, Washington and Idaho. Washington Depart Ecol WaterSuppl Bull 48 1979, 119 pp.
4. Bennett B (2009) Recharge implications of strategic pumping of the Wanapum Aquifer System in the Moscow sub-basin. University of Idaho, Moscow, ID.
5. Brooks E, Grader G (2011) Recharge from east revisited. WSU GEOL 598 Seminar, Pullman, WA, March 2011.
6. Brooks ES, Boll J (2005) A simple GIS-based snow accumulation and melt model. Proceedings of the 2005 Western Snow Conference, Great Falls, MT, p 11–14.
7. Brooks ES, Boll J, McDaniel PA (2004) A hillslope-scale experiment to measure lateral saturated hydraulic conductivity. Water Resour Res 40(4).
8. Brooks ES, Boll J, McDaniel PA (2007) Distributed and integrated response of a geographic information system-based hydrologic model in the eastern Palouse region, Idaho. Hydrol Proc 21:110–122.

9. Bush J (2005) The Columbia River Basalt Group of the Palouse Basin with hydrological interpretations, western Latah County, Idaho and eastern Whitman County, Washington. Idaho Geological Survey, Moscow, Idaho.
10. Bush J, Garwood (2005) Various maps, and materials partly published as GSA abstracts, pieces of information not included in geologic maps. GSA, Boulder, CO.
11. Candel J (2014) Identifying hydrologic recharge connections in the Moscow sub-basin. Wageningen University and Research Centre, Wageningen, The Netherlands.
12. Cappa CD, Hendricks MB, DePaolo DJ, Cohen RC (2003) Isotopic fractionation of water during evaporation. J Geophys Res Atmos 108(D16), 4526.
13. Carey L (2011) Evaluation of oxygen and hydrogen isotopes in groundwater of the Palouse Basin and Moscow sub-basin. University of Idaho, Moscow, ID.
14. Changnon SA (1987) Detecting drought conditions in Illinois, Illinois State Water Survey, Champaign, IL.
15. Craig H (1961) Isotopic variations in meteoric waters. Am Assoc AdvSci 1702–1703.
16. Dansgaard W (1964) Stable isotopes in precipitation. Tellus 16:436–468.
17. De Graaf I (2011) Spatial and temporal distribution of subsurface water in the Paradise Creek Watershed. Wageningen University, Wageningen, The Netherland.
18. Dijksma R, Brooks ES, Boll J (2011) Groundwater recharge in Pleistocene sediments overlying basalt aquifers in the Palouse Basin, USA: modeling of distributed recharge potential and identification of water pathways. Hydrogeol J 19:489–500.
19. Dillon P (2005) Future management of aquifer recharge. Hydrogeol J 13:313–316.
20. Dillon P (2009) Water recycling via managed aquifer recharge in Australia. Bol Geol Minero 120:121–130.
21. Eaton L, Melady J, Tolan T (2009) Successful implementation of ASR in basalt-hosted aquifers in the Pacific Northwest of the United States. Bol Geol Minero 120:131–156.
22. El-Naqa A, Al-Momani M, Kilani S, Hammouri N (2007) Groundwater deterioration of shallow groundwater aquifers due to overexploitation in northeast Jordan. Clean Soil Air Water 35:156–166.
23. Fairley J, Solomon M, Hinds J, Grader G, Bush J, Rand A (2006) Latah County hydrologic characterization project. Idaho Department of Water Resources, Boise, ID.
24. Foglia L, Hill M C, Mehl SW, Burlando P (2009) Sensitivity analysis, calibration, and testing of a distributed hydrological model using error-based weighting and one objective function. Water Resour Res 45:W06427.
25. Foster S, Garduno H, Evans R, Olson D, Tian Y, Zhang W, Han Z (2004) Quaternary aquifer of the North China Plain: assessing and achieving groundwater resource sustainability. Hydrogeol J 12:81–93.
26. Frankenberger JR, Brooks ES, Walter MT, Walter MF, Steenhuis TS (1999) A GIS-based variable source area hydrology model. Hydrol Proc 13:805–822
27. Froehlich K, Gibson J, Aggarwal P (2001) Deuterium excess in precipitation and its climatological significance: study of environmental change using isotope techniques. In: Proc. Int. Conf. Study ofEnvironmental ChangeUsing Isotope Techniques, Vienna, April 2001, pp 54–66
28. Gale I, Macdonald D, Calow R, Neumann I, Moench M, Kulkarni H, Mudrakartha S, Palanisami K (2006) Managed aquifer recharge: an assessment of its role and effectiveness in watershed management: final report for DFID KAR project R8169, Augmenting groundwater resources by artificial recharge: AGRAR. British Geological Survey, Keyworth, UK.
29. Gerges N, Dillon P, Sibenaler X, Martin R, Pavelic P, Howles S, Dennis K (2002) South Australian experience in aquifer storage and recovery. In: Dillon PJ (ed) Management of aquifer recharge for sustainability. Balkema, Rotterdam, The Netherlands, pp 453–458.

30. Grader G (2011) Miocene Latah Formation and subsurface geology along the Moscow Subbasin margin: "Recharge from the east" revisited. PRISEM, Spokane, WA.
31. Hansen AJ, Vaccaro JJ, Bauer HH (1994) Ground-water flow simulation of the Columbia Plateau regional aquifer system, Washington, Oregon, and Idaho. US Geol Surv Water Resour Invest Rep 91-4187.
32. Hayashi T, Tokunaga T, Aichi M, Shimada J, Taniguchi M (2009) Effects of human activities and urbanization on groundwater environments: an example from the aquifer system of Tokyo and the surrounding area. Sci Total Environ 407:3165–3172.
33. Hooper PR (1982) The Columbia River basalts. Science 215:1463–1468.
34. Hoque MA, Hoque MM, Ahmed KM (2007) Declining groundwater level and aquifer dewatering in Dhaka metropolitan area, Bangladesh: causes and quantification. Hydrogeol J 15:1523–1534.
35. Idaho Department of Water Resources (2013) Well Log Database. IDWR, Boise, ID.
36. Johnson MS, Coon WF, Mehta VK, Steenhuis TS, Brooks ES, Boll J (2003) Application of two hydrologic models with different runoff mechanisms to a hillslope dominated watershed in the northeastern US: a comparison of HSPF and SMR. J Hydrol 284:57–76.
37. Katsuyama M, Tani M, Nishimoto S (2010) Connection between streamwater mean residence time and bedrock groundwater recharge/discharge dynamics in weathered granite catchments. Hydrol Proc 24:2287–2299.
38. Kendall C, MacDonnell JJ (1998) Isotope tracers in catchment hydrology. Elsevier, Amsterdam
39. Konikow LF, Kendy E (2005) Groundwater depletion: a global problem. Hydrogeol J 13:317–320
40. Laney F, Kirkham V, Piper A (1923) Ground-water supply at Moscow, Idaho. Pamphlet 8, Idaho Bur. Min, Boise, ID.
41. Larson KR, Keller CK, Larson PB, Allen-King RM (2000) Water resource implications of 18O and 2H distributions in a basalt aquifer system. Groundwater 38:947–953.
42. Leek F (2006) Hydrogeological characterization of the Palouse Basin Basalt Aquifer system, Washington and Idaho. Washington State University, Pullman, WA.
43. Lum W, Smoot JL, Ralston DR (1990) Geohydrology and numerical model analysis of ground-water flow in the Pullman-Moscow area, Washington and Idaho. US Geol Surv Invest Rep 89-4103.
44. Macdonald D, Kulkarni H, Lawrence A, Deolankar S, Barker J, Lalwani A (1995) Sustainable groundwater development of hard-rock aquifers: the possible conflict between irrigation and drinking water supplies from the Deccan basalts of India. NERC Technical Report WC/95/52, BGS, Keyworth, UK.
45. Maliva RG, Herrmann R, Coulibaly K, Guo W (2014) Advanced aquifer characterization for optimization of managed aquifer recharge. Environ Earth Sci 73:7759–7767.
46. McDaniel P, Regan M, Brooks E, Boll J, Barndt S, Falen A, Young S, Hammel J (2008) Linking fragipans, perched water tables, and catchment-scale hydrological processes. Catena 73:166–173.
47. McDonnell JJ, Kendall C (1992) Isotope tracers in hydrology. Elsevier, Amsterdam, pp 260–261.
48. McVay M (2007) Grande Ronde aquifer characterization in the Palouse Basin: MSc Thesis, University of Idaho, Moscow, ID.
49. Mehta VK, Walter MT, Brooks ES, Steenhuis TS, Walter MF, Johnson M, Boll J, Thongs D (2004) Application of SMR to modeling watersheds in the Catskill Mountains. Environ Model Assess 9:77–89.
50. Moxley N (2012) Stable isotope analysis of surface water and precipitation in the Palouse basin: hydrologic tracers of aquifer recharge. MSc Thesis, Washington State University, Pullman, WA.

51. Nash JE, Sutcliffe JV (1970) River flow forecasting trough conceptual models, part I: a discussion of principles. J Hydrol (Amsterdam) 10:282–290.
52. Othberg KL, Breckenridge RM, Freed JS, Stanford LR (2001) Surficial geologic map of the Moscow East Quadrangle and part of the Moscow West Quadrangle, Latah County, Idaho. Idaho Geological Survey, University of Idaho, Boise, ID.
53. Pavelic P, Patankar U, Acharya S, Jella K, Gumma MK (2012) Role of groundwater in buffering irrigation production against climate variability at the basin scale in South-West India. Agri Water Manage 103:78–87.
54. Qureshi AS, McCornick PG, Sarwar A, Sharma BR (2010) Challenges and prospects of sustainable groundwater management in the Indus Basin, Pakistan. Water Resour Manage 24:1551–1569.
55. Rahman MA, Rusteberg B, Gogu R, Ferreira JL, Sauter M (2012) A new spatial multi-criteria decision support tool for site selection for implementation of managed aquifer recharge. J Environ Manage 99:61–75.
56. Reidel SP, Martin BS, Petcovic HL (2003) The Columbia River flood basalts and the Yakima fold belt: Western Cordillera and adjacent areas. In: Swanson TW (ed) Geological Society of America Field Guide. GSA, Boulder, CO, pp 87–105.
57. Robischon S (2006) 2002–2005 Palouse Groundwater Basin Use Report Moscow (ID). Palouse Basin Aquifer Committee, Moscow, ID.
58. Robischon S (2007) Palouse Basin Ground Water Management: a history. Palouse Basin Water Summit, Moscow, ID.
59. Sánchez-Murillo R, Brooks ES, Elliot WJ, Boll J (2015) Isotope hydrology and baseflow geochemistry in natural and human-altered watersheds in the Inland Pacific Northwest, USA. Isotopes Environ Health Stud 51:231–254.
60. Scanlon BR, Healy RW, Cook PG (2002) Choosing appropriate techniques for quantifying groundwater recharge. Hydrogeol J 10:18–39.
61. Shah T, Molden D, Sakthivadivel R, Seckler D (2000) Groundwater: overview of opportunities and challenges. IWMI, Colombo, Sri Lanka.
62. Soulsby C, Tetzlaff D (2008) Towards simple approaches for mean residence time estimation in ungauged basins using tracers and soil distributions. J Hydrol 363:60–74.
63. Speed M, Tetzlaff D, Soulsby C, Hrachowitz M, Waldron S (2010) Isotopic and geochemical tracers reveal similarities in transit times in contrasting mesoscale catchments. Hydrol Proc 24:1211–1224.
64. Tolan TL, Reidel SP, Beeson MH, Anderson JL, Fecht KR, Swanson DA (1989) Revisions to the estimates of the areal extent and volume of the Columbia River Basalt Group. GSA, Boulder, CO, pp 1–20.
65. US Geological Survey (2012) Discharge measurements 1935–2012. US Geological Survey, Reston, VA.
66. UNESCO (2009) The United Nations World Water Development report 3: water in a changing world. Earthscan, UNESCO, Paris.
67. Vaccaro JJ (1999) Summary of the Columbia Plateau regional aquifer-system analysis, Washington, Oregon, and Idaho. US Geological Survey, Reston, VA.
68. Van Loon A F, Gleeson T, Clark J, Van Dijk AIJM, Stahl K, Hannaford J, Di Baldassarre G, Teuling AJ, Tallaksen LM, Uijlenhoet R, Hannah DM, Sheffield J, Svoboda M, Verbeiren B, Wagener T, Rangecroft S, Wanders N, Van Lanen HAJ (2016) Drought in the Anthropocene. Nature Geosci 9:89–91.
69. Wada Y, Beek L, Bierkens M F (2012) Nonsustainable groundwater sustaining irrigation: a global assessment. Water Resour Res 48(6):W06901
70. Wassenaar L, Athanasopoulos P, Hendry M (2011) Isotope hydrology of precipitation, surface and ground waters in the Okanagan Valley, British Columbia, Canada. J Hydrol 411:37–48.
71. Whiteman K, Vaccaro J, Gonthier JB, Bauer HH (1994) The hydrogeologic framework and geochemistry of the Columbia Plateau aquifer system, Washington, Oregon, and Idaho. US Geol Surv Prof Pap 1413-B.

72. Willis KG, Garrod GD (1998) Water companies and river environments: the external costs of water abstraction. Utilities Pol 7:35–45.
73. WRCC (2013) Monthly climate averages of Idaho, 1893–2013. WRCC, Sun Valley, NV.

CHAPTER 7

Review of Soil Water Retention Characteristic (SWRC) Models between Saturation and Oven Dryness

Vitalis K. Too*, Christian T. Omuto , Elijah K. Biamah , John P.Obiero

Department of Environmental and Biosystems Engineering, University of Nairobi, Nairobi, Kenya

ABSTRACT

Soil water retention characteristic is the key soil property used in many applications in the fields of irrigation, hydrology, geotechnical engineering and soil science in general. Since the advent of digital soil mapping and digital soil assessment paradigms, there has been an upsurge of development of soil inference models and the need to increase accurate application of soil mapping products. All soils can be partially saturated with water and also near oven dryness. Therefore, constitutive models for soils should ideally represent the soil behaviour over entire range. This paper reviewed commonly used SWRC models. In order to stem potential for biasness, the models were grouped into three categories depending on the number of fitting parameters, namely, five- parameter, four-parameter and three-parameter categories. The evaluation used correlation and residual standard error statistics to choose the best overall performing model and in each category. Its results serve as a guide for selecting the models to be preferred for fitting SWRC in case there are limitations to the number of suction potential levels in the measured data.

Keywords: Water retention, Hydraulic parameters, Models

1. INTRODUCTION

Soil water characteristic is the key soil property used in many applications in the fields of irrigation, hydrology, geotechnical engineering and soil science in general. Soil water characteristic influences many physical and engineering characteristics which are key soil properties for appropriate management of the soil degradation, water resources and environment. Water content in the soil is held by two principal forces: adsorption and capillarity. Adsorption is responsible for moisture held on the surface of soil particles while capillarity holds water in the pores between the soil particles or between aggregates of the particles [1]. The relationship between combined effect of these forces (also often expressed in the form of potential head) and soil water content forms a basic property of soil referred to as the Soil Water Retention Characteristics (SWRC) [2]. Typically, soils exhibit large variations in space and occasionally in time as well. Consequently, soil water retention characteristics models have been developed that allow a physical description of soil properties with the least amount of effort.

Some of the commonly used analytical models of SWRC in the literature are Brooks-Corey, Farrell-Larson, Fredlund-Xing, Gardner, McKee-Bumb, Kosugi, van Genuchten, Campbell, Russo, Tani, Omuto, and Dexter models [3] - [5]. Many of these models have been reported successful only for certain parts of SWRC (e.g. wet or dry parts), soil textures, soil horizons, regions of the world, or limited soil stress history [6]. These models contain shape parameters known as hydraulic parameters. Applications of water retention models require knowing these hydraulic parameters. Soil hydraulic parameters are traditionally obtained by curve-fitting the water retention functions using experimental data [7].

Presently, there are many models in the literature and many more still being developed so much so that it is confusing about which model to prefer. The current choice of the appropriate model to use has not been well guided by any incisive research on the predictive performance of these models, especially between saturation and oven dryness. Consequently, SWRC model applications have been largely moved by convenience. It is worth noting that the difficulties of soil characterization increase when soil approaches oven dryness conditions a situation often encountered in arid and semi-arid areas. The cost of establishing the soil water characteristics can be relatively high in this area. The objective of this study was to review SWRC models between saturation and oven dryness so as to establish their performance and to recommend the most suitable models to be used in a given range.

2. THEORY

2.1. Features of soil water retention characteristics

Soil water retention characteristic is the relationship between amount of water held in the soil and the forces holding it. The graph giving the relation between soil water tension and soil water content is called water retention curve or soil water characteristic. If the tension is expressed as the logarithmic value of cm water, the graph is referred to as a pF-curve. Water retention curves are used:

1) To determine an index of the available water in soil and to classify soils accordingly, e.g. for irrigation purposes, construction etc.;
2) To determine the drainable pore space (effective pore space, effective porosity, specific yield) for drainage design;
3) To check changes in the structure of a soil, e.g. caused by tillage, mixing of soil layers etc.;
4) To ascertain the relation between soil water tension and other physical properties of soil (e.g. capillary conductivity, thermal conductivity, clay and organic matter content).

The maximum amount of moisture that can be held in the soil pores is denoted by the SWRC models as saturated moisture content, θ_s. A point h_a in the models where air begins to enter the soil pores (during drainage) or where all air is expelled (during wetting) is known as the air-entry potential (Figure 1). h_a and θ_s are soil hydraulic properties describing the shape of the wet-end of SWRC curves. Between h_a and the point of complete dryness, the rate of moisture release or entry in the soil is fast or slow depending on the range of pore sizes in the soil matrix, assuming that other factors such as stress history are constant [1] [5]. For example, with narrow pore-size distribution such as in sandy soil the rate is fast (and hence the part of WRC curve between h_a and point of complete dryness has a steep slope) and in broad pore-size distribution such as in clay soil the rate is slow (and WRC curve between h_a and point of complete dryness has a gentle slope). The rate of moisture release is very slow towards the end of a completely dry soil; thus, the slope is very gentle for this section of WRC. The slope of the dry section of SWRC intersects the slope of the section after h_a at a point where the corresponding moisture content is known as residual moisture content (θ_r) [5] [8] [9] (Figure 1).

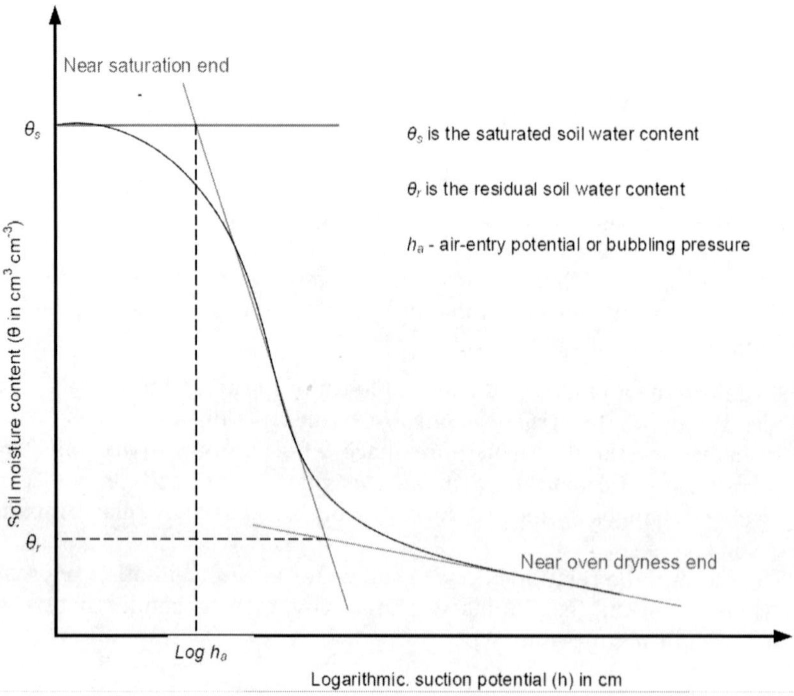

Figure 1. Soil water characteristic curve.

2.2. SWRC models

Quite a number of SWRC models have been proposed over the years to describe the soil water retention charac teristic across the complete range of soil water contents. Some of these models are new while others are modification of the existing models. Some of these models are discussed below.

2.2.1. van Genuchten water retention model

A commonly used retention model is the van Genuchten [10] closed form analytical expression. The closed form equation consists of four independent parameters which have to be estimated from observed soil water retention data. Many curve fitting and parameter optimization codes such as RETC software [7] are widely used today. These types of relationships are empirical in nature with a physical basis. The van Genuchten [10] equation is expressed as:

$$\theta(h) = \theta_r + (\theta_s - \theta_r)\left[1 + (\alpha h)^n\right]^{-m} \tag{1}$$

where, θ is the volumetric water content; h is the pressure head; θ_s and θ_r represent the saturated and residual water contents, respectively; α, n and m are empirical shape parameters.

2.2.2. Brooks-Corey water retention model

Another well-established parametric model was proposed by Brooks and Corey [11]. This is a four-parameter water retention model. A Brooks-Corey model is a type of nonlinear curve fitting model for fitting water retention characteristics using experimental data. The Brooks-Corey functions can be defined as:

$$\theta(h) = \theta_r + (\theta_s - \theta_r)(\alpha h)^{-\lambda} \tag{2}$$

Where, θ_r is the residual water content (cm^3/cm^3), θ_s is the saturated water content (cm^3/cm^3), h is the matric potential (cm), λ and α are empirical shape parameters.

2.2.3. Fredlund-Xing Water Retention Model

This is a five-parameter water retention model [8].

$$\theta = \theta_r + \frac{\theta_s - \theta_r}{\left\{\ln\left[e + (h/\alpha)^n\right]\right\}^m} \tag{3}$$

where, θ_r is the residual water content (cm^3/cm^3); θ_s is the saturated water content (cm^3/cm^3); h is the suction; cm, α, n and m are empirical shape parameters.

The Fredlund-Xing water retention model was developed based on pore size distribution of the soil. If the pore size distribution of a soil can be obtained or predicted, then the soil water characteristic curve is uniquely determined from the equation above.

2.2.4. Gardner Water Retention Model

This is a four-parameter water retention model by Gardner.

$$\theta(h)=\theta_r+(\theta_s-\theta_r)\left[1+(\alpha h)^n\right]^{-1}$$
(4)

2.2.5. Biexponential water retention model

Biexponential water retention model was developed by Omuto. This is a five-parameter water retention model contained in a bimodal pore-size distribution. The parameters are for the first and second compartments [5].

$$\theta(h)=\theta_r+\theta_1 e^{-\alpha_1 h}+\theta_2 e^{-\alpha_2 h}$$
(5)

where, θ_1 represents the difference between saturated moisture θ_{s1} and residual moisture contents θ_{r1} in

the structural pore-space; θ_2 represents the difference between saturated moisture θ_{s2} and residual moisture contents θ_{r2} in the textural pore-space; α_1 represents the inverse of air-entry potential in the structural pore-space; α_2 represents the inverse of air-entry potential in the soil textural pore-space; θ_r is the sum of residual moisture contents in the structural pore-space θ_{r2} and textural pore-space θ_{r2}.

2.2.6. Campbell water retention model

This is a three-parameter water retention model [5].

$$\theta(h)=\theta_s(\alpha h)^{\lambda}$$
(6)

where, h is suction potential/head; θ_s is the saturated moisture content. It's the moisture content when suction potential is very low (almost at the saturation point); α is the inverse of h_a air-entry potential or bubbling pressure, λ is a parameter or index for the pore-size distribution.

2.2.7. Tani water retention model

This is a three-parameter water retention model developed by Tani in 1982 [5].

$$\theta=\theta_r+(\theta_s-\theta_r)\left[1+(\alpha h)e^{-\alpha h}\right]$$
(7)

where, h Suction potential/head; θ_r is the residual moisture content. It's the moisture content when suction potential is very high (almost at the drying point); θ_s is the saturated moisture content. It's the moisture content when suction potential is very low (almost at the saturation point); α is the inverse of air-entry potential or bubbling pressure.

2.2.8. Kosugi water retention model

This is a four-parameter water retention model developed by Kosugi in 1999 [5].

$$\theta(h)=\theta_r+\frac{1}{2}(\theta_s-\theta_r)erfc\left[\frac{\ln(h/h_m)}{\sigma\sqrt{2}}\right] \tag{8}$$

where, θ_s is the saturated soil water content; θ_r is the residual soil water content; h_m is the matric potential corresponding to the median pore radius; σ is a dimensionless parameter to characterize the width of the pore-size distribution; erfc denotes the complementary error function.

2.2.9. Ruso water retention model

Ruso water retention model is a four-parameter function. This is a type of nonlinear curve fitting model for fitting water retention characteristics using experimental data [12].

$$\theta(h)=\theta_r+(\theta_s-\theta_r)\left[(1+0.5\alpha h)\mathrm{e}^{0.5\alpha h}\right]^{2/n+2} \tag{9}$$

where, h is suction potential/head as contained in the x-column of the xy water retention table or data; θ_r is the residual moisture content. It's the moisture content when suction potential is very high (almost at the drying point); θ_s is the saturated moisture content. It's the moisture content when suction potential is very low (almost at the saturation point); α is the inverse of air-entry potential or bubbling pressure; n is a parameter or index for the pore-size distribution.

2.2.10. Exponential model

This is a three-parameter water retention model [12].

$$\theta(h)=(\theta_r+\theta_s)\mathrm{e}^{-\alpha h} \tag{10}$$

where, h is suction potential/head as contained in the x-column of the xy water retention table or data; θ_r is the residual moisture content; θ_s is the saturated moisture content; α is the inverse of air-entry potential or bubbling pressure (Table 1).

3. DATA SOURCES

Water retention data used were downloaded from the ISRIC database on May 6 2013 and locally obtained data from Athi River (Kenya).The data contained soil water content (θ in cm ^{3} cm $^{-3}$) at various levels of suction potential (h) in cm (Figure 2(a) and Figure 2(b)).

The global datasets were further split into samples from Central America, South America, Africa, Europe, Middle East, North Asia and South Asia (Figure 2(c)).

4. PERFORMANCE OF SWRC MODELS USING A TEST DATA

The models' performance was assessed based on their accuracy and prediction potential between saturation and oven dryness.

The best performance of each SWRC model was the one which gave the minimum Residual Standard Error, RSE and a correlation, R^2 close to unity. The following popular statistics were used:

1) Residual standard error, RSE

$$\sum_{i=1}^{n}(y_i-\hat{y}_i)^2/(n-2) \qquad (11)$$

2) Correlation, R^2

$$\left(n\sum_{i=1}^{n}y_i\hat{y}_i-\left(\sum y_i\right)\left(\sum \hat{y}_i\right)\right)^2 \Big/ \left[\left(n\left(\sum y_i^2\right)-\left(\sum y\right)^2\right)-\left(n\left(\sum \hat{y}_i^2\right)-\left(\sum \hat{y}_i\right)^2\right)\right] \qquad (12)$$

The SWRC models were grouped based on the number of fitting parameters as shown in table 2. All the models were fitted using ISRIC global dataset and locally obtained data from Athi River (Kenya) using Hydro Me, R computer program. The performance of the models were reviewed based on the Residual Standard Error (RSE) and correlation coefficient (R^2) in the entire soil range and also at near saturation to oven dryness.

Table 1. SWRC models.

Reference	SWRC model	Parameters
Campbell (1974)	$\theta(h)=\theta_s(\alpha h)^{\lambda}$	$\theta_s, \alpha, \lambda$
Tani (1982)	$\theta(h)=\theta_r+(\theta_s-\theta_r)\left[1+(\alpha h)e^{-\alpha h}\right]$	$\theta_r, \theta_s, \alpha,$
Exponential	$\theta(h)=(\theta_r+\theta_s)e^{-\alpha h}$	$\theta_r, \theta_s, \alpha,$
van Genuchten (1980)	$\theta(h)=\theta_r+(\theta_s-\theta_r)\left[1+(\alpha h)^n\right]^{-\left(1-\frac{1}{n}\right)}$	$\theta_r, \theta_s, \alpha, n$
Gardner (1958)	$\theta(h)=\theta_r+(\theta_s-\theta_r)\left[1+(\alpha h)^n\right]^{-1}$	$\theta_r, \theta_s, \alpha, n$
Ruso	$\theta(h)=\theta_r+(\theta_s-\theta_r)\left[(1+0.5\alpha h)e^{-0.5\alpha h}\right]^{2/(n+2)}$	$\theta_r, \theta_s, \alpha, n$
Brooks-Corey (1964)	$\theta(h)=\theta_r+(\theta_s-\theta_r)(\alpha h)^{-\lambda}$	$\theta_r, \theta_s, \alpha, \lambda$
Kosugi (1999)	$\theta(h)=\theta_r+\frac{1}{2}(\theta_s-\theta_r)erfc\left[\frac{\ln(h/h_m)}{\sigma\sqrt{2}}\right]$	$\theta_r, \theta_s, \sigma, h_m$
Fredlund-Xing (1994)	$\theta(h)=\theta_r+\frac{\theta_s-\theta_r}{\left\{\ln\left[2.7183+(\alpha h)^n\right]\right\}^m}$	$\theta_r, \theta_s, \alpha, n, m$
van Genuchten (1980)	$\theta(h)=\theta_r+(\theta_s-\theta_r)\left[1+(\alpha h)^n\right]^{-m}$	$\theta_r, \theta_s, \alpha, n, m$
Biexponential (Omuto, 2009)	$\theta(h)=\theta_r+\theta_{s1}e^{-\alpha_1 h}+\theta_{s2}e^{-\alpha_2 h}$	$\theta_{s1}, \theta_{s2}, \alpha_1, \alpha_2, \theta_r$

Several attempts have been made to derive models for the purpose of analyzing the soil water characteristics in the entire range between saturation and near oven dryness. Some models have proved to perform poorly in near saturation and others in near oven dryness. This study analyzed the performance of 10 SWRC models on their performance in the entire soil range, near saturation and near oven dryness.

4.1. Between Saturation and oven dryness

Table 2 shows that Biexponential model (Omuto, 2009) had the best performance statistics for the category with five-parameter models. The Gardner (1958) model had the best performance statistics in the four-parameter models category and the Exponential model had the best performance statistics in the three-parameter categories.

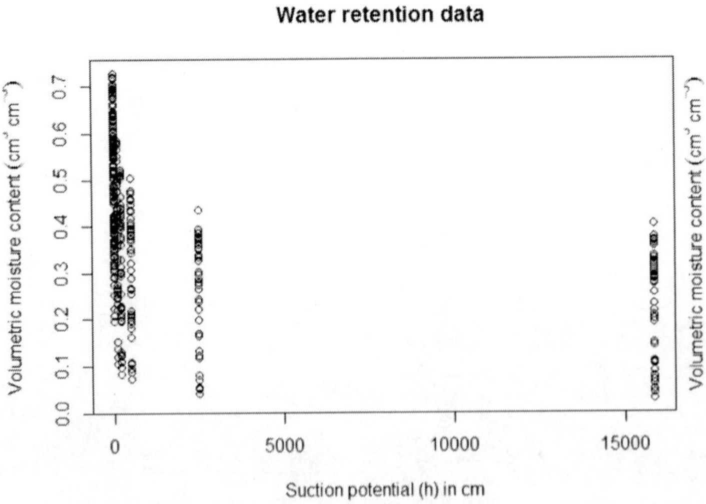
Water retention data
Volumetric moisture content (cm³ cm⁻³)
0.0
0.1
0.2
0.3
0.4
0.5
0.6
0.7
0
5000
10000
15000
Suction potential (h) in cm

(a)

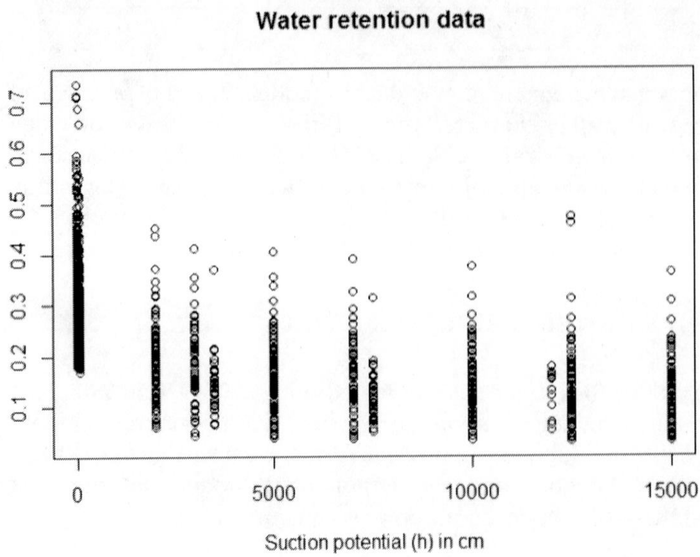
Water retention data
0.1
0.2
0.3
0.4
0.5
0.6
0.7
0
5000
10000
15000
Suction potential (h) in cm

(b)

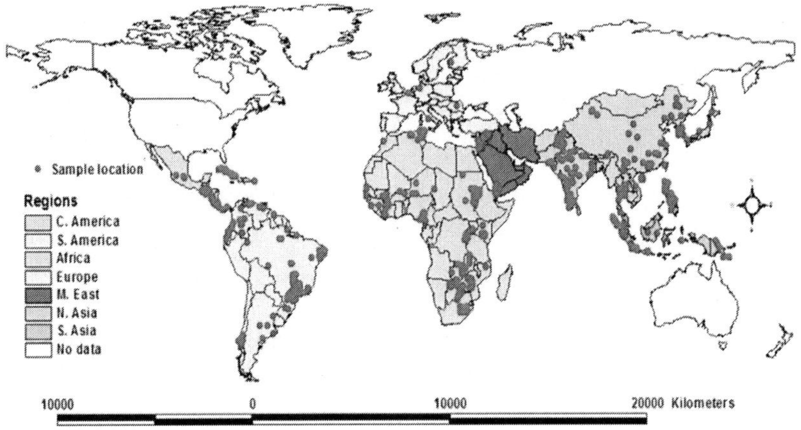

(c)

Figure 2. (a) Water retention data from Isric global data base; (b) Water retention data from Athi River, Kenya; (c) Spatial distribution of data samples.

Table 2. Performance of SWRC models.

Reference	Overall		Near saturation		Near oven dryness	
	RSE	R^2	RSE	R^2	RSE	R^2
Three-parameter models						
Campbell (1974)	0.08134	0.5246984	0.007665	0.8796749	0.003703	0.9788181
Tani (1982)	0.07873	0.5661897	0.004983	0.9509574	0.004749	0.9648825
Exponential	0.07792	0.5783668	0.003783	0.9720321	0.003793	0.9777493
Four-parameter models						
Gardner (1958)	0.016564	0.9893501	0.002568	0.991553	0.009743	0.9905466
Ruso (1998)	0.07796	0.577969	0.004351	0.9703861	0.00445	0.9770206
Brooks-Corey(1964)	0.07991	0.548148	0.008066	0.8941245	0.003813	0.9831745
Kosugi (1999)	0.07988	0.5486333	0.002266	0.9920582	0.003795	0.9833413
Five-parameter models						
Fredlund-Xing (1994)	0.07703	0.5919019	0.002646	0.9918761	0.003523	0.9904626
van Genuchten (1980)	0.07705	0.5914976	0.002568	0.992349	0.004789	0.9823130
Biexponential (Omuto, 2009)	0.017174	0.9885479	0.002463	0.9906058	0.002447	0.9931029

Review of these SWRC model performances in the entire soil range indicated that Gardner (1958) and Biexponential model had the highest R^2 value (Figure 3) compared to all other tested models. Generally, five-parameter models predicted the measured SWRC better than the others in the entire soil range. This could be true because many parameters tend to pin the model as close to the inflection points and end-points of the SWRC as possible. Hence, they allow the models to pass closest to the measured data compared to models with few parameters.

4.2. Near saturation

As shown in Table 2, van Genuchten (1980) model had the highest statistical value in the five-parameter category while Kosugi (1999) model performed better in the four-parameter category. Exponential model also performed best in the three-parameter category. Generally, the van Genuchten model predicts soil water characteristic more accurately on near saturation as compared to other SWRC models.

In Figure 4(a), the soil θ (h) curve expressed by Equation (1) (van Genuchten) does not account for the bubbling pressure but does have an inflection point, allowing the van Genuchten model to perform better than other models (for example Brooks and Corey) at near saturation soil range (Figure 4(b)).

It is worth noting that Equations (1) and (2) have considerable difficulty in representing the retention of water near oven dryness (matric suction at 15000 cm —permanent wilting point) giving an unrealistic path of the retention curve. van Genuchten and Brooks and Corey models does well when applied in wetland studies or humid region agriculture but for applications such as water flow in arid and semi-arid areas require a realistic representation for hydraulic characteristics over the all range of saturation.

4.3. Near oven dryness

Further analysis of the SWRC models' performance near oven dryness showed that Campbell (1974) performed well in the three-parameter category (Table 2). According to Omuto (2009), anchoring the SWRC at the end- points tends to improve the model performance. This is contrasted by the performance of the Campbell (1974) model in near oven dryness range as shown in Figure 5(a) and Figure 5(b). The model performed best compared to its competitors in the three-parameter model category. The study proved right the argument by some authors that elimination of the residual moisture content (q_r) can improve the predictive performance (see for example, [8]).

Gardner (1958) and Biexponential models outperformed the other models in their respective categories. In overall, Biexponential model was the best model in this soil range.

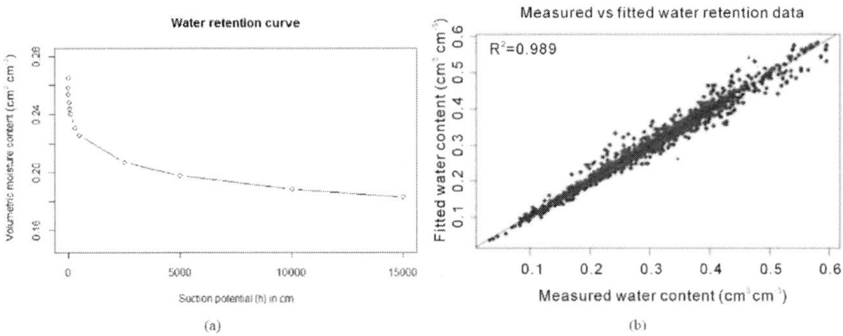

Figure 3. Gardner (1958) SWRC model.

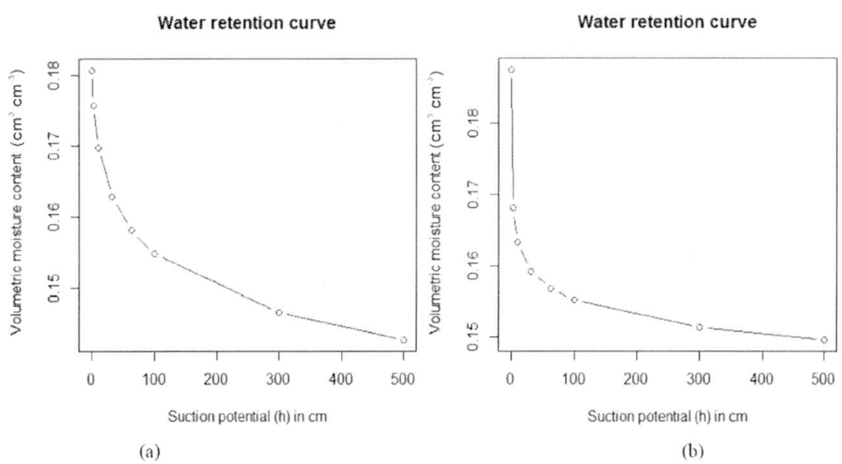

Figure 4. (a) van Genuchten model curve; (b) Brooks and Corey model curve.

It is worth noting that Biexponential model showed a lot of consistency in the three soil categories (overall, near saturation and oven dryness) by having the best fit. These results show that the model is capable of fitting the entire range of SWRC without numerical discontinuities that have been reported in other models [13]. The performance can be explained by the fact that this model is a bi-compartment model (Equation (5)) that assumes two pore-size distribution functions for the soil pores, i.e. for structural pore space and for the textural pore space unlike most SWRC models in the literature which ignore the structural pore spaces resulting in inadequate representation of the true soil water retention characteristics for field condition yet the soil pore architecture in field conditions comprises of both the textural and structural pore spaces.

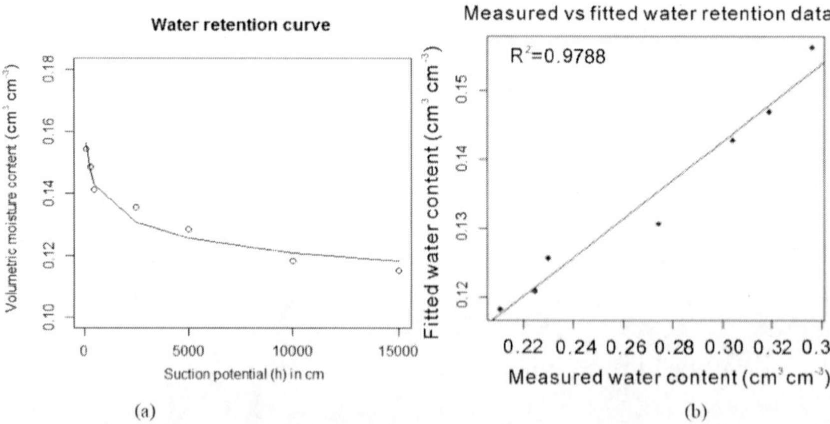

Figure 5. (a) Campbell SWRC model curve; (b) Measured vs fitted SWRC data.

5. CONCLUSIONS

It was observed that all five-parameter models performed much better than the four- and three-parameter models. Among the five-parameter models, Biexponential model (Omuto, 2009) performed better than the van Genuchten (1980), and Fredlund and Xing (1994). In the four-parameter category, Gardner (1958) performed better than Brooks and Corey (1964), Kosugi (1999) and Ruso (1998). Among the three-parameter category, Exponential model performed better than Campbell (1974) and Tani (1982).

In overall, Biexponential model performed better than other models. This is because the model is a bi-compartment model that assumes two pore-size distribution functions for the soil pores, i.e. for structural pore space and for the textural pore space unlike most SWRC models reviewed which ignore the structural pore spaces resulting in inadequate representation of the true soil water retention characteristics for field condition. Yet, the soil pore architecture in field conditions comprises of both the textural and structural pore spaces. Owing to the good performance of this model, the review also floats a suggestion for future models to develop models based on two pore-size distribution functions for the soil pores.

ACKNOWLEDGEMENTS

Vitalis K. Too, Christian T. Omuto , Elijah K. Biamah , John P. Obiero (ISRIC). The organization and countries which contributed to this dataset are highly acknowledged.

REFERENCES

1. Marshall , T.J. , Holmes , J.W. and Rose , W.C. (1996) Soil Physics. Cambridge University Press, London. http://dx.doi.org/10.1017/CBO9781139170673
2. Omuto, C.T. and Gumbe, L.O. (2009) Estimating Water Infiltration and Retention Characteristics Using a Computer Program in R. Computers and Geosciences, 35, 579-585. http://dx.doi.org/10.1016/j.cageo.2008.08.011
3. Leong, E.C. and Rahardjo, H. (1997) Review of Water Characteristic Curve Functions. Geotechnical and Geo-Environmental Engineering, 123, 1106-1117.
4. Dexter , A.R. , Czyz, E.A. , Richard , G. and Reszkowska, A. (2008) A User-Friendly Water Retention That Takes Account of the Textural and Structural Pore Spaces in Soil. Geoderma, 143, 243-253. http://dx.doi.org/10.1016/j.geoderma.2007.11.010
5. Omuto, C.T. (2009) Biexponential Model for Water Retention Characteristics. Geoderma, 149, 235-242. http://dx.doi.org/10.1016/j.geoderma.2008.12.001
6. Khlosi, M., Wim, M.C. , Ahmed , D., Martinus, T. and Donald , G. (2008) Performance Evaluation of Models That Describe the Soil Water Retention Curve between Saturation and Oven Dryness. Vadose Zone Journal, 7, 87-96.
7. van Genuchten , M.T. , Leij, F.J. and Yates , S.R. (1991)The RETC Code for Quantifying the Hydraulic Functions of Unsaturated Soils. EPA Report, 600/2-91/065, USA Salinity Laboratory, USDA.
8. Fredlung, D.G. and Xing, A. (1994) Equations for the Soil Water Characteristic Curve. Canadian Geotechnical Journal, 31, 521-532. http://dx.doi.org/10.1139/t94-061
9. Vanapalli, S.K. , Sillers, W.S. and Fredlund , M.D. (1998) The Meaning and Relevance of Residual State to Unsaturated Soils. 51st Canadian Geotechnical Conference, Edmonton, 4-7 October 1998, 1-8.
10. van Genuchten , M.T. (1980) A Closed-Form Equation for Predicting the Hydraulic Conductivity of Unsaturated Soils. Soil Science Society of America Journal, 44, 892-898. http://dx.doi.org/10.2136/sssaj1980.03615995004400050002x
11. Brooks , R.H. and Corey , A.T. (1964) Hydraulic Properties of Porous Media. Hydrology Paper, Vol. 3, Colorado State University, Fort Collins.
12. Omuto, C.T. (2007) HydroMe: Estimation of Soil Hydraulic Parameters from Experimental Data. R Comprehensive R Archive Network, USA.
13. Vogel, T., van Genuchten, M.T. and Cislerova, M. (2001) Effect of the Shape of the Soil Hydraulic Functions near Saturation on Variably Saturated Flow Predictions. Advances in Water Resources, 24, 133-144.

CHAPTER 8

Groundwater Flow into Underground Openings in Fractured Crystalline Rocks: an Interpretation Based on Long Channels

John H. Black[1] Nicholas D. Woodman[2]

[1] *In Site Hydro Ltd, East Bridgford, Nottingham NG13 8NW, UK*

[2] *Department of Engineering and Environment, University of Southampton, University Road, Southampton SO17 1BJ, UK*

ABSTRACT

Rethinking an old tracer experiment in fractured crystalline rock suggests a concept of groundwater flow in sparse networks of long channels that is supported by results from an innovative lattice network model. The model, HyperConv, can vary the mean length of 'strings' of connected bonds, and the gaps between them, using two independent probability functions. It is found that networks of long channels are able to percolate at lower values of (bond) density than networks of short channels. A general relationship between mean channel length, mean gap length and probability of percolation has been developed which incorporates the well-established result for 'classical' lattice network models as a special case. Using parameters appropriate to a 4-m diameter drift located 360 m below surface at Stripa Mine Underground Research Laboratory in Sweden, HyperConv is able to reproduce values of apparent positive skin, as observed in the so-called Macropermeability Experiment, but only when mean channel length exceeds 10 m. This implies that such channel systems must cross many fracture intersections without bifurcating. A general relationship in terms of flow dimension is suggested. Some initial investigations using HyperConv show that the commonly observed feature, 'compartmentalization', only occurs when channel density is just above the percolation threshold. Such compartments have been observed at Kamaishi Experimental Mine (Japan) implying a sparse flow network. It is suggested that compartments and skin are observable in the field, indicate sparse channel systems, and could form part of site characterization for deep nuclear waste repositories.

Keywords: Conceptual Models Crystalline Rocks Hydraulic Properties Numerical Modeling Radioactive Waste Repositories

INTRODUCTION

Many geoscientists are interested in how groundwater flows through sparsely fractured crystalline rock, none more so than those considering constructing nuclear waste repositories in such formations. During the 1970s and 1980s, several countries initiated programmes aimed at disposing of radioactive waste at depth, some of which included the construction of underground research laboratories (URLs; Witherspoon and Bodvarsson 2006). One of the major intended activities in these URLs was to study groundwater flow and radionuclide transport within poorly transmissive fractures at roughly repository depth (say ~500 m below surface). Early tracer experiments at Stripa Mine, Sweden (Abelin et al. 1985, 1991a), attempted to inject tracers into the high head gradients induced by the presence of the mine and then recover them in the underground opening. This tended to yield low rates of tracer recovery so subsequent tracer experiments in URLs were usually located some distance laterally into the rock in artificially created 'point sink' configurations (Sawada et al. 2000; Elert and Svensson 2001; Andersson et al. 2002; Einsiedl and Maloszewski 2005). In addition, these borehole to borehole tests were always sited in major fractures and fracture zones with concomitantly higher values of transmissivity compared to surrounding rock: say $\sim 10^{-7}$ relative to $\sim 10^{-10}$ m^2/s. Obviously, these experimental choices were made in order to ensure positive results since migration rates within low transmissivity fractures for even conservative tracers over small distances require many years.

However, one could reasonably question how much results from fracture and fault zones reveal about flow and transport in what is often referred to as 'background rock' (i.e. the average rock between fault and fracture zones and referred to as 'fracture domains' by Follin and Hartley 2014). The origin of such zones ought to produce a morphology and mineral content that differs from ordinary fractures in background rock. Indeed, Andersson et al. (2007) concluded after the extensive series of TRUE tracer tests at Äspö Hard Rock Laboratory, Sweden, that uncertainty concerning the tracer flowpath was greater than uncertainty about the water–rock interaction. Also that background fractures were about ten times more retentive than the main fractures involved in the tracer experiments.

Notwithstanding that the objective of the tracer tests has been to determine radionuclide transport properties, it is the premise of this paper that an early, possibly considered inconclusive, tracer test yielded a valuable insight into how groundwater flows through background rock. Based on observations from the aforementioned tracer test, a particular concept of long channels and sparse intersections is proposed. The concept is encapsulated within an innovative lattice network code that is shown to reproduce some of the features noted or measured in other URL experiments.

THE 3D (MIGRATION) EXPERIMENT

The 'Three-Dimensional (Migration) Experiment' was undertaken at the Stripa Mine between the end of 1984 and early 1987. The experiment was based on the assumption that a drift at atmospheric pressure, hundreds of metres below the water table, would provide a suitable sink for surrounding groundwater. Then, if different tracers were injected into the local groundwater system at various locations, they could all be recovered eventually as inflow to the drift (Abelin et al. 1991a, b).

An experimental drift (roughly 75 m long and 4 m diameter) was located at least 100 m from existing mine openings and connected to the old mine galleries by a 140-m-long access drift. Two 'side drifts' about 10 m long, were also excavated about 10 m before the end of the drift, thus forming a cruciform-shaped experimental gallery (Fig. 1). A major innovation of the experiment was the development of a polythene-sheet catchwater system so that inflow could be identified as emanating from individual rectangles marked on the upper walls and ceiling of the experimental drift. Three hundred and seventy-five polythene sheets, each a rectangle of 2 m^2, were attached to the upper rock surfaces using resin glue which enabled each area of the grid to yield a measurement of inflow (Fig. 1a). Inflows varied greatly from sheet to sheet; less than half of them gave any measurable inflow and 11 sheets accounted for 50% of inflow. About half of the grid-based inflow emanated from one of the side drifts. The results clearly showed that natural inflow is limited to a few areas with large dry areas between.

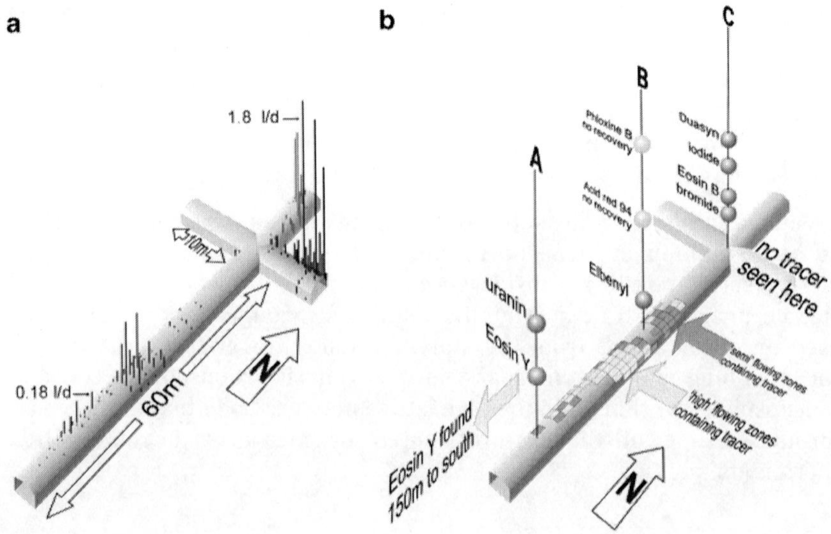

Figure. 1 The 3D (Migration) Experiment: A natural inflows of groundwater; B the measured distribution of tracer inflows from nine injection points in the three boreholes *A*, *B* and *C* (after Figs. 5 and 13 in Abelin et al. 1991a)

For the next stage of the experiment, three 70-m-long tracer injection boreholes were drilled vertically upwards into the ceiling of the main drift (Fig. 1b). Nine different tracers were injected in short (2.5 m), discrete, packer-isolated zones at distances between 10 and 56 m directly above the crown of the drift. They were injected for periods between 60 and 95 weeks followed by 25 weeks of continued monitoring. Only seven tracers were recovered in the experimental drift, of which one, bromide, incurred analysis difficulties that negated attempts to calculate cumulative recovery.

The distribution of tracer inflow was as noteworthy as the distribution of water inflow. In contrast to the water inflow that was highest into the NE arm of the crucifix-shaped excavation, all tracers that were recovered at all arrived in the middle of the experimental drift (Fig. 1b). They ignored the wettest part of the drift located in one of the 'arms' (Fig. 1a). Also surprising, was the recovery of all four tracers from borehole C given its proximity to the high-inflow 'arm'. Perhaps equally surprising was the absence of recovery of the tracers from the two more distant injection points in borehole B bearing in mind that they both appeared to show a pronounced head response to the experimental drift. A final conundrum was posed by the subsequent discovery, in significant quantities, of the tracer Eosin Y from borehole A when excavating a connecting drift about 150 m away from the injection point.

In summary, the 3D Migration Experiment revealed the extremely uneven distribution of inflow to underground excavations together with flow paths that ignore the concepts underlying groundwater flow in a porous medium. Abelin et al. (1991b) concluded that the results indicated that much flow takes place in channels that interact infrequently with neighbouring channels. Even though Abelin et al. (1991a, b) conclude that flow occurs via channels within fractures, they do not attempt to gain further insight by examining the pattern of mixing exhibited by their experiment. The pattern is enlightening because most catchwater sheets that yielded tracer, yielded mixtures from several injection points, and not always from the nearest injection points. Also, not all possible mixtures were represented. It seems reasonable to assume that mixing must occur at certain points where actively flowing channels meet others. Based on the recovery of the six tracers, it is possible to reproduce the observed mixtures based on only eleven 'mixing nodes' (Fig. 2). The incidence of mixtures containing tracers from injection points in boreholes A and C, many metres apart, suggests that the mixing nodes are relatively close to the catchwater sheets, possibly within an excavation damaged zone (EDZ). This in turn suggests that the channelized flowpaths from the injection points to the mixing nodes are long and have few 'dispersal nodes' along their length.

THE SPARSE CHANNEL NETWORK CONCEPT

The concept of channel-based flow is not new. However, since Rouleau and Gale (1985) report that fractures at Stripa Mine are seldom more than 0.5 m apart, the concept of long channels proposed here means that channels in background rock cross many fracture intersections without branching.

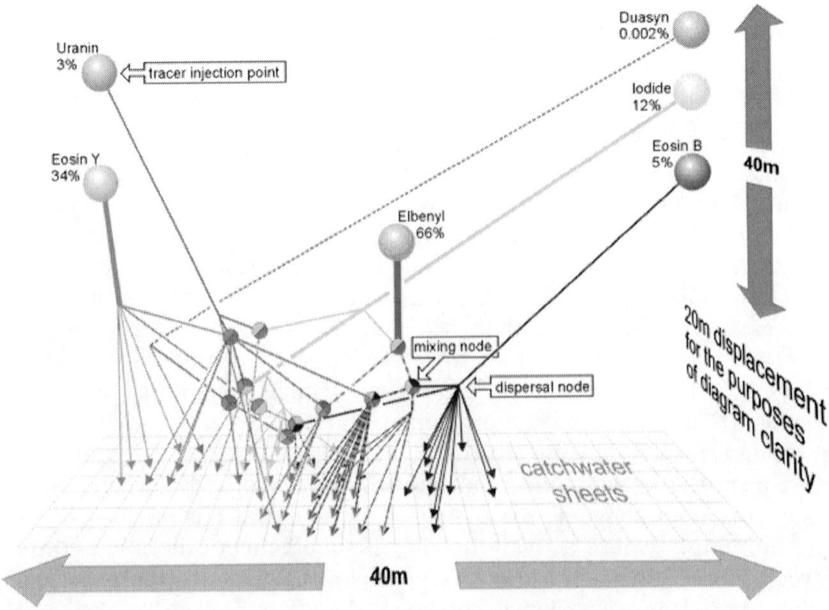

Figure. 2 Pseudo three-dimensional representation of the movement of tracers in the 3D Migration Experiment invoking imaginary 'mixing' and 'dispersal' nodes (constructed from information contained in Fig. 13 of Abelin et al. 1991a)

The concept of sparse long channels with few intersections agrees with the frequent observation in tunnels and drifts that inflows are sporadic and wet patches are separated by large regions of dry tunnel/drift surface (Tsang and Neretnieks 1998; Tsang et al. 2015). An implication of isolated points of drift inflow is that groundwater must converge to those points and that the convergence must cause head loss proportional to the 'dimension' of the convergence (Barker 1988). Figure 3a shows theoretical head distributions towards a drift based on a 4 m diameter drift in a groundwater environment with a 200 m head boundary located 50 m from the drift centreline. Figure 3a uses a logarithmic x-axis so that, based on porous medium theory, cylindrically convergent flow should plot as a straight line. If the convergent flow is of a higher dimension than cylindrical (i.e. >2D), the line of log-linear regression projected to the drift wall will intersect it at positive values of head. Most hydrogeologists would interpret this configuration as a positive skin, a region of locally reduced hydraulic conductivity relative to 'average' rock further from the drift (Fig. 3a). The general concept and mathematics of skin and its application to a drift deep below the water table is summarized in the Appendix. Recognizing and estimating skin in fractured crystalline rocks is difficult

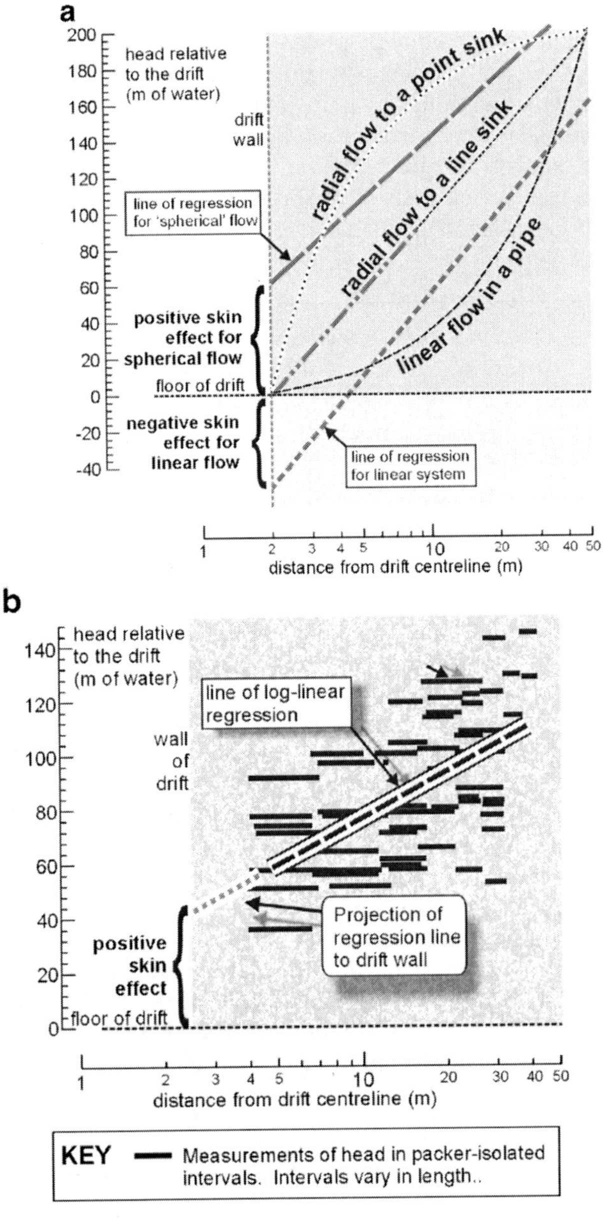

Figure. 3 Head versus distance from the centre of a drift. A Calculated values and log linear regression lines for ideal systems with zero skin. B Results from packer isolated sections in the radial boreholes of the Macropermeability Experiment at Stripa (based on Wilson et al. 1981)

because of the inherent variability of the transmissivity of adjacent fractures. Nevertheless, if enough points around a drift (actually short straddle packer intervals in nearby boreholes) are measured, then a distance–drawdown relationship is usually apparent. Figure 3b shows such a set of measurements from the Macropermeability Experiment at Stripa Mine (Wilson et al. 1981). Similar results have been reported for other experiments at Stripa, such as the Simulated Drift Experiment (Black et al. 2006), and for other URLs such as the KD90 drift at Kamaishi in Japan (Chijimatsu et al. 1996).

The idea proposed here is that observations of isolated patches of inflow on drift surfaces, positive skin effect and channelling effects in tracer experiments in 'background rock' are all linked to flow in long channels with infrequent junctions. It was decided to validate the concept by reproducing the 'apparent' skin effect in an appropriate model. The intention was to construct a numerical model of flow to limited inflow points to a drift that would, when using parameter values constrained by the Stripa experiments, yield similar values of apparent skin effect to those that were observed.

A NOVEL LATTICE NETWORK MODEL

Choice of model type

Studies of flow in fractured crystalline rocks have employed a wide variety of different conceptual and numerical models (Selroos et al. 2002). Apart from porous medium and discrete fracture network models, lattice network models have sometimes been used to understand flow and solute transport behaviour in three dimensions. Moreno and Neretnieks (1993) and Margolin et al. (1998) both used a cubic lattice model characterized by the simple probability of any single bond being present. This approach, here termed the 'classical lattice model', has the characteristic that long channels (continuous lines of bonds) are associated with short gaps and vice versa.

Since the experiments at Stripa described earlier suggest long channels separated by large gaps, the classical lattice model is considered inappropriate. A novel lattice network concept has therefore been developed that allows the mean lengths of the channels and gaps to be independent of each other. This 'two-parameter bond lattice model' has been implemented primarily as a 3D steady-state flow code named HyperConv. This name derives from the term 'hyper-convergent flow' introduced by Black et al. 2006 to represent flow converging to a point sink within cylindrical boundary conditions. It was felt that important characteristics of the code should include the ability to calculate flow and head quickly in order to run multiple realizations. HyperConv does not attempt to reflect rock or fracture structure except in as much as it is three-dimensional and channels are organized on an orthogonal grid.

Modelling approach

Hundreds of realizations are generated for two fixed parameters (e.g. mean channel and gap lengths). For all realizations, flow through each bond is determined for one of two types of boundary conditions and the key measures of bulk hydraulic conductivity and skin effect are evaluated. The measures are then averaged over all percolating realizations of a particular ensemble (i.e. of 100 realizations).

Network generation

The network is formed by creating bonds between the nodes of a 60 × 60 × 60 orthogonal lattice. A channel is conceived as a group of bonds in a continuous row. Two probabilities are used to generate channels along lines of nodes in each coordinate direction:

1. P_N (abbreviation of P_{New}) is the probability of starting a new channel if the previous bond (within a particular line of nodes) is absent.
2. P_A (abbreviation of P_{Again}) is the probability of continuing a channel to a subsequent node.

This method of network generation gives rise to channel lengths and gap lengths (in the x, y and z-directions) that both have a geometric distribution (Table 1). The mean and variance are given in Table 2. The lengths are all scaled to the length of a bond (single sub-channel) which is given unit length.

Table 1 Distributions of channel lengths, l, and gaps, g. Both follow geometric distributions on 1 to ∞ with a mode (most likely value) of 1

Distribution of lengths of:	Probability mass function	Mean	Variance
Channels, l	$\Pr\{l=n\}=P_A^{n-1}(1-P_A)=(L-1)^{n-1}/L^n$	$L=1/(1-P_A)$	$\sigma_l^2=P_A/(1-P_A)^2=L(L-1)$
Gaps, g	$\Pr\{g=n\}=P_N(1-P_N)^{n-1}=(G-1)^{n-1}/G^n$	$G=1/P_N$	$\sigma_g^2=(1-P_N)/P_N^2=G(G-1)$

Table 2 Relationships between probability generators, mean lengths and the bond probability P_{ON}

	Probability-based characterization		Length-based characterization	
Parameter	P_A	P_N	L	G
Relation to alternatives	$L = \frac{1}{1-P_A}$	$G = \frac{1}{P_N}$	$P_A = 1-\frac{1}{L}$	$P_N = \frac{1}{G}$
Probability of a bond	$P_{ON} = \frac{P_N}{1-P_A+P_N}$		$P_{ON} = \frac{L}{G+L}$	

It should be noted that any two of the parameter set $\{L, G, P_{ON}, P_N, P_A\}$ are sufficient to characterize the new lattice model and that the 'classical lattice model' is a special case where $P_{ON} = P_N = P_A$ or, equivalently, $1/L + 1/G = 1$.

Applying boundary conditions and assigning parameters

The basic channel network is produced in the form of a cube. Once this has been generated, different boundary conditions can be applied, namely face-to-face, and cylindrical (Fig. 4). Face-to-face boundary conditions are used to evaluate percolation thresholds and 'compartmentalization', whilst cylindrical boundaries are used to evaluate skin (in the context of a tunnel or drift). For face-to-face flow, four of the six faces are set as no-flow boundaries confining flow between the remaining two opposing faces at fixed head. Cylindrical boundary conditions are achieved by removing surplus nodes in the region of four parallel edges to form a cylinder and setting an inner cylindrical boundary at (zero) fixed head (Fig. 4).

The modelling was intended to simulate observations at Stripa URL so values were assigned to boundary conditions with the Stripa measurements in mind. The main choice concerns the 'equivalent' distance between adjacent nodes within the lattice and hence the length of a 'sub-channel'. A value of 1.5 m was chosen so that each 2 m × 1 m catchwater sheet of the 3D Migration Experiment (Fig. 1) could receive inflow at maximum model density ($P_{ON} = 1$). This in turn meant that the external boundary of the cylindrical version of the network was 44.25 m from the centre of the network. A value of 200 m head was chosen for the external boundary as this appeared to be the undisturbed background value near to the 3D Migration Experiment (Abelin et al. 1991a). The inner boundary was a 4-m diameter central drift connected to the atmosphere at 0 m head.

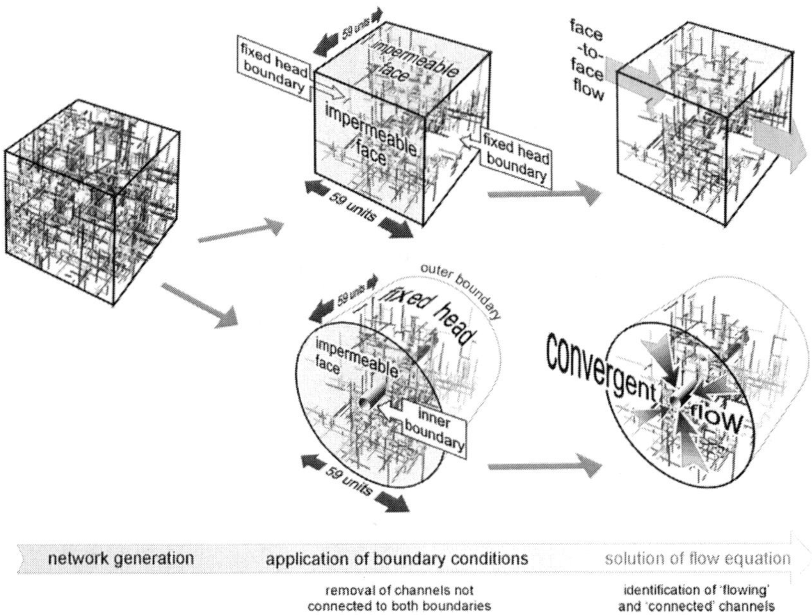

Figure. 4 Schematic of the sequence of operations involving the thinning of the HyperConv network during the application of different boundary conditions and the solution of the flow equation

Inflow to the Macropermeability Experiment, in conjunction with the near-drift hydraulic gradient, yielded an estimated value of bulk hydraulic conductivity of about 2×10^{-10} m/s. Within the HyperConv code, channel conductances (C_s) are selected randomly from a lognormal distribution. All bonds (sub-channels) comprising a channel have the same conductance (C_s).

The bond conductance values are based on a constant base (geometric mean) value of conductance, C_{sg}. So,

$$C_s = C_{sg} \times 10^{u\sigma} \tag{1}$$

where u is a randomly generated unit normal variable, and σ is the standard deviation of $\log_{10}(C_s)$.

A value of $C_{sg} = 4.5 \times 10^{-10}$ m^3/s was used to yield a target hydraulic conductivity of 2×10^{-10}m/s for a fully occupied grid. By setting the standard deviation of the log normal distribution (σ) of $\log_{10}(C_s)$ to zero, a constant conductance network can be produced. In this way it is possible to compare and therefore distinguish the significance of network structure on flow from the effect of varying channel conductance.

Solving the flow equation

The last step is to solve the flow equation for the thinned channel network. The network has mass balance applied at each node with a linear, Darcian, relationship between flow and head difference between each bond pair. The resulting linear equations are solved for the nodal heads using different routines depending on bond density. Where the network is sparse and symmetric, a pre-conditioned conjugate gradient approach is favoured. For poorly connected systems (less than 30% of the nodes connected) a general purpose sparse matrix solver with a partial Cholesky pre-conditioner is used on a system containing only the connected nodes. For fuller systems, the full lattice is solved using a specialized block-structured pre-conditioner (see Black et al. 2006 for more detail).

Finally, the channel network is assessed to identify which nodes have significant flow in them and which contain negligible flow. All the nodes containing significant flow (i.e. values greater than 10^{-16} m^3/s) are termed 'flowing' nodes. The rest are termed 'connected' nodes.

Evaluating skin

The evaluation of skin is based on finding the best-fit log-linear regression line through the heads calculated at the nodes of the network in the manner identified in Fig. 3b. The calculation is of 'skin effect' (h_{skin}), and has units of metres of water head. It comes in two versions, h_{skin}^c and h_{skin}^f that refer to the calculation based either on all connected nodes (superscript ^{c}) or only on the flowing nodes (superscript ^{f}). The values for all connected nodes, h_{skin}^c, is used in the results below because that is the more likely type of value measured in the field, albeit at less frequency at greater radial distance.

Using ensembles

HyperConv was designed to construct a relatively complex network, solve the flow equation and derive a value for skin effect quickly enough that it could be run many times. The outcomes of many realizations can then be averaged to yield a single value appropriate to the single set of parameters used to generate the ensemble (of realizations). A key question concerned how big the ensemble needed to be. Based on three runs ($\sigma = 0$, 1 and 2), each of about 4,000 realizations, and using the fact that the variance of a mean is inversely proportional to the sample size, it is estimated that for 100 realizations the relative error is typically between 2 and 5%. Therefore an ensemble size of 100 was chosen as a reasonable compromise between accuracy and computer run time (Black et al. 2007).

RESULTS

Percolation

A fundamental property of any network is whether or not it percolates (i.e. conductive elements connect across the network). Non-percolating realizations are generally of little interest for flow or transport. This study used a series of ensembles of 100 realizations for five values of mean channel length, but with different sizes of intervening gaps. Starting with large gaps (i.e. low density), only a very few realizations within an ensemble would percolate. As gap size was decreased increasing numbers of realizations would percolate until all percolated. The results show that the average number of percolating realizations is dependent on mean channel length as well as network density (Fig. 5). More importantly, systems of long channels continue to percolate at lower values of network density than short channels. Directly linked to this finding is that, of the systems examined using HyperConv, classical lattice networks require the highest values of density to percolate.

Based on this outcome, the percolation behaviour of a much larger million-node cubic lattice designed to evaluate only connectivity (i.e. without the ability to solve for flow) was evaluated. The range of behavior of a 'two-parameter bond lattice' type model is shown in Fig. 6 and spans all combinations of average channel and gap length (i.e. all P_A and P_N values). The boundary between percolating and non-percolating combinations of channel and gap length can be approximated by the equation

$$G = 2.27L^2 \tag{2}$$

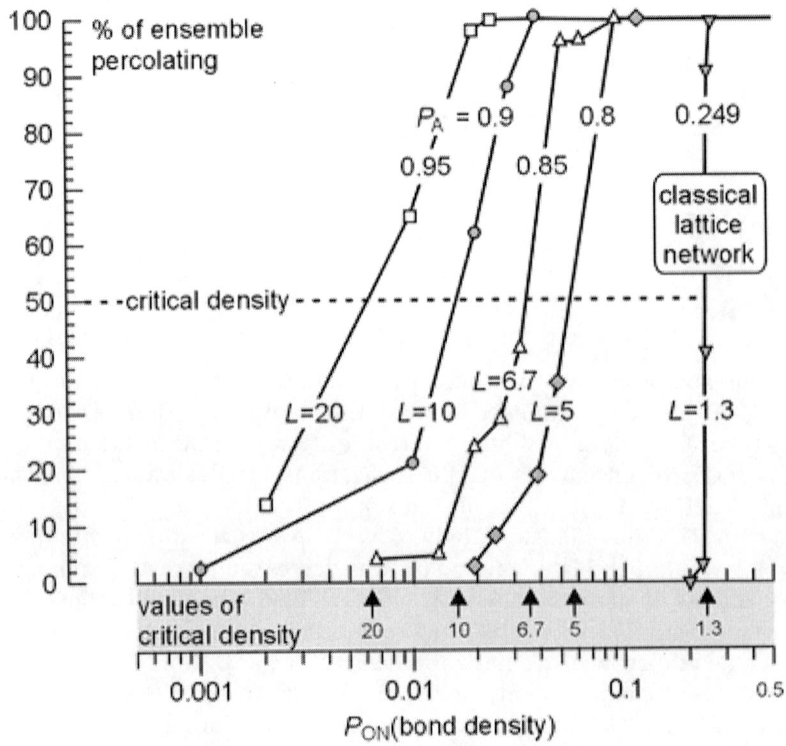

Figure. 5 The dependence of percolation probability on channel length and network density. L = average number of bonds (sub-channels) per channel. The value of probability of percolation of 50% is commonly referred to as the average percolation threshold or critical density (Harter 2005)

Figure 6 shows the two regions where percolation does and does not occur, though the boundary is only sharp for infinite networks. It also indicates that any model with a bond density, P_{ON}, greater than about 0.3 should percolate and that, as P_{ON} falls below this value, percolation only occurs for increasingly long channels and gaps. Most of the important results from the HyperConv simulations occupy this low density region. It can be seen that the possible combinations of average channel and gap lengths for a classical lattice network are limited to the single line running diagonally across Fig. 6, corresponding to $1/L + 1/G = 1$.

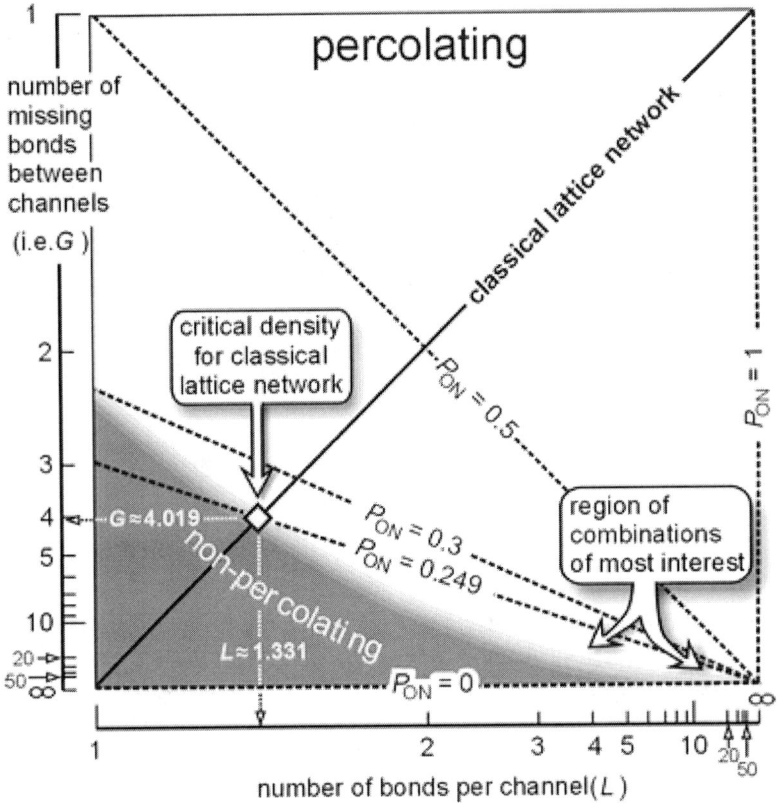

Figure. 6 The combinations of channel and gap lengths that percolate and do not percolate and their relationship to bond density (P_{ON})

Skin effect

The intention of the modelling was to investigate the effect of channel length and sparseness on the derivation of values of skin effect. Margolin et al. (1998) have previously shown that sparse flow systems can be produced by classical lattice networks based on variable bond conductance. In order to avoid confusion with effects caused by variable bond conductance, all the results reported here (except for some ensembles investigating variable conductance) concern systems comprising sub-channels that have the same value of conductance.

The variation with density in Fig. 7 is surprising. A general trend from extreme values at very low bond density to zero skin when the lattice is full (i.e. high bond density) might have been expected. Instead, while the ensemble results of some shorter channels exhibit a reasonably gradual evolution from large negative values of skin at low bond density towards zero at high density, some longer channels exhibit a peak of positive skin at intermediate values of density.

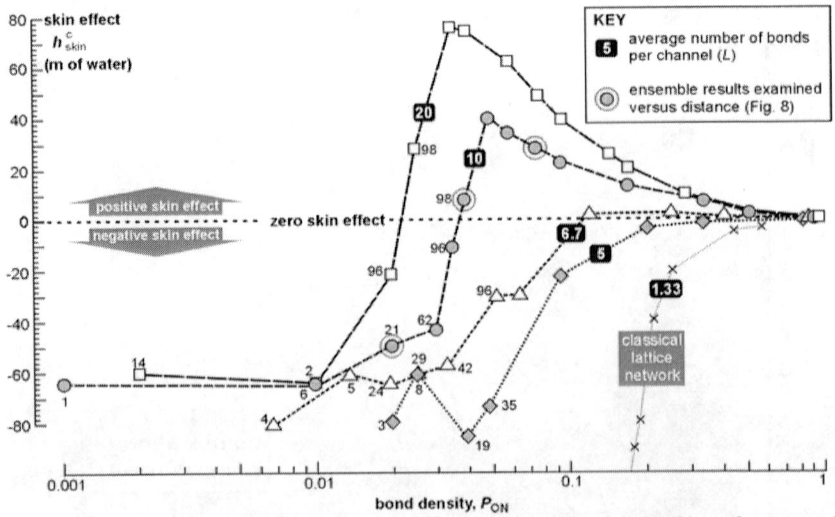

Figure. 7 The dependence of skin effect on channel length and network density. Each point represents the average value of an ensemble of 100 realizations or the number of realizations that percolated: reported as the number beside data points. Where *numbers are absent*, all 100 realizations of a given ensemble percolate

Considering the results in more detail, it can be seen that all networks yielded negative skin effect over at least some range of density values; that is they indicated a region of apparently increased permeability in the near-tunnel region. However, only some more dense ensembles of longer channels yielded a positive skin effect (i.e. apparent reduced permeability near the tunnel). The division between networks with a region of positive skin effect and those without, appears to occur around networks of 6.7 bonds average channel length (10 m). It should probably be borne in mind that the modelled system includes a 4-m diameter drift as its inner boundary and that this is of similar size to the 10-m scale dividing the two types of behaviour. Another point of note from the results is that classical lattice networks show no tendency to yield positive skin effects in the conditions applied here.

In line with the percolation results shown in Fig. 5, the expected trend of percolation percentage decreasing with bond density is maintained with a marked decline for P_{ON} less than 0.04. With decreasing numbers of realizations percolating, the variability of their average increases as evidenced by the overlapping data curves at large negative values of skin effect (Fig. 7).

The relatively sudden transition from negative to positive skin effect exhibited by networks of longer channels is a region of particular interest. It was decided to investigate this transition by examining in more detail the distance/drawdown behaviour of three ensembles of 10-bonds average length per channel ($P_{ON} \approx 0.02$, 0.04 and 0.07 as identified in Fig. 7).

Head was calculated as a function of distance from the inner boundary for all connected nodes within every percolating realization of each of the three ensembles. Values were placed in annular bins of equal thickness and an average derived (Fig. 8a). In all cases, the value of conductance of the bonds is kept constant ($\sigma = 0$) so that all outcomes are a function of geometry. It should be pointed out that these values of density represent average 'gap' distances between channels ranging between 125 and 500 bonds (i.e. beyond the boundaries of the model).

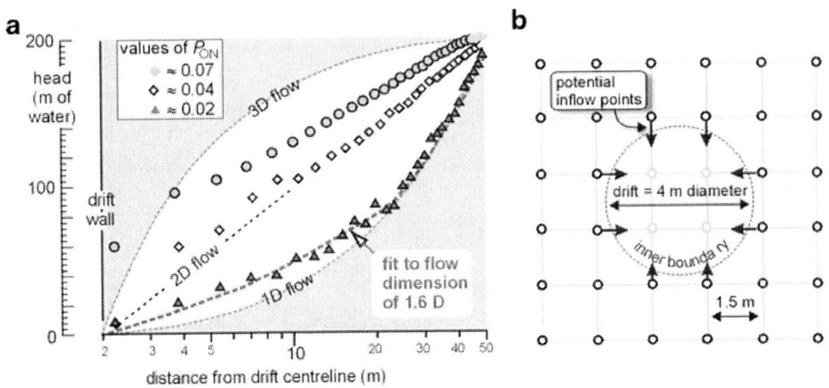

Figure. 8 A Head versus distance from the centre of a drift of 4-m diameter for percolating ensembles based on 15-m average channel length and three values of density. B Section across the axis of the inner boundary of the HyperConv model showing the configuration of potential inflow points. Note that the lattice extends out in all directions beyond the inner region depicted

The variation of head with the logarithm of distance from a drift centreline is plotted in Fig. 8aagainst a background of 1D, 2D and 3D flow in a porous medium with zero skin. The straight-line sections extending from the outer boundary for all three cases indicates reasonable adherence to cylindrically convergent flow essentially imposed by the outer boundary.

However, as the three types of flow system approach the drift wall, the least dense system diverges from cylindrical behavior at greatest distance and yields an excellent fit to the model of Barker (1988) with a flow dimension of 1.6. The most dense system of the three (i.e. $P_{ON} \approx 0.07$) adheres to cylindrically convergent flow until very close to the drift where almost half of the overall head loss occurs within 2 m of the drift wall. If the procedure summarized in the Appendix is applied, the data yields an apparent positive skin effect only for the densest of the three realizations.

It is useful to consider some practical dimensions of the lattice network at the point of inflow. Eight bonds (potential inflow points) emanate from each lattice plane perpendicular to a 4-m diameter drift (Fig 8b), so the probability of one or more of these bonds being 'on' is $(1 - P_{ON})^8$and from this it follows that the average distance between perpendicular planes with inflow points to the tunnel is

$$\Delta_{\text{inflow}} = \frac{1}{1-(1-P_{ON})^8} \tag{3}$$

For the results in Fig. 8a) this gives distances of: 2.1, 3.7 and 6.8 lattice intervals (3.3, 5.6 and 10.3 m), respectively, increasing with decreasing density. It is notable that for the two higher density results in Fig. 8a, these intervals are reasonably consistent with the radial distance at which flow ceases to be cylindrical, which is analogous to the flow pattern for a partially penetrating well.

Hydraulic conductivity

Although the objective of constructing a sparse long channel model was to simulate skin effect, it is valuable to examine the relationship between average channel length, channel density and hydraulic conductivity. The HyperConv model is capable of calculating overall network conductivity using different boundary conditions: face-to-face and radial flow to a line sink (Fig. 4).

The results shown in Fig. 9 are much as expected: as bond density decreases, hydraulic conductivity decreases. In line with skin effect, the impact of density is most pronounced where average channel lengths are shortest. It is noteworthy that the percentage of percolating realizations drops below 100% at roughly the same value of hydraulic conductivity (about 2.5 orders of magnitude less than the full lattice value) for all lengths of channel.

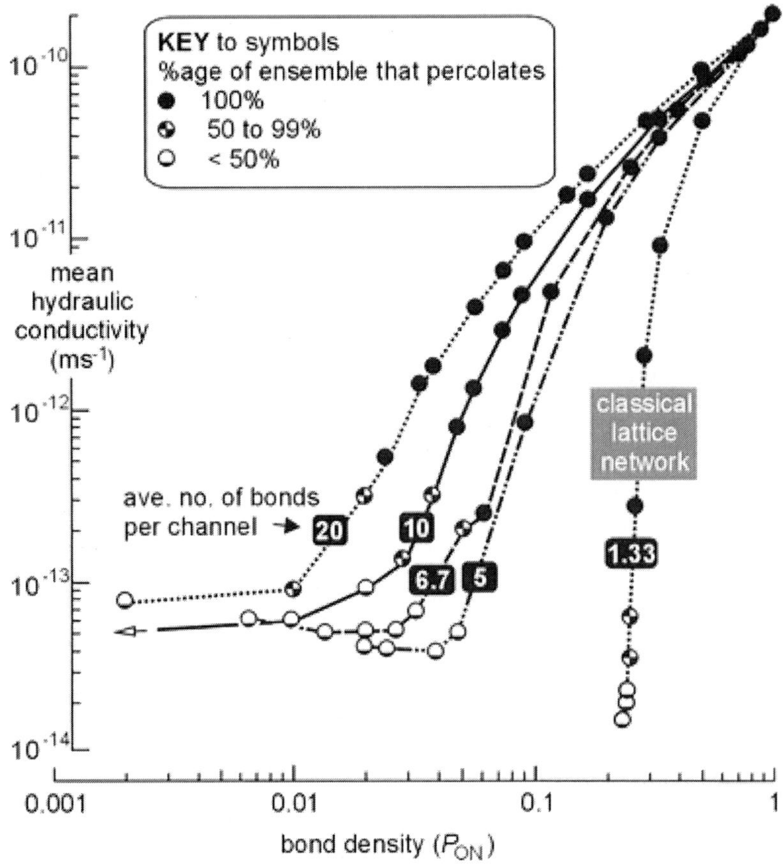

Figure. 9 The dependence of hydraulic conductivity on channel length and network density. Each point represents the average conductivity of the percolating realizations from an ensemble of 100 realizations

Effect of spatially variable channel conductance on skin effect and hydraulic conductivity

In the investigations reported in the preceding sections, the conductance of the bonds of the networks in all the realizations has the same value by setting the standard deviation of the log normal distribution of $\log_{10}$ (conductance) to zero ($\sigma = 0$). This means that all the effects observed in the model results thus far are the product of network geometry alone. Obviously, natural systems cannot be expected to be composed solely of constant transmissivity channels, so variance in conductance was introduced by setting σ greater than zero on the set of ensembles based around an average channel length of 10 bonds (i.e. average length = 15 m and the same as the investigation of transitional behaviour above) (Fig. 10).

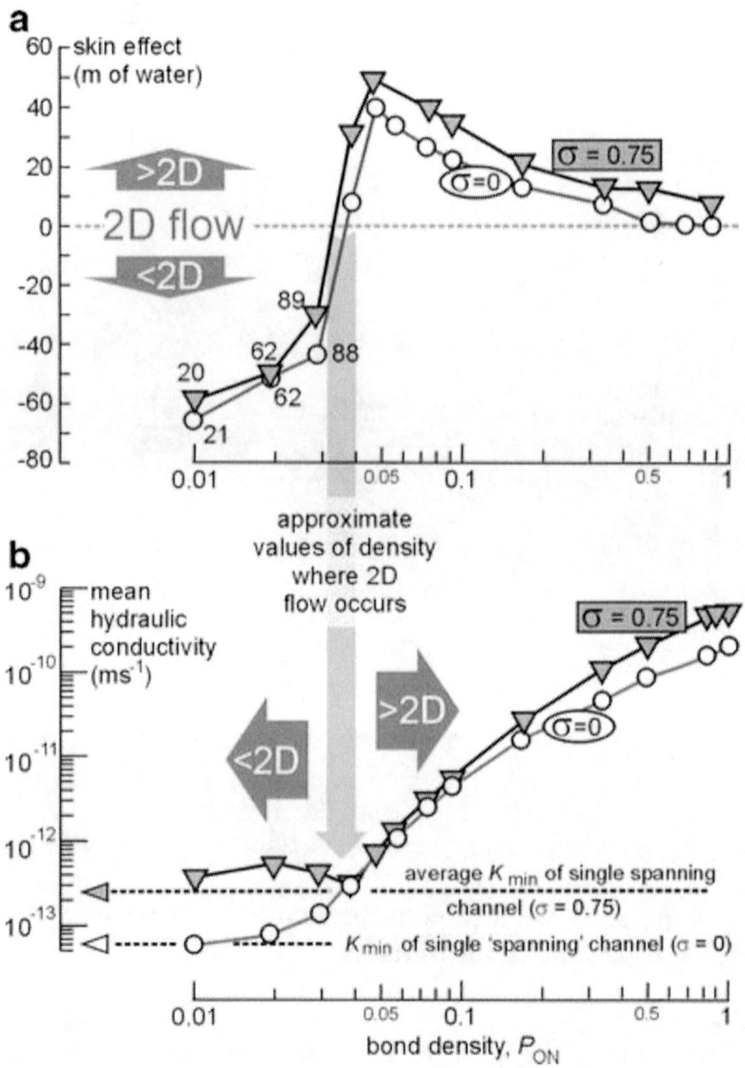

Figure. 10 Results for networks averaging 10 bonds per channel but with variable conductance: A skin effect versus density, and B hydraulic conductivity. Numbers beside data points indicate percolating percentage and apply to both figures (A). A 'spanning' channel is one that extends the entire distance between the inner and outer boundaries of the model

The skin effect values (Fig. 10a) are displaced upwards but by a relatively small amount given that the range of conductances for one standard deviation differ by a factor of 30. This suggests that the results for equal conductance channels ($\sigma = 0$) are of reasonably general applicability.

The outcome for hydraulic conductivity can be understood by reference to the analytical result obtained by Rubin (2003) for an uncorrelated, spatially stochastic, lognormal permeability field:

$$K_{\text{eff}} = K_{\text{g}} \exp\left[\left(\frac{1}{2} - \frac{1}{m}\right)\sigma_{\text{y}}^2\right] \qquad (4)$$

where K_g is the geometric mean of K, m is the geometrical (network) dimension and $\sigma_y{}^2$ is the variance of $\ln(K)$.

At high density, the network dimension, m, can be expected to be at a maximum of 3 which yields a value for K_{eff} of roughly 4×10^{-10} m/s via Eq (4) given $K_g = 2 \times 10^{-10}$ m/s. In other words, flow occurs preferentially via the more conductive channels. Thus, it must be assumed that, for any given density, the number of channels containing the bulk of flow is fewer when conductance varies than when it is held constant. This may also be the cause of the increase in skin effect at maximum density because flow continues to converge to the more conductive channels intersecting the inner boundary. Based on Eq. (4), when the network dimension $(m) = 2$, $K_{eff} = K_g$ which agrees well with Fig. 10b) for $P_{ON} \approx 1$.

The transition from positive to negative values of skin effect (at $P_{ON} \approx 0.04$) also marks the boundary between all realizations percolating and decreasing percentages percolating (Fig. 10). At network densities less than $P_{ON} \approx 0.04$, results are best understood in the context of a single individual channel extending the entire distance between inner and outer boundaries. The parameters appropriate for a single line of 'flowing' nodes (i.e. 29 bonds) are:

Bond density:29/153872 (nodes in the cylinder) = 1.88×10^{-4}

Hydraulic conductivity:5.92×10^{-14} m/s (using Thiem's equation)

Skin effect:−62.27 m (log-linear regression to 29 equally spaced head steps of 7.1 m)

It can be seen in Fig. 10b that once node density decreases to 0.01 and percolation frequency is at 21%, the value of conductivity of the equal conductance networks decreases to the value for a single ('spanning') channel intercepting the inner boundary. For the spatially variable networks where $\sigma = 0.75$, each single spanning channel can vary within the limits of the log normal distribution and an effective hydraulic conductivity given by the average conductance can be expected. For a lognormal distribution the average is $K_g \exp(\sigma_y^2/2)$ giving $K_{min} = 2.5 \times 10^{-13}$m/s for $\sigma = 0.75$ which agrees reasonably well with Fig. 10b).

BOUNDING THE MODEL BEHAVIOUR

A 'semi-generic' approach

It is possible to derive semi-generic sets of results by linking the hand-calculated minimum results (for a single spanning line of bonds) to those calculated by the lattice network model using estimated results (Fig. 11). Some of the lattice model results in the region where only small percentages of ensembles percolate (and therefore results are variable) have been manually smoothed (compare Figs. 7 and 9 with Fig. 11a, b). The resulting diagrams, Fig. 11a, bare not generic since the axes are not dimensionless. They are termed 'semi-generic' because it is considered that realizations based on different parameters from the set calculated will exhibit similar relative behaviour: in particular, the peak values of positive skin just above the percolation threshold for longer channels.

The semi-generic form of the skin effect results (Fig. 11a) indicates a division into two groups, those short channels that always yield a negative skin effect and the longer ones that exhibit an interval of positive skin effect. It is surmised that the value of this division is linked to the particular size of the inner boundary (4 m diameter) relative to the average length of the channels within each set of results (10 m in this case).

The envelope of possible values of hydraulic conductivity (Fig. 11b) is bounded by the density of a single line of bonds between the inner and outer boundaries (i.e. 1.88×10^{-4}) and the density of a full lattice (i.e. 1). The upper edge of the envelope is formed by the density–conductivity relationship of multiples of single lines of connected channels. One of the curiosities of the results set is that a single line of channels does not yield the minimum value of hydraulic conductivity. Instead, the minimum values of hydraulic conductivity are produced by a small number of very tortuous flow paths. It is not clear whether short channel networks (i.e. less than three bonds average length and including classical lattice networks) exhibit such behavior because the probability of percolation declines so rapidly with decreasing bond density that numerical results were unobtainable in this study.

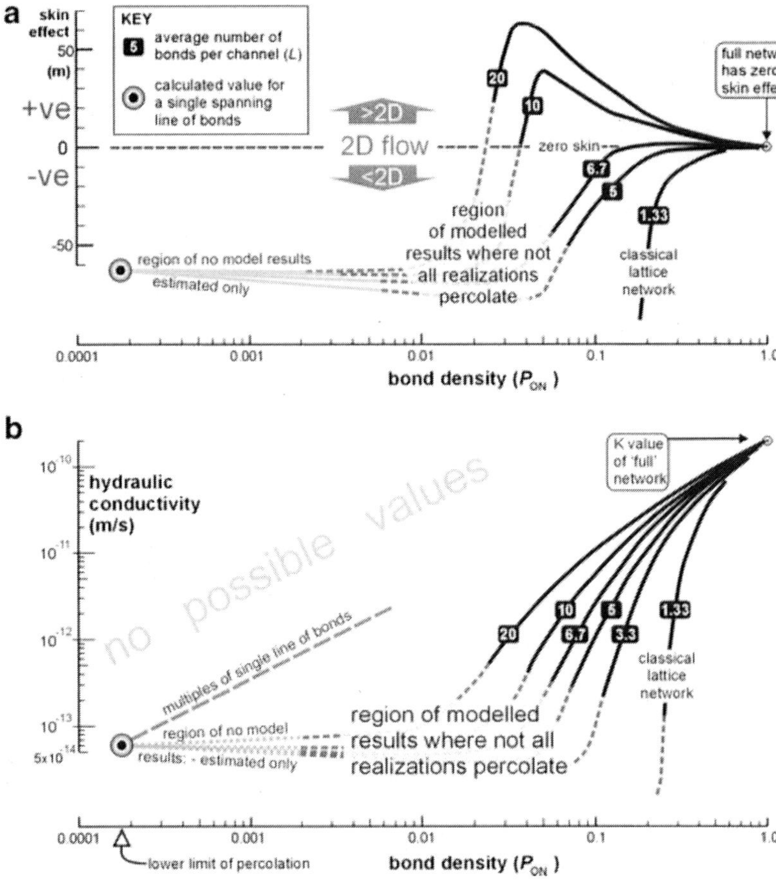

Figure. 11The relationship between A skin effect and B hydraulic conductivity, and bond (network) density shown in semi-generic form to indicate generalized behaviour. Key applies to both parts of the figure

A general approach

The skin effect results for head versus distance (Fig. 8) and the explanation of the results for spatially variable hydraulic conductivity Fig. 10b) indicate the possible usefulness of a dimension-based interpretation. A conceptual model involving three different dimensions is suggested.

The channels of the percolating lattice form a network which has a network dimension, D_{net}('*m*' in Eq. 4) which decreases from 3 at the highest densities (P_{ON} approaching unity) to 1 when the density is below about 1%. As density decreases, the number of inflow points to the tunnel decreases, which increases the flow convergence close to the tunnel wall.

This convergence yields positive values of skin and is here termed the 'boundary-condition dimension', D_{bc}, close to the tunnel. The skin factor closely relates to the flow dimension, D_{flow} (Fig. 10a).

Relating these three dimensions to two parameters of the lattice model, P_{ON} and L, requires some speculative assumptions. If it assumed that the network dimension is linearly related to the number of intersections per channel, B, which, for the two-parameter lattice network model is:

$$B = \frac{2(L+1)^2}{L+G} \tag{5}$$

Then, combining Eq. (5) with relations in Table 2, and bearing in mind that a 1D system will have zero intersections, D_{net} can be defined in terms of P_{ON} and L:

$$D_{net} = 1 + F_{net}B = 1 + F_{net}\frac{2P_{ON}(L+1)^2}{L} \tag{6}$$

where F_{net} is a constant.

Similarly, it is proposed that the boundary condition dimension is linearly related to the average interval between inflow planes along the tunnel (Δ_{inflow} as defined by Eq. 3). Since, if each plane of nodes perpendicular to the axis of the inner boundary has an inflow point to the tunnel and results in 2D flow, then a value of Δ_{inflow} greater than 1 should yield a flow dimension greater than 2; hence, Eq. (7) imposes cylindrical flow when the inflow interval is equal to the lattice grid spacing.

$$D_{bc} = 2 + F_{bc}(\Delta_{inflow} - 1) = 2 + F_{bc}\left[\frac{1}{1-(1-P_{ON})^8} - 1\right] \tag{7}$$

where F_{bc} is a constant.

A reasonable supposition is that the flow dimension is limited by both the network dimension and the boundary-condition dimension, so:

$$D_{flow} = \min(D_{net}, D_{bc}) \tag{8}$$

Since the intention is to relate apparent skin effect to flow dimension, D_{flow}, the results from HyperConv of skin effect for varying density, P_{ON}, and average channel length, L, are used for appropriate scaling. It seems reasonable that, based on Fig. 7, the maximum value of skin effect for $L = 20$ at $P_{ON} = 0.03$ should equate to a flow dimension of 3. This yields values for F_{net} of 1.5 and for F_{bc} of 0.3 and gives the variation of flow dimension with density and channel length shown in Fig. 12. A better correspondence could not be expected given the simplicity of the model, the use of averaged parameters and functions, and the fact that the results in Fig. 7 are averages over percolating realizations alone.

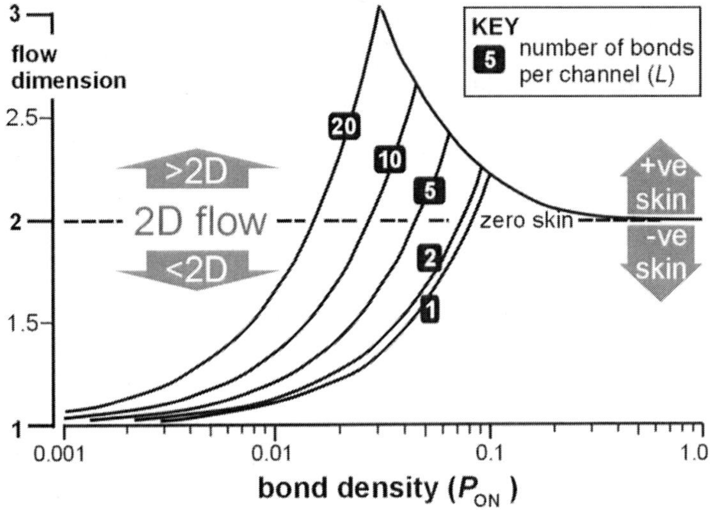

Figure. 12Variation of flow dimension with density, P_{ON}, and channel length based on Eqs. (6)–(8)

Although Eqs. (6)–(8) do not include the size of the inner boundary directly, the outcome involves Eq. (7) which is based on eight possible inflow points connecting the lattice to the inner boundary. This means that the channel lengths are a function of tunnel diameter and the interpretation may have general application.

THE IMPLICATIONS OF A LONG-CHANNEL CONCEPT

Conceptual links

The idea of long channels between bifurcations does not, at first, appear to be a particularly radical concept; however, it does have many implications that are not immediately obvious. Firstly, the HyperConv modelling clearly shows that long-channel systems are able to percolate at lower values of network density than networks consisting of shorter channels. Assuming that once water finds the shortest pathway between regions of higher and lower pressure, it will not seek extra pathways unless there is a change in boundary conditions or the original pathways alter due to external processes (e.g. seismicity, mineral precipitation, solution, etc.). This should leave most groundwater in background crystalline rock flowing within a channel network at a density just above the percolation threshold for an average channel length available within that particular rock. This should mean that the network containing flowing groundwater is sparse.

Channel sparseness has its own implications. HyperConv simulations show how, close to the percolation threshold, flow within a sparse long channel network converges to pass through a few key conductors that can be regarded as 'chokes'. The same choke effect is reported in previous variable conductance lattice network studies (Margolin et al. 1998) and in a fracture network study by Robinson (1984). Within such a flow system containing chokes, the bulk of head loss within the system will occur within the chokes, leaving the network between the chokes with little head loss as a consequence. These patches of similar head adjacent to patches of quite different head are termed 'compartments'. These ideas of sparseness, chokes and compartments are discussed in the following.

Sparseness

The main reason that channel networks such as those modelled here using the HyperConv lattice network model, emulate the skin effect behaviour observed at Stripa is their sparseness. It is difficult to appreciate the impact of sparseness without a visual cue. Figure 13 shows a single representative 2D slice through two HyperConv-like 3D lattice networks. Figure 13a is of a classical network at critical density (i.e. 50% of realizations percolate), whilst Fig. 13b is of a long channel network, also at critical density. The channels in Fig. 13b are 5× longer than those in Fig. 13a and the bond density is also 5× less. Neither slice has a direct connection across the network from face-to-face so relies for percolation on connections perpendicular to the 2D slice.

There is ample field evidence of sparseness in the form of isolated patches of inflow to drifts at Stripa URL—e.g. the Buffer Mass Test area of Pusch et al. (1985) and the site characterization and validation drift of Harding and Black (1992)—into the Kymmen (Moreno and Neretnieks 1989) and Bolmen tunnels in Sweden (Neretnieks 2006), and into tunnels in the Aar and Gotthard massifs of Switzerland (Masset and Loew 2010). At Pocos de Caldas uranium mine in Brazil, patches of oxidized pyrite associated with groundwater flow are observed together with channels of 100 m or more in length (Chapman et al. 1992).

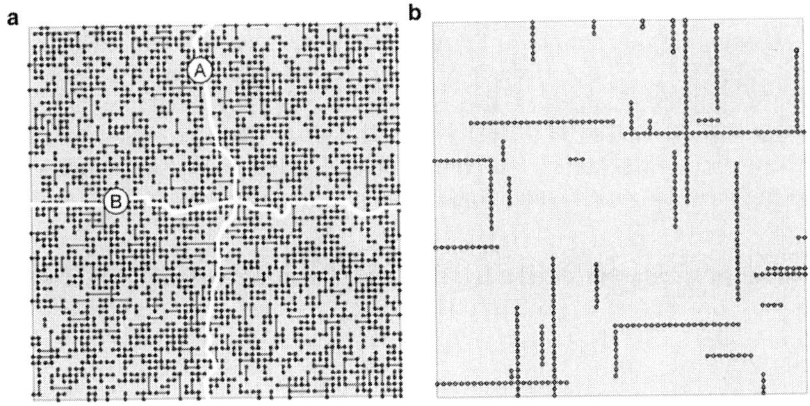

Figure. 132D slices from two HyperConv-type lattice networks at critical bond density: A a classical network (average channel length $\langle L \rangle = 1.331$ bonds, average gap length $\langle G \rangle = 4.02$ bonds), in which lines *A* and *B* identify two of the many continuous lines of gaps across the entire network; B a long channel network ($L = 10$, $G = 200$ bonds)

Chokes and compartments

The model of Margolin et al. (1998) was quite similar to HyperConv, though being of the classical lattice network type (see Fig. 6) it lacked the independent spatial correlation of bonds and gaps that HyperConv is able to simulate. Even so, Margolin et al. (1998) found that by either reducing the density of channels or increasing the variation of channel conductance they could reduce the number of percolating flow paths in a region to a small number. They also reported that chokes have a greater effect on overall network conductivity as network density decreases towards the percolation threshold (see Fig. 8 of Margolin et al. 1998). Hunt (2005), in his review of percolation theory applied to hydrogeology, stated that, "critical resistances on the optimal path control the entire field of potential drops." In other words, a sparse channel network should be associated with regions (i.e. clusters of channels) of roughly equal head (termed 'compartments') separated by head drops that reflect the head losses through the dominant chokes in the flowing system.

Almost all of the results reported here are based on constant conductance ($\sigma = 0$) networks, where any choke-like behaviour must result from the convergence of several channels towards a single channel. Whilst that behaviour appears to be exhibited by flow in long channels converging to a central opening (Fig. 8a), it does not demonstrate chokes and compartments within a more general flow regime.

It was decided therefore to search for chokes using the face-to-face boundary conditions shown in Fig. 4 and networks of long channels close to critical density. The approach required the development of an add-on code (POOLS) that searches for clusters of nodes, between which no adjacent pairs have a head drop greater than a pre-determined value (Δh). This process is repeated a number of times using increasing values of Δh until there is a small irreducible number of clusters, sometimes a single cluster.

An initial investigation of this type was performed on three realizations based on the network generation parameters used to investigate skin around tunnels. That is, a mean channel length of 10 (sub-channels/bonds) and average sub-channel densities (P_{ON}) of 0.02, 0.04 and 0.08 where P_c is just less than 0.02. The densest network did not yield any specific chokes and the number of clusters decreased gradually with increasing Δh. The two networks closer to the percolation threshold displayed behaviour associated with chokes, and heads and flows within the denser one (i.e. $P_{ON} = 0.04$) are shown in Fig. 14. The system contains five clusters of wide-ranging sizes, namely 5,301, 418, 289, 71 and 21 bonds. One of the clusters (289) is entirely separate from the rest of the network and contains flowpath 1 linking top and base directly. The largest cluster virtually coincides with the high head channels (coloured red) located in the top half of Fig. 14a and connects via several chokes to the remaining three clusters. The contrast in density between Fig. 14a, b shows how few channels provide the dominant flow pathways. Given that Fig. 14 is of a cubic system of 90 m edge, it shows how unlikely it is for boreholes in real-life tracer tests to intercept a flowing groundwater channel network.

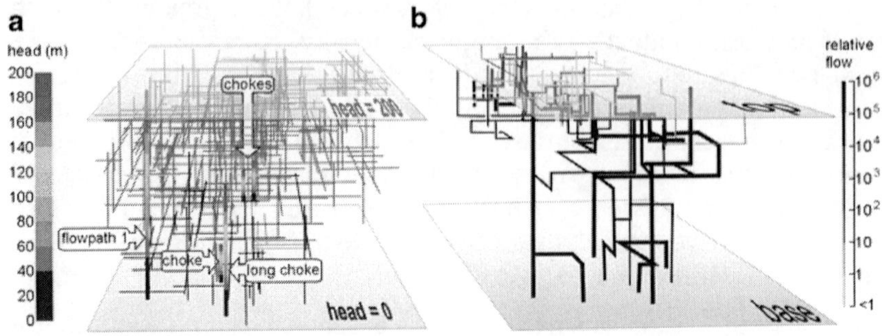

Figure. 14Heads and flows in a single realization of a HyperConv network where $L = 10$, $G = 240$ bonds and $P_{ON} = 0.04$: A head variation [note that the chokes and flowpath 1 have been visually enhanced for diagram clarity]; B Location of flowing channels (*thickness of line* is proportional to the logarithm of flow rate)

In summary, this limited investigation suggests that only networks close to the percolation threshold are likely to exhibit choke-and-compartment behavior. Hydrogeological compartments (of head) have only recently begun to be identified in fractured crystalline rocks since it requires time-consuming groundwater-based investigations (Sawada et al. 2000). Notwithstanding this difficulty, two projects around the world seeking sites for deep repositories for nuclear waste have begun to describe fractured crystalline rocks in terms of compartments (e.g. Kamaishi research mine in Japan (Sawada et al. 2000) and Bátaapáti possible repository site in Hungary (Benedek and Dankó 2009). Other repository investigations have recorded abrupt head changes in deep borehole profiles but possibly only the UK Sellafield investigation had sufficient density to warrant a compartment-based interpretation (Black and Barker 2016).

The investigation at Kamaishi research mine in granitic rock in Japan was probably the best-reported example of head compartments (Sawada et al. 2000). Seven sub-horizontal boreholes (totalling 560 m) about 350 m below surface intersected more than 3,000 fractures of which about 600 were considered open by borehole TV logging. Some of the fractures were extrapolated between boreholes based on their location, similarity of orientation and other properties such as alteration and mineralization (Fig. 15a). Hydrogeological testing for a tracer test revealed a complex arrangement of inter-borehole connections that did not seem to reflect the preceding structural analysis; however, cross-hole testing did agree with pre-existing knowledge about the disposition of head and a system of compartments was hypothesized (Fig. 15b).

Sawada et al. (2000) conclude, "the cause of the compartmentalization of flow at this site is not well understood." They also concluded that, "conductive fractures are heterogeneous internally, and a conductive fracture may not produce flow at all points of borehole penetration." The key aspect of the compartment definition at Kamaishi is the form of the boundary of the high head compartment with its complicated intrusion into its neighbours. It is obviously impossible to employ essentially planar features to define compartment boundaries of such complexity; however, it is exactly how one would expect to observe the complex interactions between the clusters of channels of Fig. 14a).

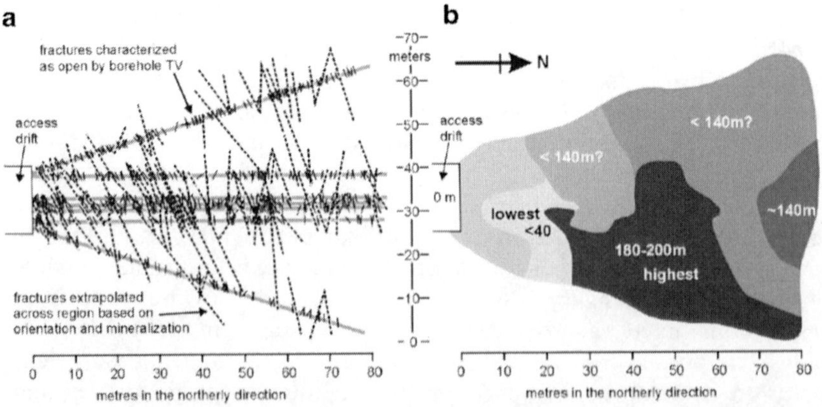

Figure. 15 Plan views of the hydrogeological and tracer testing area at Kamaishi Mine: A boreholes, open fractures and extrapolated fractures; B head compartments as derived from cross-hole testing and environmental head measurements. The compartments are coloured from *dark* equals high head to *lightest* equals low head. *Numbers* in compartments are values of head in metres above the access drift floor (figure adapted from Figs. 3, 7 and 8of Sawada et al. 2000)

PHYSICAL APPEARANCE OF LONG-CHANNEL NETWORKS

There were three phases of tracer testing performed at Stripa Mine: small-scale double fracture testing (Abelin et al. 1985), small-scale single fracture testing (Abelin et al. 1994) and the medium-scale 3D experiment described in the preceding (Abelin et al. 1991a, b). Channelling was seen to underly the results from the outset and a detailed picture emerged from the hole-to-hole measurements in the single fracture experiment (reproduced here as Fig. 16b, c).

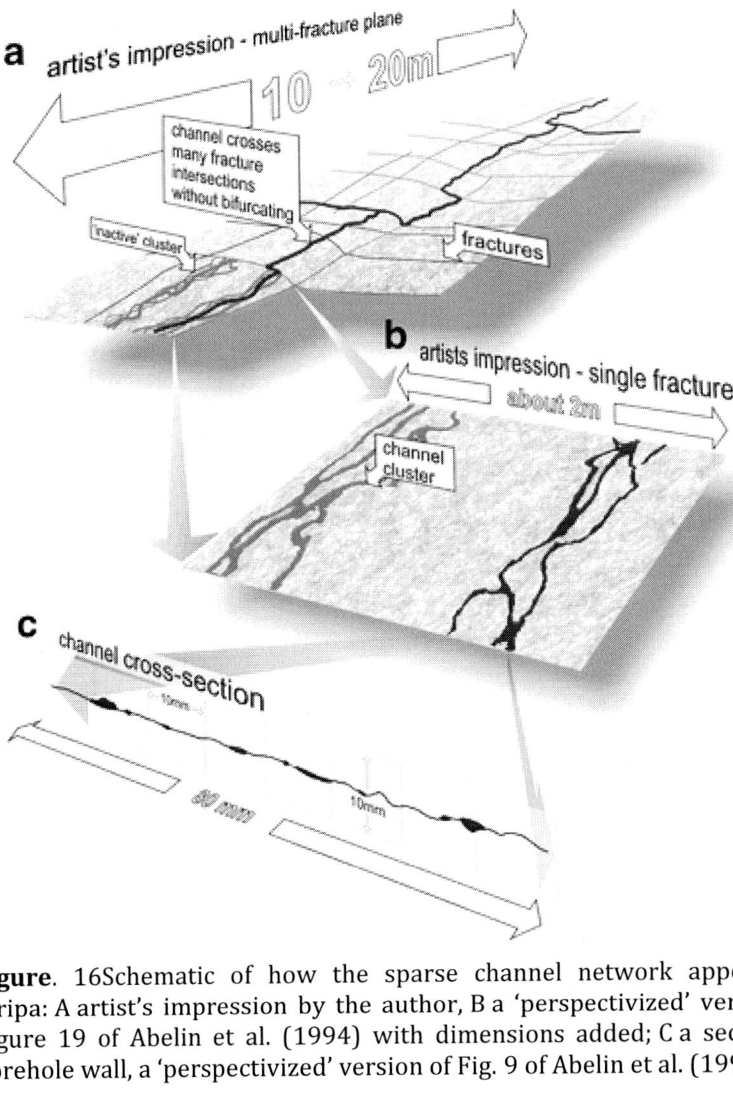

Figure. 16Schematic of how the sparse channel network appears at Stripa: A artist's impression by the author, B a 'perspectivized' version of Figure 19 of Abelin et al. (1994) with dimensions added; C a section of borehole wall, a 'perspectivized' version of Fig. 9 of Abelin et al. (1994)

The intertwined braided channels in the artist's impression in Fig. 16b are important to the overall long channel concept because they help to explain certain phenomena and are similar to the sample-based simulations of Hirano et al. (2010). The arrangement satisfies the conclusions about dead-end channels containing water that is readily accessible to tracer in actively flowing channelized groundwater (Neretnieks 2006). It also accords with the observation of 'channel swapping' in the Site Characterization and Validation Experiment at Stripa (Harding and

Black 1992), since it places 'semi-separate' channels of likely similar flow resistance in close proximity. This means that if a channel becomes clogged with either precipitate (Milodowski et al. 1998) or exsolved gas (Long et al. 1992), then there is a nearby alternative flowpath readily available. Over long periods of geological time, precipitates of very different geological age could end up in close proximity within the same fracture such as was found in the Borrowdale Volcanic Group rocks beneath West Cumbria, UK (Milodowski et al. 1998).

It should be noted that channelized flow has been proposed many times before, though mainly at the scale within a single fracture (Neretnieks et al. 1982; Hakami and Larsson 1996; Brown et al. 1998 and Vilarrasa et al. 2011). Some authors have proposed multi-fracture channel models, though most, like Cacas et al. (1990), envisage channel bifurcation at every fracture intersection. Only the model of Tsang et al. (1991) explicitly conceptualizes channels crossing fracture intersections without bifurcating. The concept visualized in Fig. 16 only differs from that of Tsang et al. (1991) in the extensiveness of the non-bifurcating channels such that the distance between major channel junctions—i.e. ignoring channel choices within clusters (Fig. 16b)—is shown to be 10–20 m.

CONCLUSIONS

A review of a decades-old tracer experiment in Stripa URL coupled with the widely noted patchiness of inflows to tunnels prompted the concept of sparse networks of long channels—that is, networks of channels that do not bifurcate at every fracture-to-fracture intersection but form long sinuous channels with only occasional junctions.

The concept has been investigated using an innovative 3D two-parameter bond lattice network model named HyperConv where the mean length of 'strings' of connected bonds (channels) and the gaps between them can be controlled by separate probability functions. It is found that networks of long channels are able to percolate at lower values of (bond) density than networks of short channels. A general relationship between mean channel length, mean gap length and probability of percolation has been developed which incorporates the well-established result for classical lattice network models as a special case.

Positive skin effects were common in the underground experiments at Stripa and were interpreted as regions of reduced hydraulic conductivity. HyperConv simulations were performed in this study using cylindrically convergent boundary conditions and yielded apparent positive skin under certain circumstances. More specifically, only realizations of channels longer than the diameter of the inner boundary (e.g. the tunnel) exhibited positive skin and reached maximum values just above critical density. Based on these numerical model results and calculations of a single channel configuration, a semi-generic pattern of behaviour is postulated that indicates that networks of short channels (relative to the size of the underground opening) are unlikely to yield positive values of apparent skin. A few ensembles based on variable conductance bonds (all other realizations used constant conductance) had only a minor effect on this unexpected behaviour.

A relatively simple analytical relationship in terms of overall channel density and the frequency of channels intersecting the inner boundary is developed that largely replicates the behaviour seen in the numerical experiments. Several researchers have suggested in the past that flow in sparse networks will be dominated by a few key connecting channels (or fractures) termed 'chokes'. A few numerical experiments were undertaken where it was found that chokes only become significant close to the percolation threshold. Values of channel density greater than about 4× critical density seem to rule out the creation of recognisable chokes. The occurrence of chokes is linked to the creation of head 'compartments' which, although the HyperConv simulations represent conditions at the Stripa URL, appear very similar to actual head compartments reported at Kamaishi experimental mine in Japan.

It is concluded that, based on this relatively limited numerical study, the long-channel concept is capable of reproducing many groundwater-related effects observed in underground research laboratories in fractured crystalline rocks. Networks of long channels are intimately linked to sparseness which is itself linked to 'apparent skin', chokes and head compartments. Where these effects are observed would seem to indicate that the groundwater is flowing in a long-channel network close to critical density. These links between sparse long-channel networks and observable features in the field present the possibility of recognising and characterising such networks in future site characterization projects for nuclear repositories in fractured crystalline rock.

ACKNOWLEDGEMENTS

The model 'HyperConv' was developed by Peter Robinson of Quintessa Ltd. This study was supported in part by the Swedish Nuclear Fuel and Waste Management Company, SKB.

APPENDIX: THE INTERPRETATION OF (HYDRAULIC) SKIN AROUND TUNNELS BASED ON HEAD MEASUREMENTS

'Skin' is a concept used to describe poor performance of groundwater and oil production boreholes relative to what would be expected under 'ideal' conditions. It is a catchall term for effects due to permeability variations close to the borehole wall. Within nuclear repository projects, the concept has been adapted to apply to tunnels and drifts (Rhén et al. 1997), even though tunnels are horizontal rather than vertical and time-scales, length-scales and rocks are significantly different.

In groundwater hydraulics, skin is usually quantified in terms of a dimensionless 'skin factor', s_f, additional to any well function describing head loss involved in flow through an aquifer towards a sink. Normally, this factor would be added to the well function evaluated at the well, but to adopt the convention of the tunnel being at zero head, the skin factor is added to the head in the formation outside the well:

$$h(r) - h(r_t) = \frac{q}{2\pi K}\left[W(r_t, r_t) + s_f\right] \quad r > r_t \tag{9}$$

where the mathematical terms are as follows:

h(r) head at radial distance *r* from the centre of the tunnel

q inflow rate per unit length of tunnel

K average hydraulic conductivity of the rock surrounding the tunnel

W(r_t,r_t) well function describing steady-state head during cylindrically convergent flow to a tunnel. This function would normally be formulated to be zero at the tunnel wall, $W(r_t,r_t) = 0$

***s_f*'** skin factor' (dimensionless)

Using the well function representing cylindrically symmetrical flow in an infinite system, Eq. (9) can be re-arranged to give:

$$h(r) - h(r_t) = \frac{q}{2\pi K}\ln\left(\frac{r}{r_{\mathrm{eff}}}\right) \tag{10}$$

where an effective radius is introduced given by:

$$r_{\text{eff}} = r_{\text{t}} e^{-s_{\text{f}}} \tag{11}$$

Thus, a positive skin factor (equating to a skin of reduced permeability) corresponds to an effective radius of the tunnel which is less than the actual radius, and vice versa (see Fig. 17e, f).

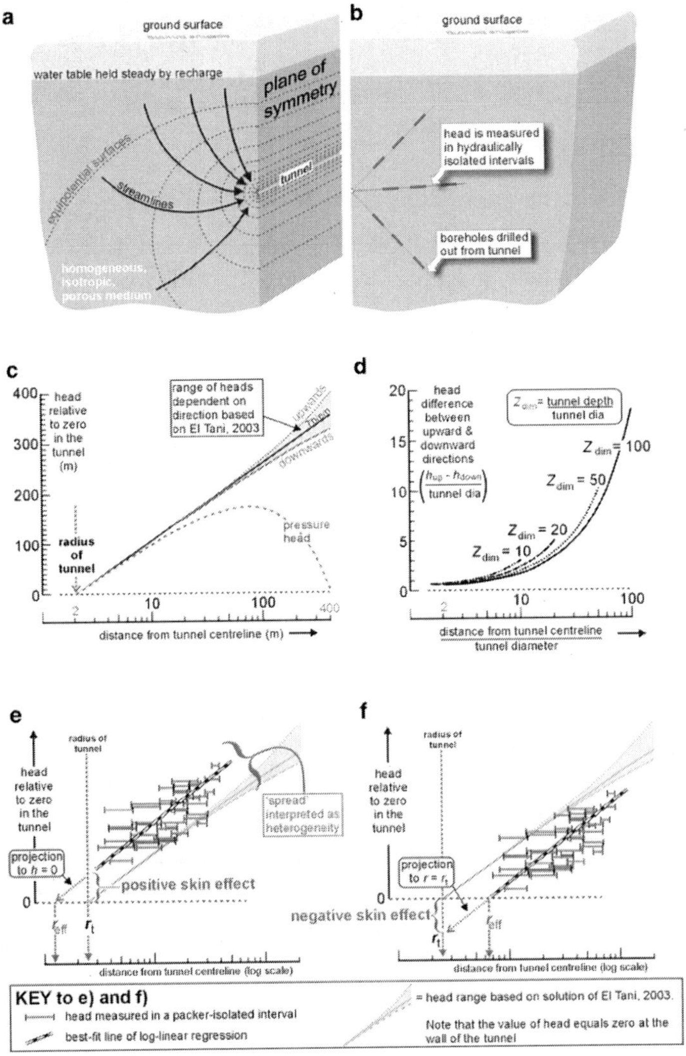

Figure. 17Steady-state inflow to a deep tunnel and the interpretation of skin

A somewhat more real-world solution than Eq. (9) is obtained by assuming a horizontal water table (at a constant head) combined with constant pressure conditions at the tunnel wall (like an air-filled tunnel) to yield a solution of more general application. Equations for the head distribution were derived by El Tani (2003). The full solution is too complex to reproduce here but a key result is the flux to the tunnel:

$$q = \frac{2\pi K z_t}{\left(\frac{\varsigma^2+1}{\varsigma^2-1}\right)\ln\varsigma} \qquad \varsigma = \frac{z_t}{r_t} - \sqrt{\frac{z_t^2}{r_t^2} - 1} \tag{12}$$

where z_t = depth of the tunnel centre below the water table.

When z_t/r_t is large, the El Tani (2003) solution differs from the Thiem equation (with zero head at radius $2r_t$) by less than 1% when $z_t/r_t > 7.2$.

As can be seen in Fig. 17a, the system is symmetrical about the vertical plane intersecting the axis of the tunnel but not about the horizontal plane, since all the inflow is sourced from the water table above. As the ratio of depth below the water table relative to tunnel diameter increases, the head distribution becomes more radially symmetrical about the tunnel axis. Cylindrically symmetrical flow at great depth is the assumption underlying the well-known approximate solution of Goodman et al. (1965). The asymmetry of the solution is seen as divergent heads dependent on direction from the tunnel centreline (Fig. 17c).

The maximum amount of divergence occurs between vertically upward and vertically downward directions and can be generalized by expressing depth and distance in terms of tunnel diameters (Fig. 17d). For example, a 4-m-diameter tunnel 400 m below the water table (i.e. 100 tunnel diameters) would have upward and downward measurements of head that differed by 4 m (i.e. 1 tunnel diameter) at 6.1 tunnel diameters distance from the drift centreline (i.e. 24 m; from Fig. 17d).

It can be concluded that for either a water-filled tunnel or a tunnel at atmospheric pressure, groundwater flow can be approximated as cylindrically convergent flow to a line source when at distances of more than a few times the tunnel diameter. This can be represented by the Thiem equation with skin:

$$h(r) - h(r_t) = \frac{q}{2\pi K}\left(\ln\frac{r}{r_t} + s_f\right) \tag{13}$$

Hence, a plot of head versus the log of distance from the tunnel centreline should yield a straight line passing through zero at the tunnel wall (i.e. Fig. 17c). The deviation of real measurements from this ideal gives rise to interpretations of skin for tunnels.

Most commonly, head variation around tunnels is measured in boreholes drilled 'out' from the tunnel itself so that measurements are readily expressed in terms of distance from the tunnel (Fig. 17b). In nuclear repository projects, boreholes are instrumented with packers to hydraulically isolate individual sections so that resultant head measurements apply to specific borehole intervals (e.g. as in Fig. 17b, e, f).

All of the formulae given earlier are based on the assumption of flow in a homogeneous isotropic porous medium. In real fractured crystalline rock, measurements usually vary much more widely than that caused by the asymmetry of the theoretical solution. This variation is usually interpreted as resulting from the basic heterogeneity of fractured rock and the regression line is assumed to be indicative of the average behaviour of the groundwater system interacting with the underground-opening. In many cases, the regression line does not intercept the tunnel wall at zero head (Fig. 17e, f).

REFERENCES

1. Abelin H, Neretnieks I, Tunbrant S, Moreno L (1985) Final report of the migration in a single fracture: experimental results and evaluation. OECD/NEA, Stripa Project TR 85–03, Swedish Nuclear Fuel and Waste Management Co., Stockholm, 103, pp.
2. Abelin H, Birgersson L, Gidlund J, Moreno L, Neretnieks I (1991a) A large-scale flow and tracer experiment in granite: 1. experimental design and flow distribution. Water Resour Res 27:3107–3117. doi:10.1029/91WR01405.
3. Abelin H, Birgersson L, Moreno L, Widén H, Ågren T, Neretnieks I (1991b) A large-scale flow and tracer experiment in granite: 2. results and interpretation. Water Resour Res 27:3119–3135. doi:10.1029/91WR01404.
4. Abelin H, Birgersson L, Widén H, Ågren T, Moreno L, Neretnieks I (1994) Channeling experiments in crystalline fractured rocks. J Contaminant Hydrol 15:129–158. doi:10.1016/0169-7722(94)90022-1.
5. Andersson P, Byegård J, Winberg A (2002) TRUE Block Scale Project Final Report 2: tracer tests in the block scale. Technical report TR-02-14, Swedish Nuclear Fuel and Waste Management Co., Stockholm.
6. Andersson P, Byegård J, Billaux D, Cvetkovic V, Dershowitz W, Doe T, Hermanson J, Poteri A, Tullborg E-L, Winberg A (2007) TRUE Block Scale Continuation Project final report. TR-02-14, Technical report, Swedish Nuclear Fuel and Waste Management Co., Stockholm.
7. Barker JA (1988) A generalised radial flow model for hydraulic tests in fractured rock. Water Resour Res 24:1796–1804. doi:10.1029/WR024i010p01796.
8. Benedek K, Dankó G (2009) Stochastic hydrogeological modelling of fractured rocks: a generic case study in the Mórágy Granite Formation (South Hungary). Geol Carpath 60/4:271–281. doi:10.2478/v10096-009-0019-y.

9. Black JH, Barker JA (2016) The puzzle of high heads beneath the West Cumbrian coast, UK: a possible solution. Hydrogeol J. 24:439–457. doi:10.1007/s10040-015-1340-4.
10. Black JH, Robinson PC, Barker JA (2006) A preliminary investigation of the concept of hyper-convergence. Research report R-06-30, Swedish Nuclear Fuel and Waste Management Co., Stockholm, 66 pp.
11. Black JH, Barker JA, Woodman N (2007) An investigation of 'sparse channel networks': characteristic behaviours and their causes. Research report R-07-35, Swedish Nuclear Fuel and Waste Management Co., Stockholm, 123 pp.
12. Brown S, Caprihan A, Hardy R (1998) Experimental observation of fluid flow channels in a single fracture. J Geophys Res 103(B3):5125–5132. doi:10.1029/97JB0354.
13. Cacas MC, Ledoux E, de Marsily G, Tillie B, Barbreau A, Durand E, Feuga B, Peaudecerf P (1990) Modeling fracture flow with a stochastic discrete fracture network: calibration and validation—the flow model. Water Resour Res 26/3:479–489. doi:10.1029/WR026i003p00479.
14. Chapman NA, McKinley IG, Penna Franca E, Shea MJ, Smellie JAT (1992) The Pocos de Caldas project: an introduction and summary of its implications for radioactive waste disposal. J Geochem Explor 45:1–24. doi:10.1016/0375-6742(92)90120-W
15. Chijimatsu M, Fujita T, Sugita Y, Ishikawa H (1996) Coupled thermo-hydro-mechanical experiment at Kamaishi Mine: initial data around T-H-M experiment area. PNC report 03-95-03, JAEA, Ibaraki, Japan.
16. Einsiedl F, Maloszewski P (2005) Tracer tests in fractured rocks with a new fluorescent dye: pyrene-1,3,6,8-tetra sulphonic acid (PTS). Hydrol Sci J 50(3):543–554. doi:10.1623/hysj.50.3.543.65026.
17. El Tani M (2003) Circular tunnel in a semi-infinite aquifer. Tunn Undergr Space Technol 18:49–55. doi:10.1016/S0886-7798(02)00102-5.
18. Elert M, Svensson H (2001) Evaluation of modelling of the TRUE-1 radially converging tests with conservative tracers. The Äspö Task Force on Modelling of Groundwater Flow and Transport of Solutes, Tasks 4E and 4F. Technical report TR-01-12, Swedish Nuclear Fuel and Waste Management Co., Stockholm.
19. Follin S, Hartley L (2014) Approaches to confirmatory testing of a groundwater flow model for sparsely fractured crystalline rock, exemplified by data from the proposed high-level nuclear waste repository site at Forsmark, Sweden. Hydrogeol J 22:333–349. doi:10.1007/s10040-013-1079-8.
20. Goodman RE, Moye DG, van Schalkwyk A, Javandel I (1965) Ground water inflows during tunnel driving. Eng Geol 2(1):39–56
21. Hakami E, Larsson E (1996) Aperture measurements and flow experiments on a single natural fracture. Int J Rock Mech Min Sci Geomech Abstr 33(4):395–404. doi:10.1016/0148-9062(95)00070-4.
22. Harding W, Black J (1992) Site characterization and validation: inflow to the validation drift. OECD/NEA technical report 92-18 of the Stripa Project, Swedish Nuclear Fuel and Waste Management Co., Stockholm, 84 pp.
23. Harter T (2005) Finite-size scaling analysis of percolation in three-dimensional correlated binary Markov chain random fields. Phys Rev E 72:026120. doi:10.1103/PhysRevE.72.026120.
24. Hirano N, Ishibashi T, Watanabe N, Okamoto A, Tsuchiya N (2010) New concept discrete fracture network model simulator, GeoFlow, and three dimensional channeling flow in fracture network. Proceedings World Geothermal Congress 2010 Bali, Indonesia, 25–29 April 2010.
25. Hunt A (2005) Percolation theory and the future of hydrogeology. Hydrogeol J 13:202–205. doi:10.1007/s10040-004-0405-6.
26. Long JCS, Olsson O, Martel S, Black JH (1992) Effects of excavation on water inflow to a drift. In: Myer LR, Tsang C-F, Cook NGW, Goodman RE (eds) Fractured and jointed rock masses. Proc. of Conf. on Fractured and Jointed Rock Masses, Lake Tahoe, California. Balkema, Rotterdam, The Netherlands, pp 543–549.

27. Margolin G, Berkowitz B, Scher H (1998) Structure, flow, and generalized conductivity scaling. Water Resour Res 34:2103–2121. doi:10.1029/98WR01648.
28. Masset O, Loew S (2010) Hydraulic conductivity distribution in crystalline rocks, derived from inflows to tunnels and galleries in the Central Alps, Switzerland. Hydrogeol J 18(4):863–892. doi:10.1007/s10040-009-10569-1.
29. Milodowski AE, Gillespie MR, Naden J, Fortey NJ, Shepherd TJ, Pearce JM, Metcalfe R (1998) The petrology and paragenesis of fracture mineralization in the Sellafield area, west Cumbria. Proc Yorkshire Geol Soc 52(2):215–242.
30. Moreno L, Neretnieks I (1989) Channeling in fractured zones and its potential impact on the transport of radionuclides. Symposium on Scientific Basis for Nuclear Waste Management XII, Berlin, 10–13 October 1988, pp 779–786.
31. Moreno L, Neretnieks I (1993) Fluid flow and solute transport in a network of channels. J Contaminant Hydrol 14:163–192. doi:10.1016/0169-7722(93)90023-L.
32. Neretnieks I (2006) Channeling with diffusion into stagnant water and into a matrix in series. Water Resour Res 42, W11418, 15 pp. doi:10.1029/2005WR004448.
33. Neretnieks I, Eriksen T, Tähtinen P (1982) Tracer movement in a single fissure in granitic rock: some experimental results and their interpretation. Water Resour Res 18:849–863. doi:10.1029/WR018i004p00849.
34. Pusch R, Nilsson J, Ramqvist G (1985) Final report of the Buffer Mass Test, vol I: scope preparative fieldwork and test arrangement. Tech. Report no. 85-11, Report of the Stripa Project, Swedish Nuclear Fuel and Waste Management Co., Stockholm.
35. Rhén I, Gustafson G, Wikberg P (1997) Äspö HRL: geoscientific evaluation 1997/4—results from pre-investigations and detailed site characterization—comparison of predictions and observations—hydrogeology, groundwater chemistry and transport of solutes. Technical report no. 97-05, Swedish Nuclear Fuel and Waste management Co., Stockholm, 334 pp.
36. Robinson P (1984) Connectivity, flow and transport in network models of fractured media. PhD Thesis, Oxford.
37. Rouleau A, Gale JE (1985) Statistical characterization of the fracture system in the Stripa Granite, Sweden. Int J Rock Min Sci Geomech Abstr 22/6:353–367. doi:10.1016/0148-9062(85)90001-4.
38. Rubin Y (2003) Applied stochastic hydrogeology. Oxford University Press, Oxford, 390 pp.
39. Sawada A, Uchida M, Shimo M, Yamamoto H, Takahara H, Doe TW (2000) Non-sorbing tracer migration experiments in fractured rock at the Kamaishi Mine, northeast Japan. Eng Geol 56(1–2):75–96. doi:10.1016/S0013-7952(99)00135-0.
40. Selroos J-O, Walker DD, Strom A, Gylling B, Follin S (2002) Comparison of alternative modelling approaches for groundwater flow in fractured rock. J Hydrol 257:174–188. doi:10.1016/S0022-1694(01)00551-0.
41. Tsang C-F, Neretnieks I (1998) Flow channeling in heterogeneous fractured rocks. Rev Geophys 36(2):275–298. doi:10.1029/97RG03319.
42. Tsang C-F, Tsang YW, Hale FV (1991) Tracer transport in fractures: analysis of field data based on a variable aperture channel model. Water Resour Res 27(12):3095–3106. doi:10.1029/91WR02270.
43. Tsang C-F, Neretnieks I, Tsang Y (2015) Hydrologic issues associated with nuclear waste repositories. Water Resour Res 51:6923–6972. doi:10.1002/2015WR017641.
44. Vilarrasa V, Koyama T, Neretnieks I, Jing L (2011) Shear induced flow channels in a single rock fracture and their effect on solute transport. Transp Porous Media 87:503–523. doi:10.1007/s11242-010-9698-1.
45. Wilson CR, Long JCS, Galbraith RM, Karasaki K, Endo HK, DuBois AO, McPherson MJ, Ramqvist G (1981) Geohydrological data from the Macropermeability Experiment at Stripa, Sweden. Report no. LBL-12520, Lawrence Berkeley Natl. Lab., Berkeley, CA, 194 pp.

46. Witherspoon PA, Bodvarsson GS (2006) Geological challenges in radioactive waste isolation, fourth worldwide review, Rep. LBNL-59808, Lawrence Berkeley Natl. Lab., Berkeley, CA.

CHAPTER 9

Delineating Groundwater Vulnerability and Protection Zone Mapping in Fractured Rock Masses: Focus on the DISCO Index

Helen Meerkhan [1], José Teixeira [1,2,3], Jorge Espinha Marques [4], Maria José Afonso [1,2] and Helder I. Chaminé [1,2,*]

[1] *Laboratory of Cartography and Applied Geology (LABCARGA), Department of Geotechnical Engineering, School of Engineering (ISEP), Polytechnic of Porto, 4200-072 Porto, Portugal*

[2] *Centre GeoBioTec|UA—Georesources, Geotechnics and Geomaterials group, University of Aveiro, 3810-193 Aveiro, Portugal*

[3] *Department of Geography, Faculty of Arts, University of Porto (FLUP), 4099-002 Porto, Portugal*

[4] *Institute of Earth Sciences (ICT|UP), Department of Geosciences, Environmental and Land Planning, Faculty of Sciences, University of Porto (FCUP), 4099-002 Porto, Portugal*

ABSTRACT

Hard-rock catchments are considered to be source of valuable water resources for water supply to inhabitants and ecosystems. The present work aims to develop a groundwater vulnerability approach in the Caldas da Cavaca hydromineral system (Aguiar da Beira, Central Portugal) in order to improve the hydrogeological conceptual site model. Different types of information were overlaid, generating several thematic maps to achieve an integrated framework of key sectors in the study site. Thus, a multi-technical approach was used, encompassing field and laboratory techniques, whereby different types of data were collected from fields such as geology, hydrogeology, applied geomorphology and geophysics and hydrogeomechanics, with the fundamental aim of applying the so-called DISCO index method. All of these techniques were successfully performed and an intrinsic groundwater vulnerability to contamination assessment, based on the multicriteria methodology of GOD-S, DRASTIC-Fm, SINTACS, SI and DISCO indexes, was delineated. Geographic Information Systems (GIS) provided the basis on which to organize and integrate the databases and to produce all the thematic maps. This multi-technical approach highlights the importance of groundwater vulnerability to contamination mapping as a tool to support hydrogeological conceptualization, contributing to improving the decision-making process regarding water resources management and sustainability.

Keywords: hard-rock hydrogeology; fractured media; GIS mapping; vulnerability assessment; DISCO index; hydrogeological conceptual model

1. INTRODUCTION

Groundwater in fractured aquifers is considered an important source for irrigation, water supply, and industrial purposes [1]. Moreover, fissured rock mass hydrosystems are considered essential sources of groundwater resources in large areas of the Earth. Fractures have a dual role, since they may act as hydraulic systems, providing preferential flow paths, as well as barriers that prevent flow across them, contributing to control of the transport of contaminants through and into the subsurface [2,3,4,5]. Many factors control the groundwater flow paths and the impact on the potential leakage of contaminants into groundwater, which in turn lead to the vulnerability of groundwater to contamination in varying degrees. These factors include lithology, geostructure, topography, fracture network, slope, weathering grade, permeability, drainage pattern, climate and land use [6].

The term "groundwater vulnerability" was first mentioned in the 1970s [7] and became used on a large scale in the 1980s [8]. Groundwater vulnerability is not a property that can be directly measured in the field. It is a concept that is based on the existence of areas that are more vulnerable to groundwater contamination than others. In fractured aquifers, extreme vulnerability is associated with a shallow water table, where fractures provide low contaminant decrease [9,10,11]. In this case, fractures increase vertical permeability, which affects the recharge rate and decreases the amount of contaminant attenuation. Thus, in fractured hydrosystems where groundwater flow is relatively easy and rapid, contamination may become widely dispersed [12,13].

The assessment of groundwater vulnerability to contamination aims to identify zones with high contamination risk in order to assess the possibility of an aquifer being contaminated [14,15]. Groundwater vulnerability assessments are usually represented by means of a map displaying zones where the resource is vulnerable to contamination from a number of sources. Vulnerability mapping is a suitable technique for assessing hydrogeological factors, among others, with the groundwater for potential contamination in a specific region shown on a map [16]. Displaying zones in different colors is an easy and intuitive way to classify, distinguish and interpret groundwater vulnerability, and it can also be used for delineating protection zones [17]. In short, hydrogeological mapping, groundwater vulnerability mapping and Geographic Information System (GIS) water-related mapping are excellent tools for supporting the description, assessment, modelling and communicating of groundwater resources [9,10,11,16,18,19, 20,21,22,23,24].

The main objective of this study is to contribute to the development of a methodology for assessing groundwater vulnerability to pollution and the delineation of protection zones in fractured media. To achieve this goal, groundwater vulnerability indexes were applied to the Caldas da Cavaca site to point out the most vulnerable zones. The research is based on a hydrogeological system case study of Caldas da Cavaca [25,26,27]. The fractured rock media were characterized with GIS-based vulnerability mapping with a multi-technique approach involving geological, morphotectonic, hydrological, hydrogeological, geophysical and geotechnical data. Finally, the collected data allowed the generation of thematic maps, encompassing lithology, weathering grade, tectonic lineaments density, rock hydrogeomechanics, land cover, drainage network density, slope and rainfall.

Due to the great diversity of geological and hydrogeological conditions in such fractured media, it is not possible to delineate protection zones for these groundwater systems by means of a single approach. Thus, several methods were applied, such as GOD-S [10], DRASTIC-Fm [8,28], SINTACS [29,30], SI [31,32] and DISCO [33,34]. Moreover, this study focused on the application and discussion of the importance of the DISCO methodology on wells and spring areas. Finally, this study highlights the role of groundwater vulnerability mapping in developing environmental hydrogeology tools, contributing to the development of water resource management, environmental sustainability and groundwater protection.

2. REGIONAL FRAMEWORK: CALDAS DA CAVACA HYDROMINERAL SYSTEM

The Caldas da Cavaca site (40°46′ N–40°47′ N latitude and 7°34′ W–7°35′ W longitude) is an area which has had a thermal spa tradition since late nineteenth century [35,36,37,38]. It is situated in Central Portugal, in the municipality of Aguiar da Beira, covering an area of about 10 km^2 and belonging to the River Dão catchment, which is part of the basin of Mondego River (Figure 1).

The study region is located in the Central-Iberian Zone and is also considered part of the Dão granitic complex from the Beiras granitic belt [39], close to the western boundary of the Bragança–Vilariça–Manteigas major fault zone [31], following a general trend towards NNE-SSW. The main tectonic structure is the Dão fault zone with a general trend towards NE-SW and fracture network systems, each one affecting the regional drainage network and controlling the thermal water circulation [39]. At the local scale, the Ribeira de Coja fault zone is a main morphotectonic feature, trending NNE–SSW, with steep slopes and altitude difference of about 170 m [40].

The geology of the site comprises thin alluvial deposits (argillaceous silty sandy deposits) and granitic rocks which encompass coarse grained granites occasionally interrupted by quartz veins, pegmatite-aplite veins or dolerite dykes, where most of the basic rocks are exposed and fresh to weathered [39]. Regarding the landscape surrounding the site, it is mainly characterized by the existence of agricultural areas, *Pinus pinaster* forest and a large area of granitic rock outcrops.

Generally, the climate in Caldas da Cavaca region is moderate, which means a temperate Mediterranean climate with a dry and warm summer (Csb according to Köppen classification [42]). The average annual temperature is 13 °C, ranging from 6.2 °C in January to 20.1 °C in July. The region suffers from water deficit from June to September, particularly in July and August. The average annual rainfall is 1252.4 mm/year, reaching 189 mm in January (the wettest month) and 16 mm in July (the driest month) [40].

The Caldas da Cavaca site is characterized by the presence of shallow groundwaters and hydromineral waters, where the shallow waters correspond to weathered or fractured granitic zones and surface water/groundwater from alluvia. The hydromineral waters are related to deep circulation in rock masses [40,41]. Recently, hydrogeological, applied geomorphological and hydrogeomechanical studies were carried

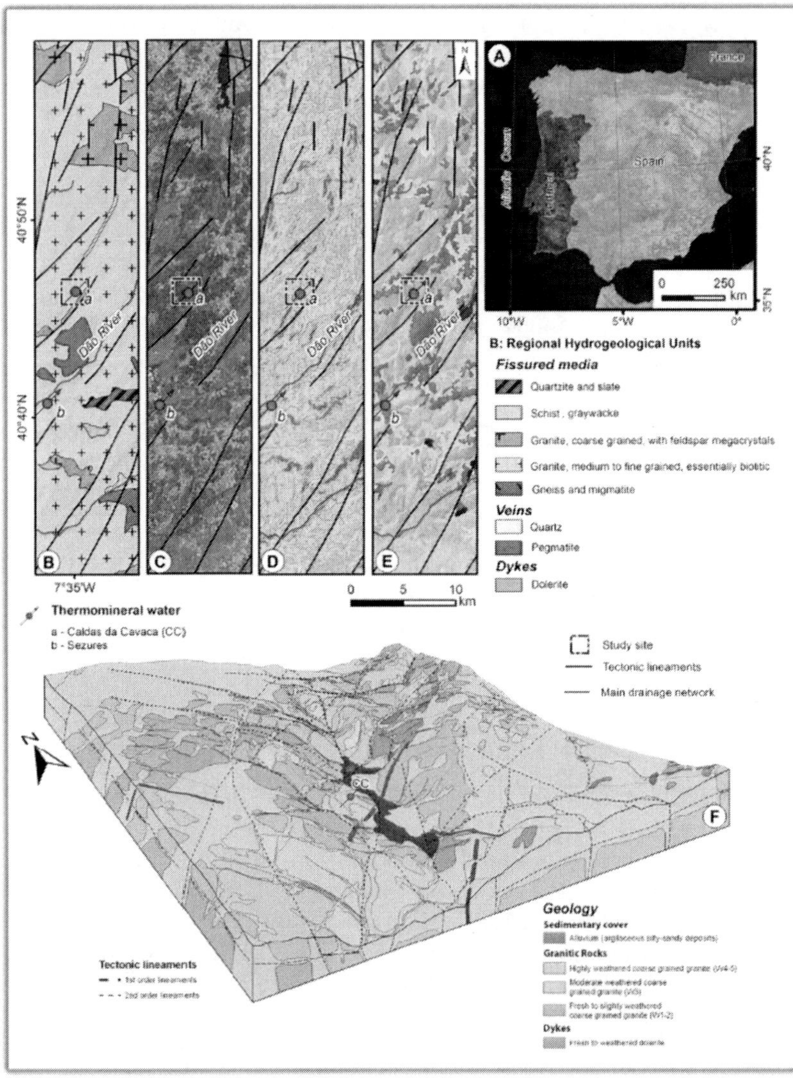

Figure 1. Regional framework of the study area (Caldas da Cavaca hydromineral system, Aguiar da Beira), based on [25,39,40,41]: (**A**) Regional location of the study site (Central Portugal); (**B**) shaded relief and regional hydrogeology; (**C**) Satellite image and tectonic lineaments; (**D**) Slope; (**E**) Land cover; (**F**) Block diagram illustrating the basic geological framework of the study site.

out in this site as a result of the need to improve the management of the thermal water resources (e.g., [25,26,27,41,43]). In the past, water withdrawal relied on a spring and a well which were not able to assure the required thermal water volume and quality. Presently, the thermomineral aquifer is exploited by means of two new wells, constructed according to the best technical procedures [44,45,46].

3. MATERIALS AND METHODS

This study was successfully undertaken through the combination of several major steps, where all the data collection techniques can be classified into two main categories (Figure 2) [21,22,26,47,48]: (i) mapping data: which include topography, remote sensing, morphotectonics, structural geology, land use and climatology; (ii) field and laboratory data: which comprise rock mass geotechnics, hydrogeological inventory, and hydrogeochemical and isotopic analysis.

The field and laboratory data were systematized and assessed in a GIS environment, where the GIS technique mapping provided the geology (i.e., lithology, structure, and weathering grade), drainage, land cover, slope, and rainfall, in order to get a comprehensive framework of the study site and the assessment of groundwater vulnerability areas. These maps were classified into four main categories: (i) geological description of rock masses; (ii) geographical description; (iii) hydrogeomorphological features and (iv) hydrological features.

The infiltration potential is a key parameter in some vulnerability indexes [26,40,48]. Thus, the identification and calculation of the infiltration potential zoning index was very important in this study. The internal scores were evaluated using fieldwork data and the identification of the relative weight and score for each factor was achieved using the Analytical Hierarchy Process (AHP) method to ensure objective weight assignment [24,26,49]. The grid data structure consists of a pixel of 1 m × 1 m.

The importance of GIS-based mapping comes from its ability to produce geodatabases and to create vulnerability maps. Thus, the evaluation of groundwater vulnerability in this study was carried out using several methods, each one adopting a specific set of parameters:

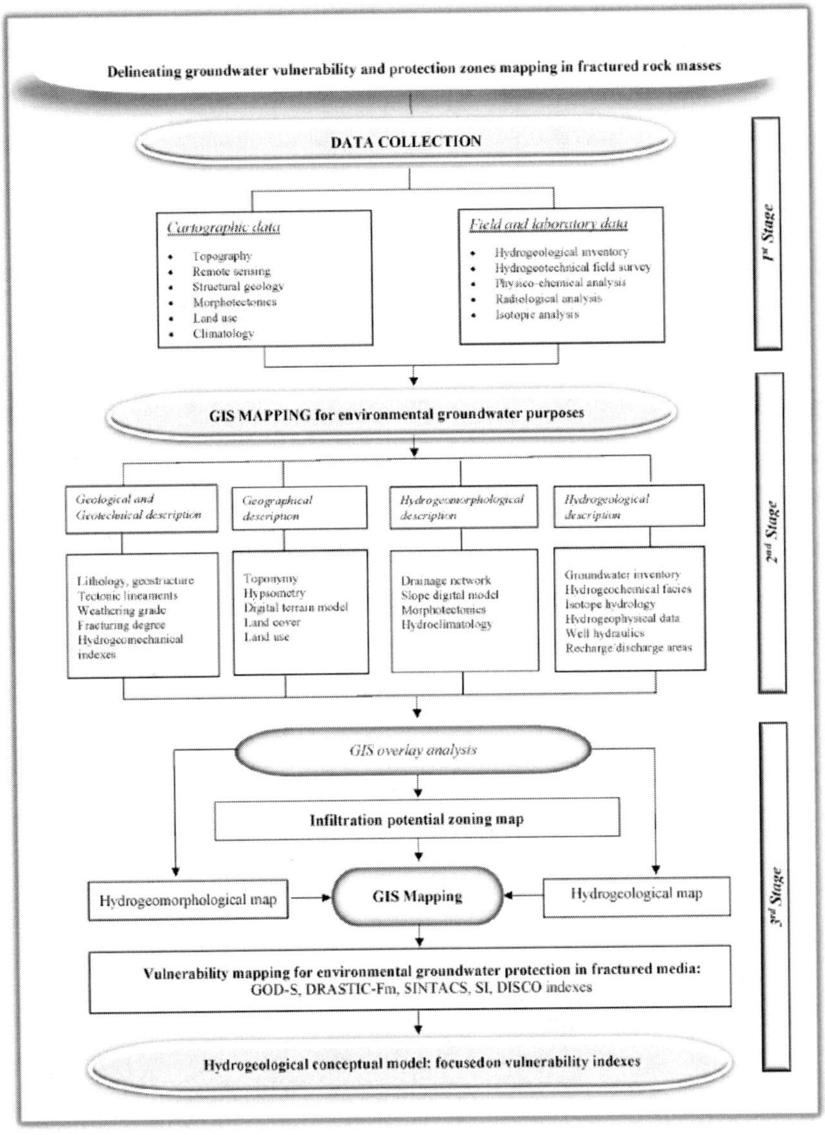

Figure 2. Conceptual flowchart representing the methodology used in the study.

- GOD-S is an evolution of GOD index, considering soil media properties [10];
- DRASTIC-Fm is a modified version of DRASTIC, taking into consideration the characteristics of fissured hard-rock aquifers.

The fractured media parameter (Fm) was derived from the map of tectonic lineaments density, and collected into four classes, with ratings ranging from 4 to 10, depending on the tectonic lineament density [8,26,28,48,50];

- SINTACS was created for vulnerability assessments and mapping in medium and large scale maps, and the multiplier weights of normal string were used [29,30];
- SI, where the Land Use (LU) parameter was derived from land cover maps [32];
- DISCO is carried out in the vulnerable springs and in the mineral water wells related to a highly heterogeneous and fractured media aquifer [33,34].

The GOD-S, DRASTIC-Fm, SINTACS and SI multi-criteria indexes are often used for local and regional mapping purposes (medium to small-scale), and the DISCO index [34] is very useful for in situ environmental hydrogeological investigations (large-scale maps). The most efficient methodology is an integrative approach encompassing all previous multi-criteria indexes [22].

Finally, a comprehensive approach was achieved in order to provide knowledge of the groundwater systems of the Caldas da Cavaca site and to improve the hydrogeological conceptual model and thus develop environmental sustainability and groundwater protection.

4. RESULTS AND DISCUSSION

4.1. Local Hydrogeological, Hydrogeomorphological and Hydrogeomechanical Framework

The hydrogeological inventory has proven to be an essential methodology for evaluating the Caldas da Cavaca hydromineral system, and several fieldwork campaigns have been performed. During the fieldwork campaigns, electrical conductivity, yield, water temperature and pH were measured (Figure 3A). It was found that water points are distributed unevenly, located at small depths (less than 5 m), and are connected to the unconfined aquifer [25,26,40]. The springs are sited at 650–700 m, show water temperatures of below 17 °C and discharge between 0.01 and 0.05 $L \cdot s^{-1}$, as well as having electrical conductivities below 50 $\mu S \cdot cm^{-1}$.

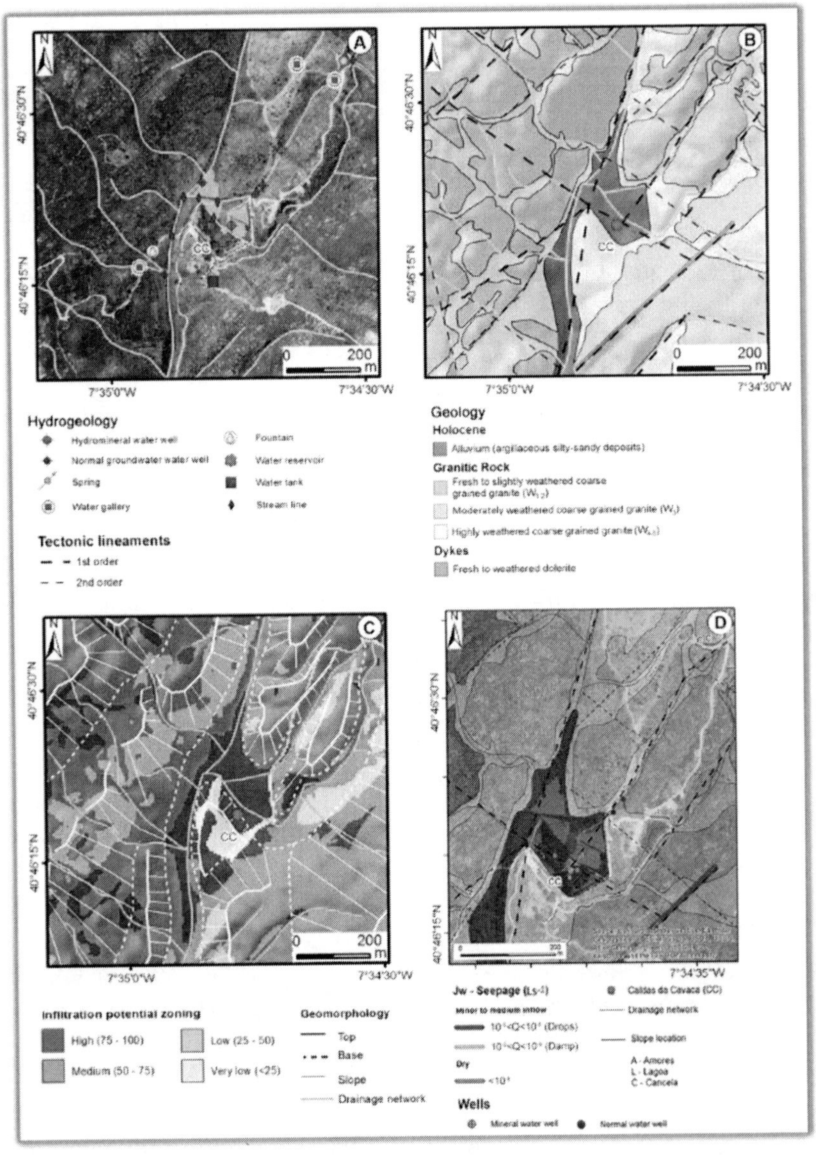

Figure 3. Local hydrogeological framework of Caldas da Cavaca site, revised and updated from [26,27,40]. Hydrogeological inventory (**A**); geology and geotechnical framework (**B**); hydrogeomorphology and infiltration potential zoning (**C**); hydrogeomechanical assessment based on seepage parameter (Jw—joint water reduction) on selected rock masses (**D**).

Two types of waters are recognized in Caldas da Cavaca [26,40]: (i) sodium chloride to sodium bicarbonate waters related to the water table aquifer and (ii) hydromineral waters (sodium bicarbonate facies) related to the confined aquifer.

The shallow groundwaters have two origins, the first from water circulating in sedimentary deposits (with electrical conductivity up to 20 $\mu S \cdot cm^{-1}$ and pH ranging from 5.5 to 6) and the second from groundwater circulating in weathered or fractured granitic zones, with electrical conductivity of 20–50 $\mu S \cdot cm^{-1}$ and pH ranging from 5 to 6.5. However, this aquifer has long-term well yields below 1 $L \cdot s^{-1}$ and transmissivity values below 1 m^2/day. The springs in the site also have low yield values (around 0.1 $L \cdot s^{-1}$) [26,40].

Caldas da Cavaca hydromineral waters are very different to shallow waters due to the following features [43]: (i) high temperature, around 29.8 °C; (ii) relatively high pH values, around 8.5; (iii) high silica contents, around 55 mg/L; (iv) electrical conductivity ranges from 353 to 427 $\mu S \cdot cm^{-1}$; (v) Total Dissolved Solids (TDS) contents in the range of 262–272 mg/L; (vi) high fluoride concentrations, reaching 14 mg/L; and (vii) the presence of reduced Sulfur species (HS–Hydrogen Sulfide, *c.* 0.9 mg/L). The hydromineral water wells are located at the bottom of the valley of Ribeira da Coja, and are characterized by transmissivities ranging from 27 to 136 m^2/day, with yields between 1 and 4 $L \cdot s^{-1}$. These wells intersect the alluvia deposits in the first meters, and reach a maximum depth of 220 m.

1. The main geological features of the study site (lithology, structure, weathering grade and fracturing degree) are presented in Figure 3B. The dominant rock is a coarse grained granite which, depending on its weathering degree, is categorized into three main groups ([26,27,40]):

2. Fresh to slightly weathered coarse grained granite ($W_{1\text{-}2}$), which dominates in the higher altitudes, around 600–700 m, and shows moderate fracturing degree (F_3) with very close to close fracturing degree ($F_{4\text{-}5}$);

3. Moderately weathered coarse grained granite (W_3), which dominates in the lower altitudes, around 500–650 m, particularly in a wide corridor between 500 and 1000 m length that is compatible with a general NE–SW trend;

4. Highly weathered coarse grained granite ($W_{4\text{-}5}$), which basically prevails in plateau regions.

The doleritic dykes show NW–SE and NE–SW trending. These basic rocks usually exhibit green to orange color as a result of different degrees of weathering. Locally, the weathering grade in the dominant granitic lithology is very strong, sometimes reaching depths of 50 m, particularly in the trend towards NNE–SSW, where the fault zone of Ribeira de Coja is outcropped and characterized by the presence of sedimentary cover (silty-sandy deposits) at the bottom of the valley, with thickness ranging from 3 to 5 m [40].

The Caldas da Cavaca area shows high infiltration potential (Figure 3C). The majority of the water points examined in the area are compatible with the high infiltration potential zones or with the transition between the moderate and high infiltration potential zones, because of the low slopes, the moderately to highly weathered granite and the moderate to high density of tectonic lineaments. Thus, the bottom of the Ribeira da Coja valley has high infiltration potential, as a result of the alluvial cover and the very low slope of the valley bottom. However, the recharge area of the hydromineral aquifer of Caldas da Cavaca is located at higher altitudes, at more than 675 m. Moreover, the most effective infiltration zones are characterized by highly weathered granite and high thickness arenization, which are found in the NW and SE strips of the Caldas da Cavaca site. To the contrary, the less effective infiltration zones are characterized by less weathered granite and higher slopes, which are found in NE–SW bands [26,40]. The drainage network density is an important factor which helps evaluate the properties of the groundwater infiltration zones, where it reflects important characteristics of the surface and groundwater flows. The Caldas da Cavaca site shows a high drainage network density with the general trend of stream lines towards the NE to SW, especially in the SW sector (Figure 3C).

In order to characterize the shallow hydrogeomechanical behavior in the Caldas da Cavaca site, a detailed study was performed on selected rock slopes (Amores slope, A; Lagoa slope, L; and Cancela slope, C) in the study area [27]. The total length of the rock slopes has an extension of about 484 m, with an altitude that ranges between 1 and 7 m above the road. In addition, the two main orientations of the slopes studied were: N 80°–110° E for the Cancela slope and N 45°–60° E for the Amores and Lagoa slopes (Figure 3D). The rock hydrogeomechanical parameters for the site are summarized ([27,40]):

1. *Type of discontinuities*: joints are the dominant discontinuities in the rock mass. They are grouped into two sets of joints: the first set has a direction of N 120°–150° E and dip ranges from 75° to 90° NE/SW; and the second set has a general direction towards N 20°–80° E and dips between 55° SE and 90° SE;
2. *Weathering grade:* In all of the slopes in the area, a moderately weathered rock (W_3) prevails and, in addition, a highly weathered rock (W_{4-5}), as well as some fresh to slightly

weathered rock (W_{1-2}) on the Amores and Lagoa slopes (Figure 3D);

3. *Water seepage content*: This is related to flow through the rock mass joints. Although the Lagoa slope displayed small sections classified as wet (10–100 drop per minute) or damp (1–10 drop per minute), most of the slope was classified as dry. As for the Amores slope, small sections were classified as damp ($1 \leq$ drops/min < 10), while the Cancela slope was completely dry;
4. *Fracturing degree:* the parameter trend exhibited higher occurrence of wide spacing discontinuities (F_{1-2}) and discontinuities with moderate spacing (F_3), as well as a small percentage of close spacing discontinuities (F_{4-5}).

4.2. Geophysical Survey

Two geophysical surveys using geoelectrical methods were analyzed in the discharge zone of the hydromineral aquifer [40,45,46,51,52], (Figure 4): (a) electrical resistivity tomography [44,45]; (b) electromagnetic methods, especially electromagnetic conductivity (using the geophysical device Geonics EM34-3 model) [51,52]. These surveys were used to define potential geostructures, which were determined by high electrical conductivity values through electric conductivity maps and vertical dipole configuration, in addition to the geoelectrical surveys in the study site (Figure 4).

In general, the depths reached in electrical resistivity and electromagnetic surveys did not exceed 50 m, and the electromagnetic survey was carried out with a set of electromagnetic profiles on an area of about 3184 m^2 (Figure 4). The trend of the geostructure was determined through the interpretation of structural geology and geomorphological maps, encompassing digital elevation model and field maps, showing that the trend of the geostructure is compatible with the main tectonic lineament systems in the area ([40,44,51]). However, the location of boreholes was determined by coupling hydrogeology and structural geology information from earlier studies [35,37] with new data from the latest groundwater engineering operations carried out in the area [40,45,46,51,52].

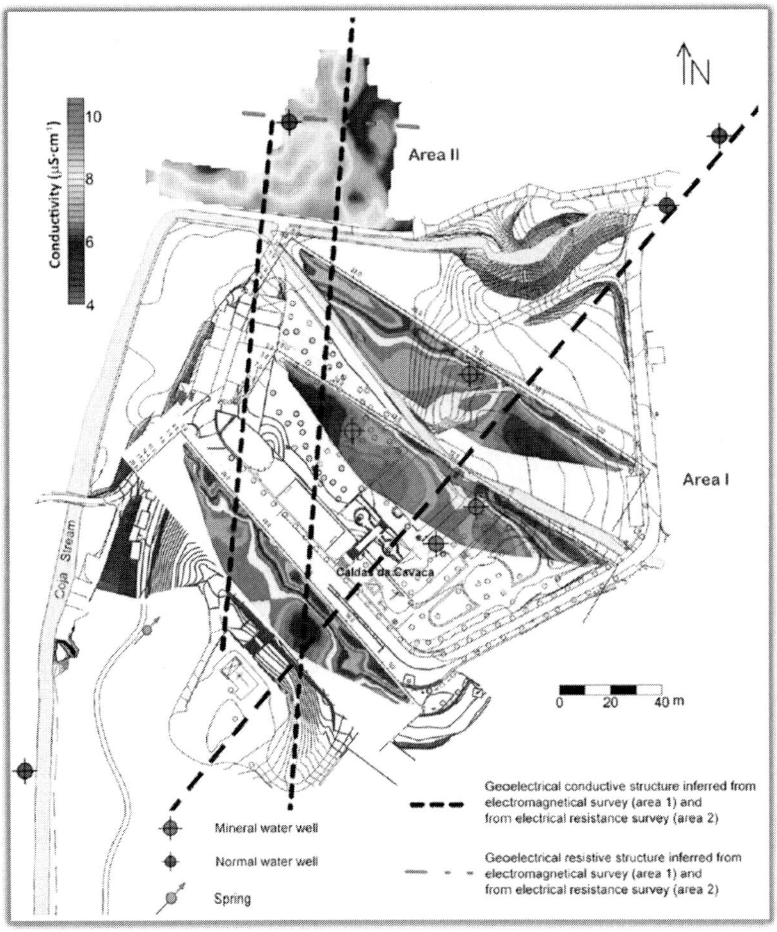

Figure 4. Synthetic interpretation of geophysical surveys for hydrogeological exploration (geoelectrical methods) in the area of Caldas da Cavaca, updated from [45,46,51,52].

The existence of mineral water in boreholes permitted the interpretation of a major geostructure with a N30° E trend, which is considered as a potential hydrogeological trap, sometimes compatible with a higher permeability zone in the rock mass that enables the discharge of mineral water. Permeability increases according to the presence of deep tectonic nodes that develop by the intersection of the main categories of fracture network [26,40,44]: (i) fractures with N–S trend dipping 70° W–80° W; and (ii) fractures with NW–SE trending with subvertical dip. Moreover, the emergence of mineral water is probably related to three factors: (i) the presence of a deep tectonic node; (ii) the occurrence of strong hydraulic heads at depth and (iii) the lower mineral water density as a result of higher temperature and gases in solution.

4.3. Multicriteria Intrinsic Vulnerability Assessment

Prior to this study, a variety of methods such as the GOD-S, DRASTIC-Fm, SINTACS and SI indexes (Figure 5) were successfully applied in a larger area by the authors of [26] to assess groundwater vulnerability to contamination for the fractured media in Caldas da Cavaca. Table 1 shows a summary of the parameter description, along with the classification adopted for intrinsic groundwater vulnerability methods, for a more detailed assessment of the present study. Figure 6 presents schematic block diagrams illustrating ground conditions and vulnerability index inputs (GOD-S, DRASTIC-Fm, SINTACS and SI).

The assessment of groundwater vulnerability according to the GOD-S index (Figure 5A) shows that most of the Caldas da Cavaca area fits in a moderate vulnerability category. This category corresponds to the highly weathered granite ($W_{4\text{-}5}$). The high vulnerability category corresponds to the alluvia cover through a narrow strip along the valley bottom. Furthermore, the low and very low vulnerability categories match the moderately weathered (W_3) to slightly weathered ($W_{1\text{-}2}$) granite and the dolerite dykes.

The DRASTIC-Fm index (Figure 5B) shows that the flat parts of the valley bottom are associated with the moderate-high and high vulnerability categories, due to the presence of alluvia cover. The NE–SW corridors correspond to the moderate vulnerability category, due to the fracture density, the slope and less weathered granite. Over 50% of the site, especially the NW and SE zones of the study site, near the Caldas da Cavaca site, fits into the low-moderate and low vulnerability categories.

Regarding the SINTACS index (Figure 5C), the surroundings of the Caldas da Cavaca site are comprised of very high and extremely high vulnerability categories due to the low slope values and the presence of alluvia and highly weathered granite. Moreover, the large NE–SW strip along the valley slopes belongs to moderate-high vulnerability categories, as a result of the high slope values and less weathered granite. The low vulnerability category corresponds to the dolerite rocks, and is related to the argillaceous weathering of these dykes. When comparing the SI index results (Figure 5D), an identical pattern to that of SINTACS is revealed. Land use can be clearly seen to be an important parameter, namely around the settlements, where the buildings and agricultural areas are concentrated. These areas have higher slope values

(20°–40°) and correspond to high or very high vulnerability categories. On the contrary, the high slope values are related to the less weathered granite and are associated with moderate and low-moderate vulnerability. The SI index shows only a very small area in the high vulnerability category.

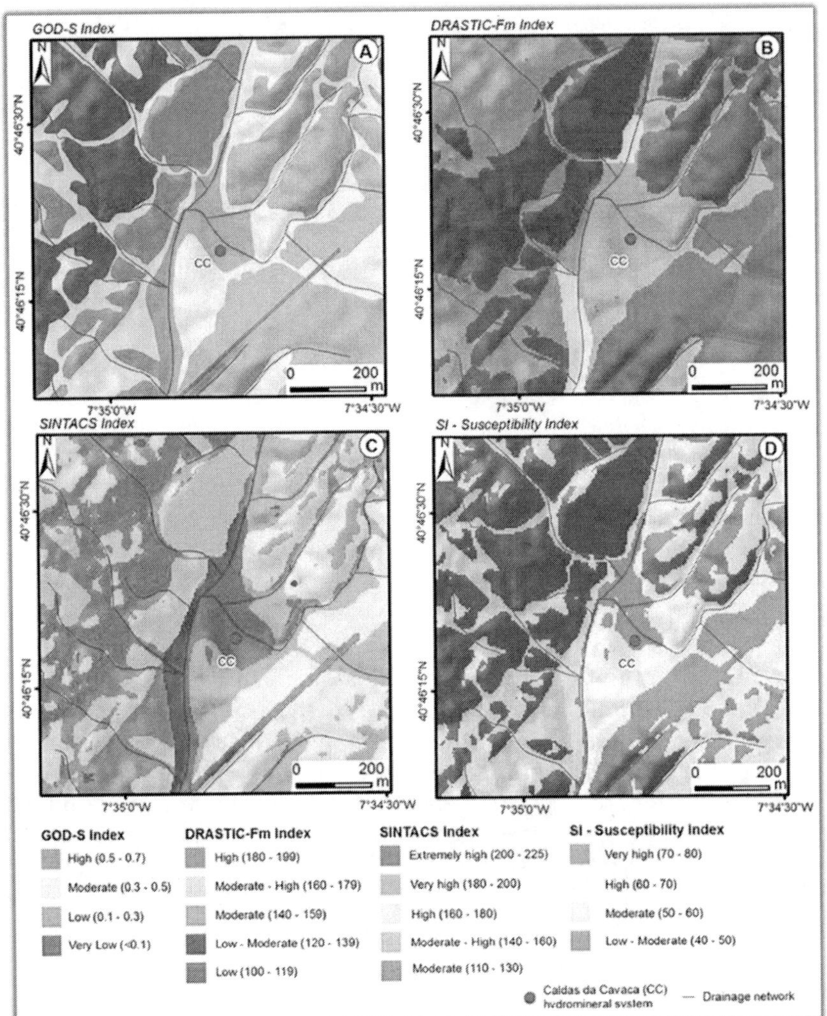

Figure 5. Multicriteria intrinsic vulnerability indexes from Caldas da Cavaca groundwater systems and surrounding area: (**A**) GOD-S; (**B**) DRASTIC-Fm; (**C**) SINTACS; (**D**) SI, revised and updated from [26].

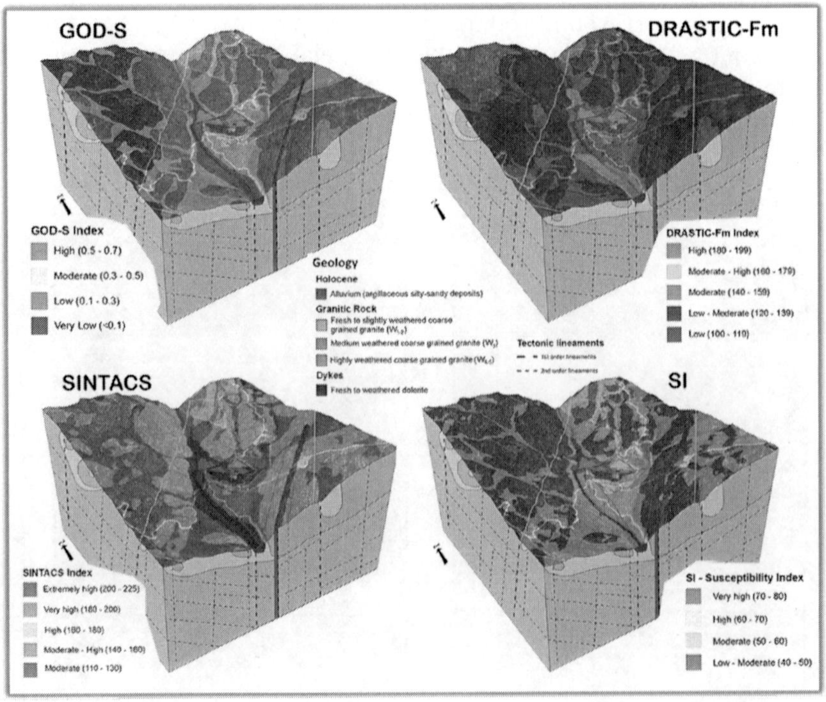

Figure 6. Ground conditions and intrinsic vulnerability indexes mapping: GOD-S, DRASTIC-Fm, SINTACS, SI.

In this study a groundwater vulnerability assessment was carried out by means of the DISCO index [34]. The delineation of groundwater protection zones in the study site was applied for two hydromineral water wells (Figure 7) and for an important shallow groundwater spring (0.5–1 $L \cdot s^{-1}$) (Figure 8). The location of the wells was determined according to the geological, geomorphological and hydrogeological studies carried out previously [25,26,40]. In addition, the location of the springs was controlled by the rock discontinuities network in the groundwater watershed. Locally, both mineral water wells and normal groundwater springs are located along major deep fractures, or near the intersections between tectonic lineaments.

Table 1. GOD-S, DRASTIC-Fm, SINTACS and SI (DRAT-LU) parameters used in the surrounding area of the Caldas da Cavaca site, adapted from [26].

	Hydrogeological Units				
	Sands and Gravels	Highly Weathered Coarse Grained Granite (W_{4-5})	Moderately Weathered Coarse Grained Granite (W_3)	Fresh to Slightly Weathered Coarse Grained Granite (W_{1-2})	Fresh to Weathered Dolerite Dyke (W_{1-2} to W_{4-5})
G	unconfined	confined/ semi-confined	confined/ semi-confined	confined/ semi-confined	confined
O	alluvial sands/gravels	igneous formations	igneous formations	igneous formations	igneous formations
D	<5	5–20	5–20	<5	5–20
IS	silty-sand	silty-clay	silt	thin/absent	silty-clay
D	1.5–4.6	4.5–9	4.5–9	1.5–4.6	4.6–9.1
R	250	175	175	175	175
A	sand and gravel	igneous/weathered igneous	igneous/weathered igneous	igneous	igneous/weathered igneous
S	sandy loam	clay loam	silty-loam	sandy loam	clay loam
T	0%–60%	0%–60%	0%–60%	0%–60%	0%–60%
I	sand and gravel with significant clay	sand and gravel with significant clay	igneous	igneous	igneous
C	<4.1 m/day	<4.1 m/day	<4.1 m/day	<4.1 m/day	<4.1 m/day
Fm	0–25 km of lineament/km^2	0–25 km of lineament/km^2	0–5 km of lineament/km^2	0–25 km of lineament/km^2	0–25 km of lineament/km^2
S	1.5–4.6	4.5–9	4.5–9	1.5–4.6	4.6–9.1
I	250	175	175	175	175
N	coarse alluvial deposit	fissured plutonic rock	fissured plutonic rock	fissured plutonic rock	fissured plutonic rock
T	sandy-loam	clay-loam	silty-loam	sandy-loam	clay-loam
A	coarse alluvial deposit	fissured plutonic rock	fissured plutonic rock	Fissured plutonic rock	fissured plutonic rock
C	5×10^{-5}–1×10^{-4}	$<5 \times 10^{-5}$	$<5 \times 10^{-5}$	$<5 \times 10^{-5}$	$<5 \times 10^{-5}$
S	*The rating for this parameter is variable since slope values are calculated for each pixel of the ArcGIS raster dataset*				
D	1.5–4.6	4.5–9	4.5–9	1.5–4.6	4.6–9.1
R	250	175	175	175	175
A	sand and gravel	igneous/weathered igneous	igneous/weathered igneous	igneous	igneous/weathered igneous
T	*The rating for this parameter is variable since slope values are calculated for each pixel of the ArcGIS raster dataset*				
LU	*The rating for this parameter is variable since land use is calculated for each pixel of the ArcGIS raster dataset*				

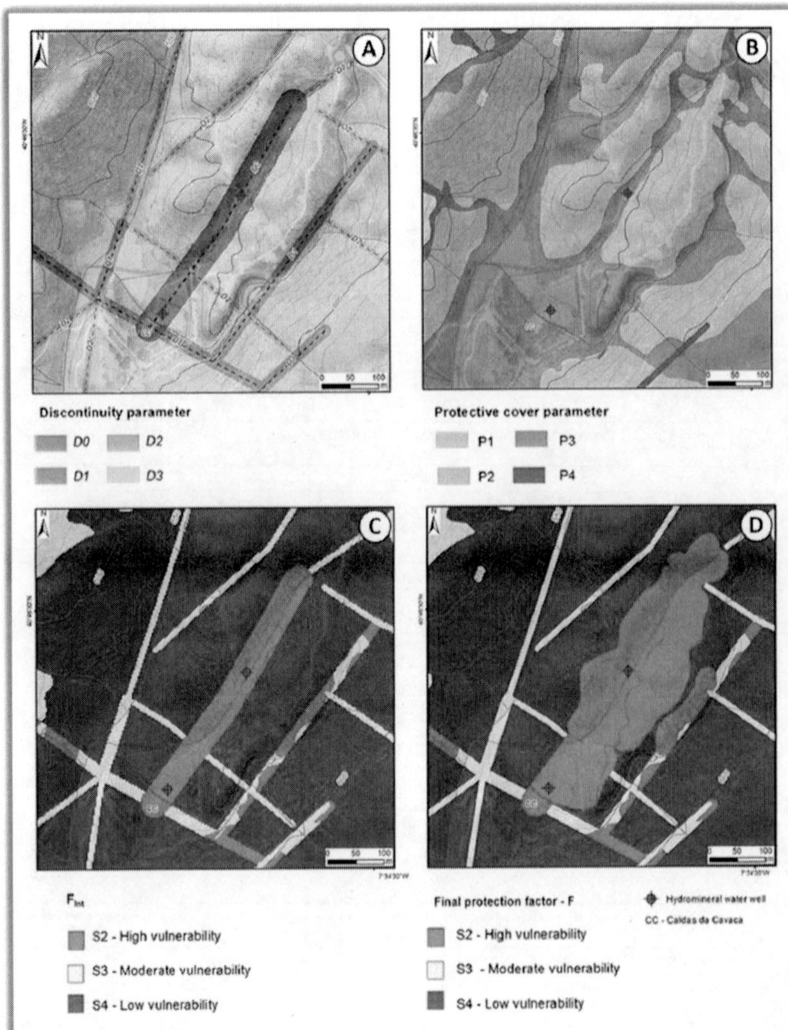

Figure 7. DISCO vulnerability index from Caldas da Cavaca aquifer systems: (**A**) discontinuities parameter for the hydromineral water wells; (**B**) protective cover parameter for the hydromineral water wells; (**C**) intermediate protection factor for the hydromineral water wells; (**D**) final protection factor for the hydromineral water wells.

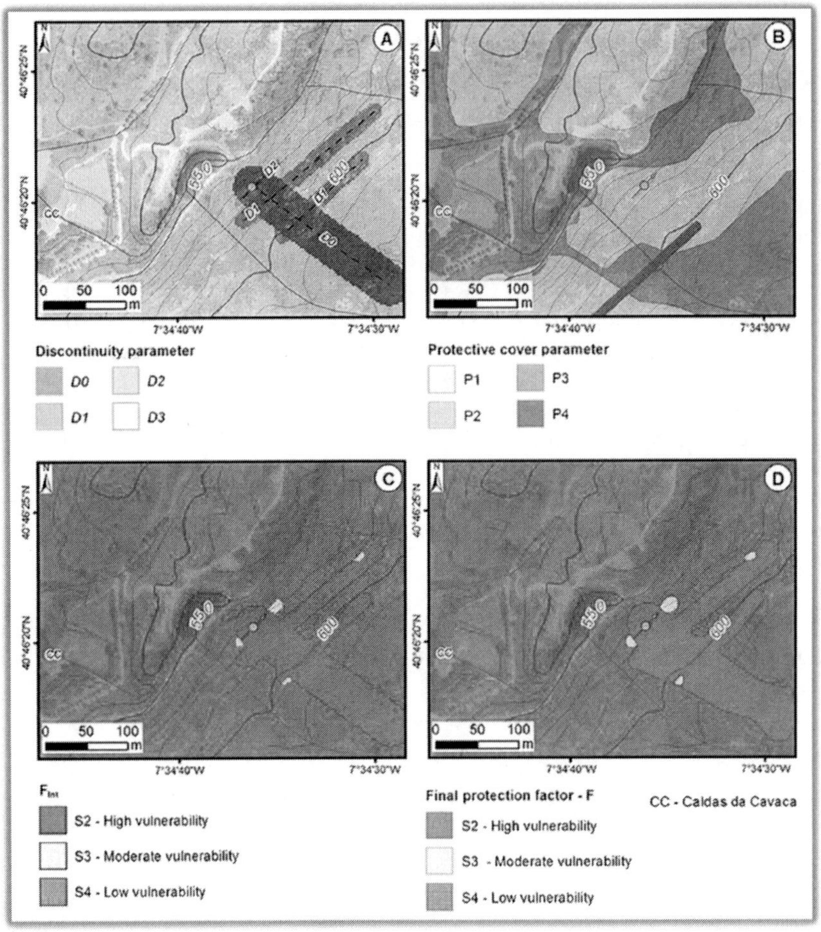

Figure 8. DISCO method applied to a spring at the Caldas da Cavaca site: (**A**) discontinuity parameter; (**B**) protective cover parameter; (**C**) intermediate protection factor (*[Math Processing Error]*); (**D**) final protection factor (*F*).

The application of the DISCO method includes three main steps (details in [33,34]):

1. ASSESSMENT OF THE DISCONTINUITIES (DIS) AND PROTECTIVE COVER (CO) PARAMETERS

The discontinuities parameter (DIS) is based on the tectonic lineaments analysis and on field geostructural and hydrogeomechanical surveys, as well as on hydrogeological and hydrogeophysical data. The delineation of protection zones at the site (D_0, D_1, D_2 and D_3) was defined according to the rating and criterion defined by Pochon et al. [34] (Figure 7A). Therefore, buffers were assigned for each category: 25 m for D_0, 10 m for D_1 and 5 m for D_2, while D_3 was assigned to the rest of the area.

The protective cover parameter (CO) is based on soil analysis, geomorphological mapping, drilling and hydrogeophysics, and was defined according to the rating and criterion defined by Pochon et al. [34], taking into account soils and geological formations overlying the aquifer (Table 2).

Table 2. Protective cover parameter evaluation taking into consideration geological formations and soils overlying aquifers in the Caldas da Cavaca site, adapted from [53].

		Hydrogeological Units				
		Alluvia (Silty-Sandy Deposits)	Highly Weathered Coarse Grained Granite ($W_{4\text{-}5}$)	Moderately Weathered Coarse Grained Granite (W_3)	Fresh to Slightly Weathered Coarse Grained Granite ($W_{1\text{-}2}$)	Fresh to Weathered Dolerite Dyke ($W_{1\text{-}2}$ to $W_{4\text{-}5}$)
Protective cover (CO): pedological soil	Description	High permeability soil (sand, pebbles)	Moderate permeability soil (silt, loam)	Moderate permeability soil (silt, loam)	Moderate permeability soil (silt, loam)	Low permeability soil (loam, clay)
	Thickness (m)	>5	>1	0.2–0.5	0–0.2	>1
	Class	P_1	P_1	P_0	P_0	P_3
Protective cover (CO): geological formations	Description	Combined with P_1 soil	Combined with P_1 soil	Combined with P_0 soil	Combined with P_0 soil	Combined with P_3 soil
	Thickness (m)	>2	>2	1–2	< 1	>2
	Class	P_3	P_3	P_2	P_1	P_4
Rating		3	3	2	1	4

Regarding this approach, the rating values of "P" range from 1 to 4, with increasing values being associated with thick protective cover and low permeability of the formations. In this way, "P" areas correspond to Figure 7B: (i) P1—fresh to slightly weathered coarse grained granite, covered by moderate permeability soils; (ii) P2—moderately weathered coarse grained granite, covered by moderate permeability soils; (iii) P3—highly weathered coarse grained granite and alluvia deposits, covered by moderate to high permeability soils; (iv) P4—fresh to weathered dolerite, covered by low permeability soils.

2. DETERMINATION OF THE INTERMEDIATE PROTECTION FACTOR (*[MATH PROCESSING ERROR]*)

It is possible to calculate the intermediate protection factor (*[Math Processing Error]*) using Equation (1) [34]. This factor ranged from 2 (low protective effect) to 7 (moderate protective effect).

$$F_{int} = 2 \times D + P \ (1)$$

Depending on the conversion between the protection factor F and the groundwater protection zones defined by Pochon et al. [34], the relation between F_{int} and groundwater protection zones (S) is the following (Figure 7C): (i) Zone S2 is compatible with F_{low} (2, 3, 4) where the values 2, 3 and 4 are related to lineaments and to the moderately to highly weathered coarse grained granite W_3 to $W_{4\text{-}5}$ respectively (ii) Zone S3 is compatible with $F_{Moderate}$ (5, 6, 7) where the values 5 and 6 are related to some lineaments and value 7 is related to the fresh to slightly weathered coarse grained granite $W_{1\text{-}2}$.

3. PROTECTION ZONE DELINEATION (*F*)

Because of the influence of slope topography and the low permeability cover over some parts of the study site, it is important to modify the F_{int} map, taking the runoff into consideration. Consequently, the protection zone map (Figure 7D) shows that: (i) zones S2 (high vulnerability) were assigned to the first order lineaments; (ii) zones S3 (moderate vulnerability) were assigned to the second order lineaments in the site; and (iii) zones S4 (low vulnerability) were assigned to the rest of the site.

A similar approach, presented in Figure 8, was applied to a spring of normal groundwater located on the slope to east of the spa. The discontinuity parameter is mapped in Figure 8A, the protective cover parameter corresponds to Figure 8B, the intermediate protection factor (F_{int}) is represented in Figure 8C and the final protection factor (F) is mapped in Figure 8D. In the F factor map once again, the close control of the spatial distribution of the high and moderate vulnerability classes (S2 and S3) by the first order and second order tectonic lineaments is clear. The remaining area, without any hydraulic connection to this spring, shows the low vulnerability class (S4).

4.4. Wellhead Protection Zones: An Updated Proposal

According to the Portuguese law (D.L. 382/99; L. 58/2005; P. 702/2009), a wellhead protection area is to be established considering the hydrogeological conditions and comprises the following: (i) the immediate protection zone, where all activities not related to groundwater exploitation are forbidden (the corresponding area is normally protected by a fence); (ii) the intermediate protection zone, where all activities or facilities which may contaminate the groundwater, whether by the infiltration of contaminants or by changing the flow paths, are forbidden or at least strictly controlled; (iii) the extended protection zone, where the constraints to activities and facilities are still applied but in a less severe way.

At Caldas da Cavaca, the existing wellhead protection area was defined for the former mineral water well in 1996 using simple criteria, (Figure 9). Since then, new wells have been drilled and knowledge of the hydrogeological system has improved significantly, e.g., [25,26,40,44]. The results of the DISCO methodology provide a solid basis for the review of the wellhead protection area in order to adjust it to the present exploitation conditions and to the updated hydrogeologic conceptual model. Given the results of the DISCO index, the intermediate protection zone is the most critical. This zone should be expanded to NE and to NW (Figure 9) with the purpose of encompassing all the high vulnerability zones (S2), as well as the most important moderate vulnerability zones (S3). The existing extended vulnerability zone should retain its limits because it is large enough to include all the relevant deep-crustal geostructures and the related vulnerability zones. The immediate protection zone should be restricted to the vicinity of the new mineral water wells.

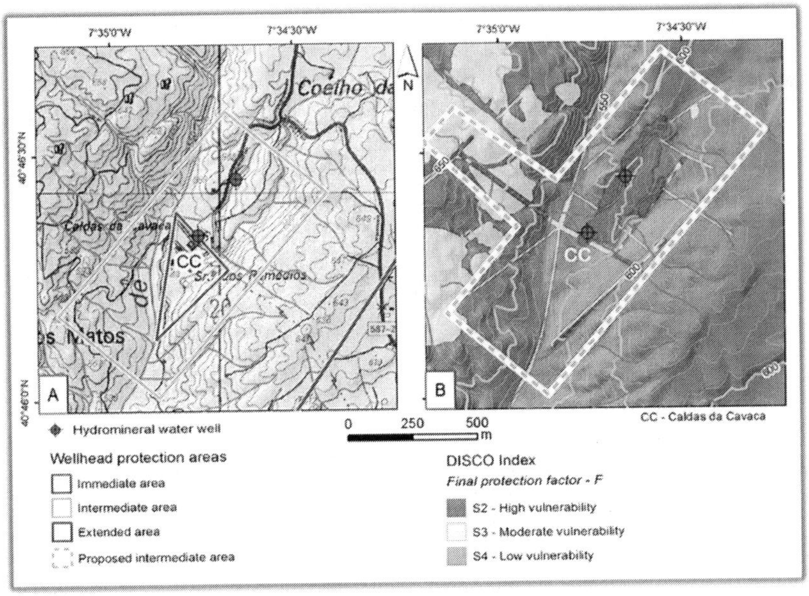

Figure 9. Wellhead protection areas for Caldas da Cavaca hydromineral system: (**A**) wellhead protection areas defined in 1996; (**B**) proposed intermediate wellhead protection area.

4.5. Hydrogeology Conceptual Site Model: Inputs from Intrinsic Vulnerability Mapping

The discharge zone of the Caldas da Cavaca hydromineral system is characterized by three aquifer types represented in the hydrogeological conceptual model of Figure 10, as follows [26,40,44,53]:

- **Shallow and unconfined aquifer:** this is located in the bottom of the valley, associated with the alluvial deposits. The groundwater (sodium chloride to sodium bicarbonate facies) of this aquifer is characterized by pH of around 5–6.5 and very low mineralization, with electrical conductivity under 20 μS/cm; the temperature of the water is similar to the mean monthly air temperature. The intrinsic vulnerability to contamination of this aquifer varies from high to very high.
- **Unconfined to semi-confined aquifer:** this is located in the fractured and weathered granitic with transmissivity lower than 1 m^2/day and very low water yields (less than 0.05 $L \cdot s^{-1}$). The groundwater from this aquifer is characterized by a sodium chloride facies, with pH around 5–6.5 and low mineralization, with electrical

conductivity between 20 and 50 μS/cm. The vulnerability to contamination is moderate to high.

- **Deep confined hydromineral aquifer:** this is located in the fresh granite dominated by a deep fault zone with transmissivity ranging from 27 to 136 m^2/day. The Ribeira de Coja fault zone is considered the main structure of the hydromineral water reservoir. These waters are characterized by pH of around 8.4–8.6, temperatures around 29 °C and intermediate mineralization with electrical conductivity ranging from 350 to 430 μS/cm. Moreover, the hydromineral waters of this aquifer have an alkaline reaction, are fluoridated and sulfurous, and have sodium bicarbonate hydrogeochemical facies. The intrinsic vulnerability to contamination is high in the discharge area (wellhead site), but very low in the deep hydromineral reservoir.

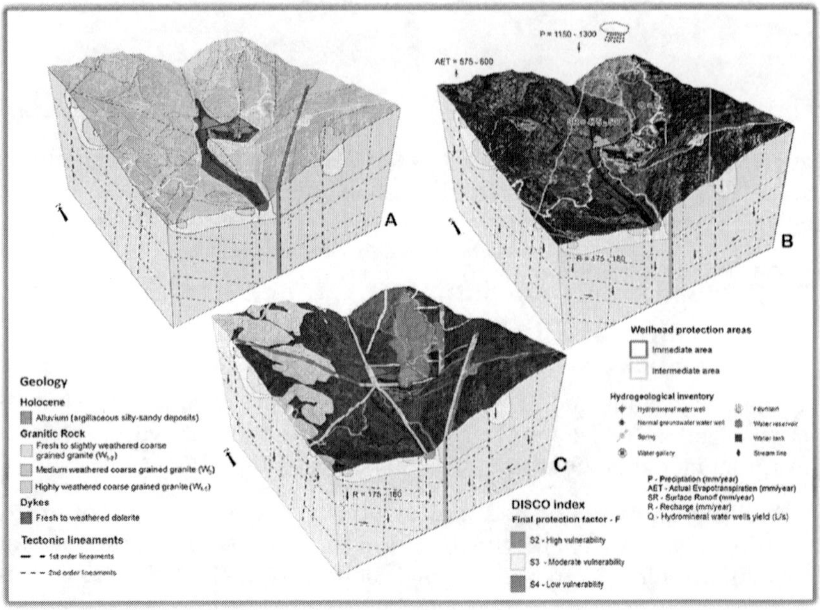

Figure 10. Hydrogeological conceptual model from the Caldas da Cavaca hydromineral system: vulnerability DISCO index assessment input: (**A**) general geology and surface rock mass conditions; (**B**) hydrogeological conceptual site model, updated from [26]; (**C**) vulnerability DISCO index inputs. In general, the hydromineral system of Caldas da Cavaca is characterized by a precipitation range of 1150–1300 mm/year; actual evapotranspiration ranges from 575 to 600 mm/year; surface runoff from 475 to 500 mm/year and recharge from 175 to 180 mm/year, while the hydromineral water wells yield [Q] varies from 1 L·s^{-1} to 4 L·s^{-1} [26].

The results of the application of the DISCO methodology (the final protection factor, *F*) make it possible to improve the Caldas da Cavaca hydrogeological conceptual model. Figure 10C shows the hydrogeological conceptual site model depending on the vulnerability DISCO index inputs. In a broad sense, the DISCO index results are compatible with the other vulnerability indexes studied, especially GOD-S, DRASTIC-Fm, SINTACS and SI (see Figure 6). However, the DISCO map, particularly regarding the high and moderate vulnerability areas, reflects much more effectively the role of the main tectonic structures in the groundwater circulation. Thus, high vulnerability was assigned to the first order lineaments, moderate vulnerability was assigned to the second order lineaments in the site (related fractures and joints) and low vulnerability was assigned to the rest of the site.

5. CONCLUSIONS

Vulnerability assessment mapping is an essential step in assessing intrinsic groundwater vulnerability to contamination and in delineating wellhead protection areas. In addition, this approach provides a visual analysis and a practical tool for helping planners and decision makers in order to achieve the sustainable management of water resources.

Intrinsic groundwater vulnerability studies are often based on multi-technical studies to achieve a comprehensive model of groundwater vulnerability to contamination. GIS-based mapping methodology provides an accurate approach for evaluating groundwater vulnerability, as well as for delineating the hydrogeological conceptual site model. Furthermore, comparative and detailed studies applying different scale mapping and methodologies are a valuable approach to the effective evaluation of the intrinsic groundwater vulnerability. The more frequent multicriteria indexes, GOD-S, DRASTIC-Fm, SINTACS and SI, are often used in local to regional-scale study areas. The DISCO index [34] is more suitable to local-scale site hydrogeological investigations related to wellhead protection areas and springs. The integrated multi-technical approach has been demonstrated to be suitable for fractured media, mainly in contexts of data scarcity.

This comprehensive study highlights the need to perform a merging balance of the most common multicriteria intrinsic vulnerability evaluation techniques, used generally in larger areas, along with the DISCO index, more suitable for local-scale investigations. In addition, the design of a methodological GIS mapping approach, particularly focused on supporting the local-scale site management and protection of groundwater resource areas, is an essential task which provides valuable output. Finally, the integrative approach generates more realistic hydrogeological conceptual site models, which are of greater interest in the scope of water resource management, especially for mineral water exploitation, encompassing mineral water bottling, thermal spa activities and the related tourism with respect to health and wellness.

ACKNOWLEDGMENTS

This study was carried out under the framework of the projects LABCARGA|ISEP re-equipment program (IPP-ISEP|PAD'2007/08), Centre GeoBioTec|UA (UID/GEO/04035/2013), and HydroSPOT|LABCARGA (ESA-ID5750). Helen Meerkhan was supported by GPSS|Foundation Casa do Regalo (Lisbon), particularly Jorge Sampaio and Helena Barroco, as well as by P. Porto-Polytechnic of Porto (Rosário Gâmboa) and ISEP-School of Engineering of Porto (João Rocha). Jorge Espinha Marques was supported by the IES (UID/GEO/04683/2013), through the project ICT (POCI-01-0145-FEDER-007690). Special thanks are given to our colleagues José Martins Carvalho, Rui Moura, Ricardina Moura, Liliana Freitas, João Meirinhos and Robin Nagano. The authors are grateful to the Aguiar da Beira Council, Caldas da Cavaca SA Spa, TARH Lda and GeoSonda Lda for their support in some stages of the research. We acknowledge the anonymous reviewers for the constructive comments that helped to improve the focus of the manuscript.

REFERENCES

1. Richter, H.E.; Gungle, B.; Lacher, L.J.; Turner, D.S.; Bushman, B.M. Development of a shared vision for groundwater management to protect and sustain baseflows of the Upper San Pedro River, Arizona, USA. *Water* **2014**, *6*, 2519–2538.
2. Committee on Fracture Characterization and Fluid Flow (CFCFF). *Rock Fractures and Fluid Flow: Contemporary Understanding and Applications*; National Academy of Sciences: Washington, DC, USA, 1996.
3. Scesi, L.; Gattinoni, P. *Water Circulation in Rocks*; Springer: Dordrecht, The Netherlands, 2009.
4. Singhal, B.S.; Gupta, R.P. *Applied Hydrogeology of Fractured Rocks*, 2nd ed.; Springer: Heidelberg, Germany, 2010.
5. Gustafson, G. *Hydrogeology for Rock Engineers*; BeFo and ISRM: Stockholm, Sweden, 2012.
6. Jaiswal, R.K.; Mukherjee, S.; Krishnamurthy, J.; Saxena, R. Role of remote sensing and GIS techniques for generation of groundwater prospect zones towards rural development—An approach. *Int. J. Remote Sens.* **2003**, *24*, 993–1008.
7. Albinet, M.; Margat, J. Cartographie de la vulnérabilité à la pollution des nappes d'eau souterraine. *Bull. BRGM (Paris)* **1970**, *2*, 13–22.
8. Aller, L.; Bennet, T.; Lehr, J.H.; Petty, R.J. *DRASTIC: A Standardized System for Evaluating Groundwater Pollution Potential Using Hydrologic Settings*; Robert, S.,

Ed.; Technical Report; U.S. EPA Report, 600/2-87/035; Kerr Environmental Research Laboratory: Ada, OK, USA, 1987.
9. Vrba, J.; Zaporozec, A. *Guidebook on Mapping Groundwater Vulnerability*; International Contributions to Hydrogeology; Velrag Heinz Heise: Hannover, Germany, 1994; Volume 16.
10. Foster, S.; Hirata, R.; Gomes, D.; Delia, M.; Paris, M. *Groundwater Quality Protection: A Guide for Water Utilities, Municipal Authorities, and Environment Agencies*; The International Bank for Reconstruction and Development; The World Bank: Washington, DC, USA, 2002.
11. Witkowski, A.; Kowalczyk, A.; Vrba, J. (Eds.) *Groundwater Vulnerability Assessment and Mapping*; Taylor & Francis Group: London, UK, 2007.
12. Morris, B.; Lawrence, A.; Chilton, J.; Adams, B.; Klinck, B.; Calow, R. *Groundwater and Its Susceptibility to Degradation: A Global Assessment of the Problem and Options for Management*; United Nations Environment Programme: Nairobi, Kenya, 2003.
13. Robins, N.S.; Chilton, P.J.; Cobbing, J.E. Adapting existing experience with aquifer vulnerability and groundwater protection for Africa. *J. Afr. Earth Sci.* **2007**, *47*, 30–38.
14. Gemitzi, A.; Petalas, C.; Tsihrintzis, V.; Pisinaras, V. Assessment of groundwater vulnerability to pollution: A combination of GIS, fuzzy logic and decision making techniques. *Environ. Geol.* **2006**, *49*, 653–673.
15. Pisinaras, V.; Polychronis, C.; Gemitzi, A. Intrinsic groundwater vulnerability determination at the aquifer scale: A methodology coupling travel time estimation and rating methods. *Environ. Earth Sci.* **2016**, *75*, 85.
16. Daly, D.; Warren, W.P. *Mapping Groundwater Vulnerability: The Irish Perspective*; Geological Survey of Ireland: Dublin, Ireland, 1998.
17. González, A.; Sánchez, A.L.; Requena, P.M.; Varela, M.S. Assessment of the microbiological quality of groundwater in three regions of the Valencian Community (Spain). *Int. J. Environ. Res. Public Health* **2014**, *11*, 5527–5540.
18. Jha, M.K.; Chowdhury, A.; Chowdary, V.; Peiffer, S. Groundwater management and development by integrated remote sensing and geographic information systems: Prospects and constraints. *Water Resour. Manag.* **2007**, *21*, 427–467.
19. Saaty, T.L. Decision making with the analytic hierarchy process. *Int. J. Serv. Sci.* **2008**, *1*, 83–98.
20. Margat, J.; van der Gun, J. *Groundwater around the World: A Geographic Synopsis*; Taylor & Francis Group: Boca Raton, FL, USA, 2013.
21. Chaminé, H.I. Water resources meet sustainability: New trends in environmental hydrogeology and groundwater engineering. *Environ. Earth Sci.* **2015**, *73*, 2513–2520.
22. Chaminé, H.I.; Carvalho, J.M.; Teixeira, J.; Freitas, L. Role of hydrogeological mapping in groundwater practice: Back to basics. *Eur. Geol. J.* **2015**, *40*, 34–42.
23. Junior, R.F.; Varandas, S.G.P.; Fernandes, L.F.S. Multicriteria analysis for the monitoring of aquifer vulnerability: A scientific tool in environmental policy. *Environ. Sci. Policy* **2015**, *48*, 250–264.
24. Sahoo, S.; Dhar, A.; Kar, A.; Chakraborty, D. Index-based groundwater vulnerability mapping using quantitative parameters. *Environ. Earth Sci.* **2016**, *75*, 522.
25. Teixeira, J.; Chaminé, H.I.; Espinha Marques, J.; Gomes, A.; Carvalho, J.M.; Pérez-Alberti, A.; Rocha, F. Integrated approach of hydrogeomorphology and GIS mapping to the evaluation of groundwater resources: An example from the hydromineral system of Caldas da Cavaca, NW Portugal. In *Global Groundwater Resources and Management: Selected Papers from the 33rd International Geological Congress*; Paliwal, B.S., Ed.; Scientific Publishers: Jodhpur, India, 2010; pp. 227–249.
26. Teixeira, J.; Chaminé, H.I.; Espinha Marques, J.; Carvalho, J.M.; Pereira, A.J.; Carvalho, M.R.; Fonseca, P.E.; Pérez-Alberti, A.; Rocha, F. A comprehensive

analysis of groundwater resources using GIS and multicriteria tools (Caldas da Cavaca, Central Portugal): Environmental issues. *Environ. Earth Sci.* **2015**, *73*, 2699–2715.
27. Meirinhos, J. Hydrogeomechanics for Rock Engineering: Coupling Subsurface Hydrogeomechanical Assessment and Hydrogeotechnical Mapping on Fractured Rock Masses. Master's Thesis, School of Engineering, Polytechnic of Porto, Porto, Portugal, 2015.
28. Denny, S.C.; Allen, D.M.; Journeay, M. DRASTIC-Fm: A modified vulnerability mapping method for structurally-controlled aquifers. *Hydrogeol. J.* **2007**, *15*, 483–493.
29. Civita, M.V.; de Maio, M. Assessing and mapping aquifer vulnerability to contamination: The Italian "combined" approach. *Geofísica Internacional* **2000**, *43*, 513–532.
30. Civita, M.V. *Le Carte Della Vulnerabilità Degli Acquiferi All'inquinamento*; Teoria & Practica Pitagora: Bologna, Italy, 2010.
31. Ribeiro, A.; Munhá, J.; Dias, R.; Mateus, A.; Pereira, E.; Ribeiro, L.; Fonseca, P.E.; Araújo, A.; Oliveira, J.T.; Romão, J.; et al. Geodynamic evolution of the SW Europe Variscides. *Tectonics* **2007**, *26*, 1–24.
32. Stigter, T.Y.; Ribeiro, L.; Dill, A.M. Evaluation of an intrinsic and a specific vulnerability assessment method in comparison with groundwater salinisation and nitrate contamination levels in two agricultural regions in the south of Portugal. *Hydrogeol. J.* **2006**, *14*, 79–99.
33. Pochon, A.; Zwahlen, F. *Délimitation des Zones de Protection des Eaux Souterraines en Milieu Fissuré*; Technical Report; Office Fédéral de L'environnement: Berne, Switzerland, 2003.
34. Pochon, A.; Tripet, J.-P.; Kozel, R.; Meylan, B.; Sinreich, M.; Zwahlen, F. Groundwater protection in fractured media: A vulnerability-based approach for delineating protection zones in Switzerland. *Hydrogeol. J.* **2008**, *16*, 1267–1281.
35. Freire Andrade, C. *Projecto de Modificação de Captagem das Águas das Caldas da Cavaca*; Unpublished Technical Report; Serviços Geológicos: Lisboa, Portugal, 1935; p. 11.
36. Freire Andrade, C. *Os Vales Submarinos Portugueses e o Diastrofismo das Berlengas e da Estremadura*; Serviços Geológicos: Lisboa, Portugal, 1937; p. 236. [Google Scholar]
37. Freire Andrade, C. *Nota Acerca dos Trabalhos Realizados para a Modificação da Captagem das Águas Medicinaes das Caldas da Cavaca*; Unpublished Technical Report; Serviços Geológicos: Lisboa, Portugal, 1938; p. 7.
38. Acciaiuoli, L. *Le Portugal Hydromineral*; Direcção Geral, Serviços Geológicos: Lisbonne, Portugal, 1952; Volume 2.
39. Boorder, H. Petrological Investigations in the Aguiar da Beira Granite Area, Northern Portugal. Ph.D. Thesis, University of Amsterdam, Amsterdam, The Netherlands, 1965.
40. Teixeira, J. Hidrogeomorfologia e Sustentabilidade de Recursos Hídricos Subterrâneos. Ph.D. Thesis, University of Aveiro, Aveiro, Portugal, 2011.
41. Carvalho, J.M. Prospecção e Pesquisa de Recursos Hídricos Subterrâneos no Maciço Antigo Português: Linhas Metodológicas. Ph.D. Thesis, University of Aveiro, Aveiro, Portugal, 2006.
42. Peel, M.C.; Finlayson, B.L.; McMahon, T.A. Updated world map of the Köppen-Geiger climate classification. *Hydrol. Earth Syst. Sci.* **2007**, *11*, 1633–1644.
43. Brum Ferreira, A. Neotectonics in Northern Portugal: A geomorphological approach. *Z. Geomorphol.* **1991**, *82*, 73–85.
44. Carvalho, J.M.; Chaminé, H.I.; Afonso, M.J.; Espinha Marques, J.; Teixeira, J.; Cerqueira, A.; Coelho, A.; Gomes, A.; Fonseca, P.E. Prospecção hidrogeológica da área do sistema hidromineral das Caldas da Cavaca (Aguiar da Beira, Portugal Central): Implicações na gestão dos recursos hídricos subterrâneos. In *Actas del I Foro Ibérico Sobre Aguas Envasadas y Balnearios*; Fernández Rubio, R., Ed.;

Asociación Nacional de Empresas de Aguas de Bebida Envasadas: Madrid, Portugal, 2005; pp. 109–121.
45. Espinha Marques, J. *Revisão do Plano de Exploração da Concessão Hidromineral HM-11 Caldas da Cavaca (Dezembro)*; Unpublished Technical Report; Faculdade de Ciências da Universidade do Porto: Porto, Portugal, 2008; p. 24.
46. Espinha Marques, J. *Revisão do Plano de Exploração da Concessão Hidromineral HM-11 Caldas da Cavaca (Maio)*; Unpublished Technical Report; Faculdade de Ciências da Universidade do Porto: Porto, Portugal, 2008; p. 61.
47. Chaminé, H.I.; Carvalho, J.M.; Afonso, M.J.; Teixeira, J.; Freitas, L. On a dialogue between hard-rock aquifer mapping and hydrogeological conceptual models: Insights into groundwater exploration. *Eur. Geol. J.* **2013**, *35*, 26–31.
48. Teixeira, J.; Chaminé, H.I.; Carvalho, J.M.; Pérez-Alberti, A.; Rocha, F. Hydrogeomorphological mapping as a tool in groundwater exploration. *J. Maps* **2013**, *9*, 263–273.
49. Saaty, T.L. A scaling method for priorities in hierarchical structures. *J. Math. Psychol.* **1977**, *15*, 234–281.
50. Narany, T.S.; Ramli, M.F.; Aris, A.Z.; Sulaiman, W.N.; Fakharian, K. Spatial assessment of groundwater quality monitoring wells using indicator kriging and risk mapping, Amol-Babol Plain, Iran. *Water* **2014**, *6*, 68–85.
51. Terra, Ambiente & Recursos Hídricos Lda. *Estudo Hidrogeológico da Concessão Hidromineral HM-11 Caldas da Cavaca e Terrenos Envolventes*; Unpublished Technical Report; TARH—Terra, Ambiente & Recursos Hídricos Lda.: Sacavém, Portugal, 2005; p. 42.
52. GeoSonda Lda. *Prospecção Electromagnética nas Caldas da Cavaca*; Unpublished Technical Report; GeoSonda Lda.: Santa Maria da Feira, Portugal, 2004; p. 7.
53. Meerkhan, H. Delineating Environmental Groundwater Vulnerability and Protection Zones Mapping in Fractured Rock Masses. Master's Thesis, School of Engineering, Polytechnic of Porto, Porto, Portugal, 2015.

CHAPTRER 10

Inventory of Managed Aquifer Recharge Sites in Europe: Historical Development, Current Situation and Perspectives

C. Sprenger[1] N. Hartog[2] M. Hernández[3] E. Vilanova[4] G. Grützmacher[5] F. Scheibler[6] S. Hannappel[6]

[1]*Kompetenzzentrum Wasser Berlin GmbH, Cicerostr. 24, 10709 Berlin, Germany*

[2]*KWR - Watercycle Research Institute, Groningenhaven 7, 3433 PE Nieuwegein, The Netherlands*

[3]*Cetaqua, Water Technology Center, Carretera d'Esplugues 75, Cornellà de Llobregat, 08940 Barcelona, Spain*

[4]*Amphos21 Consulting SL, Passeig de Garcia Fària, 49, 08019 Barcelona, Spain*

[5]*Berliner Wasserbetriebe, Neue Jüdenstraße 1, 10179 Berlin, Germany*

[6]*HYDOR Consult GmbH, Am Borsigturm 40, 13507 Berlin, Germany*

ABSTRACT

Different types of managed aquifer recharge (MAR) schemes are widely distributed and applied on various scales and for various purposes in the European countries, but a systematic categorization and compilation of data has been missing up to now. The European MAR catalogue presented herein contains various key parameters collected from the available literature. The catalogue includes 224 currently active MAR sites found in 23 European countries. Large quantities of drinking water are produced by MAR sites in Hungary, Slovakia, the Netherlands, Germany, Finland, Poland, Switzerland and France. This inventory highlights that, for over a century, MAR has played an important role in the development of European water supply and contributes to drinking-water production substantially. This development has occurred autonomously, with "trial-and-error" within the full range of climatically and hydrogeologically diverse conditions of the European countries. For the future, MAR has the potential to facilitate optimal (re)use and storage of available water resources and to take advantage of the natural purification and low energy requirements during MAR operations. Particularly with respect to the re-use of wastewater treatment-plant effluent and stormwater, which is currently underdeveloped, the use of MAR can support the public acceptance of such water-resource efficient schemes. Particularly for the highly productive and urbanized coastal zones, where the pressure on freshwater supplies increases by growing water demand, salinization and increased agricultural needs for food production (such as along the Mediterranean and North Sea coasts), MAR is expected to be increasingly relied on in Europe.

Keywords: Artificial recharge History of hydrogeology Future Europe

INTRODUCTION

Managed aquifer recharge (MAR) describes the intentional recharge and storage of water into an aquifer for subsequent recovery or for environmental benefits. MAR can be used to store and treat water in an appropriate aquifer from a variety of source waters, including river water, treated effluent, stormwater, desalinated seawater, rainwater, and even groundwater from other aquifers. MAR is a cross-cutting technology applicable for the drinking water supply, for processing water for industry, for irrigation and for sustaining groundwater dependent ecosystems. MAR relies on naturally occurring processes in the subsurface such as mechanical filtering, sorption and biodegradation, and is often applied in combination with engineered treatment systems, as pre-treatment prior to recharge and post-treatment after recovery.

The European MAR catalogue presented herein contains data from the literature about the MAR type, coordinates, year of first operation and closure, reason for closure, operational scale, aquifer properties (hydraulic conductivity, target aquifer thickness), horizontal aquifer passage, residence time during subsurface passage, recovery rate/share of bank filtrate, source water type and end-use. This catalogue is aimed at (1) outlining the historical origins, (2) describing the current situation and (3) giving perspectives of future developments of MAR in Europe.

MATERIALS AND METHODS

MAR classification

There is a large variety of different MAR types, which can be classified based on the recharge and storage technique into four major groups (Table 1).Enhanced infiltration techniques rely on gravitational infiltration and percolation, and include different surface-spreading methods (areal recharge), point/line recharge and in-channel modifications. Surface-spreading methods are among the simplest and most widely applied MAR techniques. In such methods, the source water is spread over a permeable land surface to enhance the infiltration and percolation to the groundwater body. Most of the existing large-scale MAR sites in Europe make use of this technique, and typically utilise infiltration ponds to increase the groundwater availability. Soil aquifer treatment (SAT) describes the infiltration and percolation of reclaimed water (treated wastewater) through soil and aquifer passage. Due to the lower quality of source water, SAT systems are often operated in wet and dry cycles to allow for maintenance (e.g., mechanical removal of clogging layer) or to restore aerobic conditions in the infiltration zone. During point or line recharge, source water is infiltrated either in elongated (e.g., shafts, drains) or punctual (e.g., abandoned dug wells, bore holes) recharge structures.

Table 1 Classification and overview of MAR types included in the catalogue

Recharge technique and main MAR type	Specific MAR type
Enhanced infiltration	
Surface-spreading methods (areal recharge)	Infiltration ponds
	Soil aquifer treatment (SAT)
	Excess irrigation, ditches, trenches, sprinkler irrigation
Point or line recharge	Well/borehole infiltration
	Reverse drainage, shaft recharge
In-channel modifications	Check dams
	Riverbed scarification
	Sand dams
Induced bank filtration (IBF)	Riverbank filtration
	Lake bank filtration
Well injection	Aquifer storage and recovery (ASR)
	Aquifer storage, transfer and recovery (ASTR)
	Aquifer storage as hydraulic barriers (AS)
Enhanced storage	Subsurface dams

In-channel modifications are structures or measures applicable to stream or channel beds. Riverbed scarification is a measure to enhance the riverbed recharge by mechanical removal of the impermeable top layer in the riverbed. Sand dams are constructed in non-perennial rivers where during periods of high flow, the suspended solids in the stream flow accumulate upstream of the dam over time. Runoff water can easily infiltrate these highly permeable deposits, creating a small-scale artificial aquifer upstream of the dam. Sand dams are often found in Sub-Saharan countries.

Induced bank filtration (IBF) describes the infiltration of surface water induced by pumping from a nearby well. In opposite to bank filtration, which is occurring due to losing river conditions, induced bank filtration emphasizes the purposeful abstraction of surface water by wells. During bank filtration, water quality improvement (treatment) of the induced surface water is commonly observed. IBF sites often consist of several wells situated in a line parallel or perpendicular to the bank of the surface water body. Induced bank infiltration sites are typically installed near perennial streams and lakes that are in hydraulic contact with the adjacent aquifer.

Well injection techniques are used where thick, low-permeability strata overlie the target aquifer(s). Well injection methods include aquifer storage and recovery (ASR), aquifer storage transfer and recovery (ASTR) and aquifer storage (AS). ASR is defined as "the storage of water in a suitable aquifer through a well during times when water is available, and the recovery of water from the same well during times when it is needed" (Pyne 2005), whereas ASTR includes an aquifer passage between the injection and abstraction well. ASTR serves to bridge seasonal gaps in source water by storing water in times of excess and recovering it in times of demand. The stored water is flowing according to the local hydraulic gradient toward the point of abstraction. ASTR is more appropriate in dry climates with pronounced seasonal water availability, but is also found in moderate climatic zones, e.g., the Netherlands. The increased transport component for all injected water as compared to ASR, enhances the removal potential for, e.g., bacteriological parameters. Also, enhanced mixing with native groundwater will occur, which can be seen as a positive or negative aspect, depending on location-specific conditions.

Water storage to bridge seasonal gaps in water supply is often the primary goal when applying this technique. Aquifer storage (AS) primarily aims at groundwater replenishment in the target aquifer, e.g., to counteract seawater intrusion. Sub-surface dams, which are rarely used in Europe, do not lead to additional recharge but enhanced groundwater storage where required. This type is therefore classified as enhanced storage.

Other techniques such as rooftop water harvesting (also called rainwater harvesting), is a method for collecting source water (in this case rain) in the capture zone. Rooftop water harvesting can be combined with injection or infiltration techniques as per local conditions and requirements, and therefore, is not considered as a MAR technique.

Data availability, coverage and limitations

The data presented herein were compiled mostly from scientific publications (i.e., peer-reviewed papers, textbooks, PhD theses) accounting for 42% of all literature sources. Web pages of MAR operators account for 25%, presentations (both talks and posters) for 14%, technical reports and documents for 10%, reports from governmental and non-governmental projects 4%, and personal communication with specialists and operators, and newspaper articles account for 5%.

The European MAR catalogue does not claim to be a comprehensive database including all existing MAR sites in Europe (which is virtually impossible). The lack of data for some countries does not necessarily mean the lack of MAR sites, but can rather be attributed to the fact that language barriers restricted the literature research to languages spoken by members of the author team (i.e., English, Spanish, German, Polish, Dutch and French).

The database contains 278 MAR sites, out of which 56 sites were shut down before 2013. In most cases, the reason for closure is not reported, but many of the closed sites were used as pilots for a limited period of time. At other sites, operation has been suspended temporarily or was shut down entirely due to economic, technical or political reasons.

Data presented herein build on reports from the DEMEAU project (Hannappel et al. 2014), but have undergone substantial changes. Selected data from the European MAR catalogue are available on a web-based geographic information system (GIS) platform and incorporated into a global inventory of MAR sites developed by the International Groundwater Resources Assessment Centre (IGRAC 2016). Maps presented herein were created with the public domain software QGIS 2.8.3; the coordinate reference system is ETRS89, Lambert Azimuthal Equal Area.

Quality assurance and plausibility control

Several persons with various professional backgrounds contributed to data acquisition. Besides the risk of typing error during data entry, many other factors, e.g., false interpretation, translation errors, outdated information sources, will challenge the quality of the collected data. Thus, following the data acquisition period, various quality control measures were carried out to ensure a high level of data integrity. Identified outliers and conspicuous extreme values of the database's numerical fields were double-checked using the respective references. Besides these relatively simple statistical tests on individual fields, logical checks were performed between related parameters in order to identify data gaps. Implausible or unlikely combinations of parameter values were also checked and corrected if necessary.

RESULTS AND DISCUSSION

Historical development of MAR in Europe

Information on the year of first operation and the year of closure allows outlining the historical development of MAR in Europe. The modern history of what is called MAR today begins with two techniques which are most prominently represented in the MAR catalogue: (1) induced bank filtration and (2) surface-spreading methods (Fig. 1).

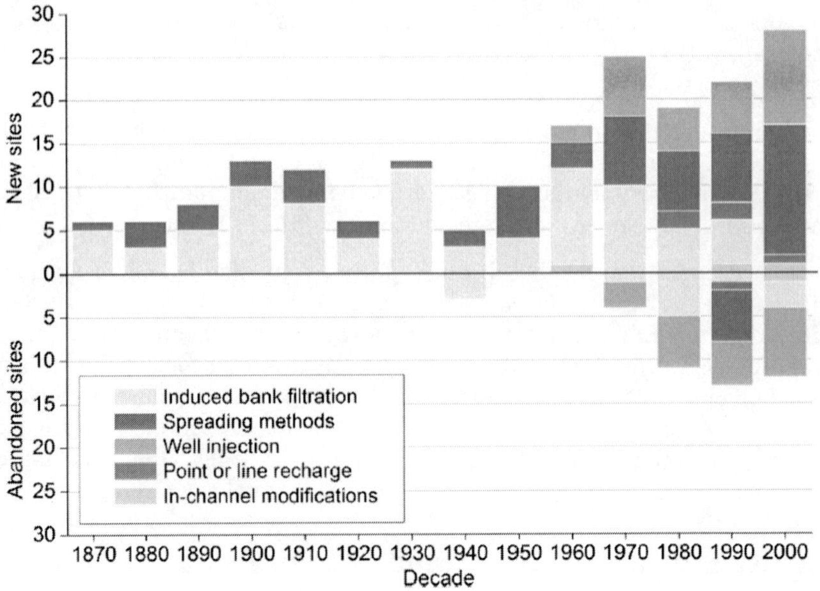

Figure. 1 Outline of the historical development of MAR in Europe showing the number of MAR sites opened or closed per decade between the 1870s and 2000s

The first reported MAR site in Europe was in Glasgow (UK) where in the year 1810 the Glasgow Waterworks Company constructed a perforated collector pipe parallel to the Clyde River (Ray et al. 2002) and abstracted bank filtrated water (BMI 1985; note this site is not shown in Fig. 1 because it seems to be historically isolated). This method was successful at the beginning and many other cities in the UK (e.g., Nottingham, Perth, Derby, Newark; Ray et al. 2002) adopted the idea; thus, the 1860s became the first heyday of "naturally filtered water" in the UK (BMI 1985). However, many of these early sites experienced problems with decreasing well performance and had to be abandoned in later years (BMI 1985); nevertheless, the idea of "naturally filtered underground water" was born and spread to continental Europe, where it was soon adopted by cities in the Netherlands, Belgium, Sweden, France, Austria and Germany. The scientific investigation of "artificial underground water" in Europe began with the water engineers Dupuy, Belgrand, Salbach, Thiem and Richert in the late 19th century (Richert 1900). The progressing industrialization in the 19th century and growing population in European cities presented the water suppliers with new challenges. The traditional water supply based on surface water was impaired by increasing contamination from the new industries and improper sanitation. At that time, based on the experiences in the UK, Thiem proposed the application of riverbank filtration to cope with degrading hygienic surface-water quality and increasing water demand; thus, the pioneers of MAR in Germany can be found at the industrialization hubs close to the Rhine River (e.g., Water Works Düsseldorf 1870), the Ruhr River (e.g., WW Essen 1875), the Elbe River (e.g., WW Saloppe 1875, WW Hosterwitz 1908) around Dresden, and in the Berlin area (e.g., WW Müggelsee, switched to groundwater in 1904–1909, WW Tegel 1901–1903). To increase the abstracted water quantity, many water works constructed infiltration ponds which were often situated on the land side of the abstraction well galleries. Similar to the development in Germany, riverbank filtration (RBF) and infiltration ponds found application in the Netherlands, Sweden and Switzerland—for example, in the Netherlands, the first known RBF-based water supply was reported to have started its operation in 1890 (Stuyfzand 1989). The first MAR site in Switzerland started its operation in Basel "Langen Erlen" in 1912. Eastern European cities then followed and in Hungary the first RBF site was installed north of Budapest on a Danube island (Szentendre Island) in the 1920s (Homonnay 2002); to date, this MAR system is the main drinking water source for Budapest (Homonnay 2002). Additional RBF sites have been developed on other Danube islands (e.g., Csepel) and nowadays several RBF sites exist along the rivers of Raba, Drava, Ipoly, Sajo and Hernad (Homonnay 2002). In Romania the MAR history starts with the operation of the Iasi water supply system at the Moldova River in 1911 and the cities of Cluj Napoca followed in 1935 with conjunctive use of RBF and infiltration ponds and Bacau in 1961 (Rojanschi et al. 2002). In Finland the first water plant using groundwater replenishment by infiltration ponds started its operation in 1929 in Vaasa (Tapio et al. 2006). A few other plants were developed before and after World War II, but the systematic development of MAR in Finland started in the 1960s (Tapio et al. 2006). It is reported that in the year 1992 about 20 water suppliers relied on different MAR types mainly constructed in the 1970s and by 2002, already 25 operating water works utilized MAR in Finland (Tapio et al. 2006). Finally, Tapio et al. (2006) report that after several decades of experience with MAR, this technique is continuously favored by water suppliers.

Research and development of well injection methods began in the 1960s. These early sites were mostly situated in the Netherlands where pilot-scale trials began. Stuyfzand et al. (2012) report that many of these early ASR sites have been closed due to clogging problems. An exception is located in Barcelona, Spain, where a dozen ASR wells were constructed in the early 1970s and are still active today (Hernández et al. 2011). Here, the high transmissivity of the target aquifer (up to 40,000 m^2/day) and low turbidity (<1 NTU) of source water are key parameters associated with the good long-term system performance (Hernández et al. 2015). Cleaning cycles consist of pumping episodes of 15 days with a flow rate four times higher than the injection flow (Azcon and Dolz 1978). Maintenance strategies and clogging aspects are known to be important to consider for MAR practices, but were only rarely reported in the available literature for the European sites. However, in addition to turbidity, as illustrated by the example given, redox mixing and clay swelling are some of the additional factors to consider with respect to clogging risks, which may require pre-treatment prior to injection.

In-channel modifications have been practiced since the 1960s in the Llobregat River in the Barcelona area by riverbed scarifications. Recently this technique experienced decreasing infiltration rates and it is planned to stop or to modify current practices in the near future. Currently this old technique is operated together with other complementary MAR techniques (ASR, infiltration ponds) in the Llobregat area of Spain (Hernández et al. 2015).

Until today, the expression "artificial recharge" has often been used. In Europe, this term dates back to the early investigators such as Richert (1900) and describes "underground" water recharged by human activities. Later on in the late 1990s and the beginning of the 21st century the term 'managed aquifer recharge' was introduced. Some authors reason that "artificial recharge" falsely implies that a somewhat artificial process occurs (Dillon 2005), which can be misleading because the purification in the subsurface relies on natural processes. Moreover, the term MAR refers to the management of aquifer recharge, which implies that risks are managed in a quantitative way.

Current situation of MAR in Europe

The catalogue includes 224 MAR sites active in the year 2013, found in 23 European countries (Fig. 2).Most of the active MAR sites are found in Germany ($n = 64$) followed by the Netherlands ($n = 41$), France ($n = 21$), Finland ($n = 14$), Sweden ($n = 11$), Switzerland ($n = 10$) and Spain ($n = 10$), while in the other countries less than 10 active MAR sites have been found. The most widespread MAR type is induced bank filtration with 127 sites (57% of total active sites); surface-spreading methods rank second among all MAR types with 77 sites (34% of total active sites). Well injection schemes form the third largest group of MAR types with 11 active sites (5% of total active sites) and 23 abandoned sites. Active-point or line-recharge and in-channel modification sites have been found 7 and 1 time(s), respectively. Enhanced storage MAR types, i.e., sub-surface dams, were not found in the literature for Europe.

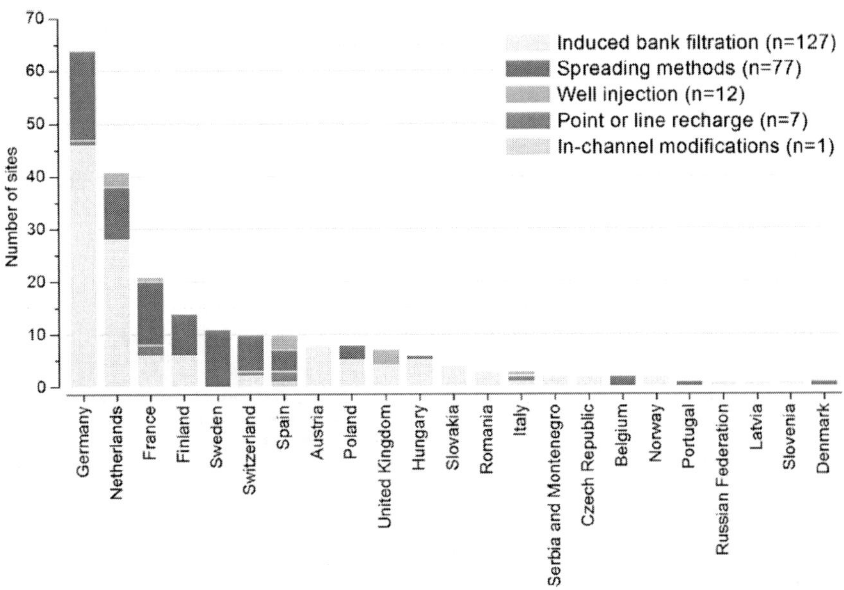

Figure. 2 Number of MAR sites and types for European countries (only MAR sites active in 2013 are shown)

The spatial occurrences of MAR sites and aquifer properties are shown in Fig. 3; these were derived from the International Hydrogeological Map of Europe ('IHME 1500') as reported in BGR & UNESCO (2014). IHME 1500 is a generalized hydrogeological map series covering the European continent. Aquifer properties are displayed by their hydraulic productivity and dominant rock type.

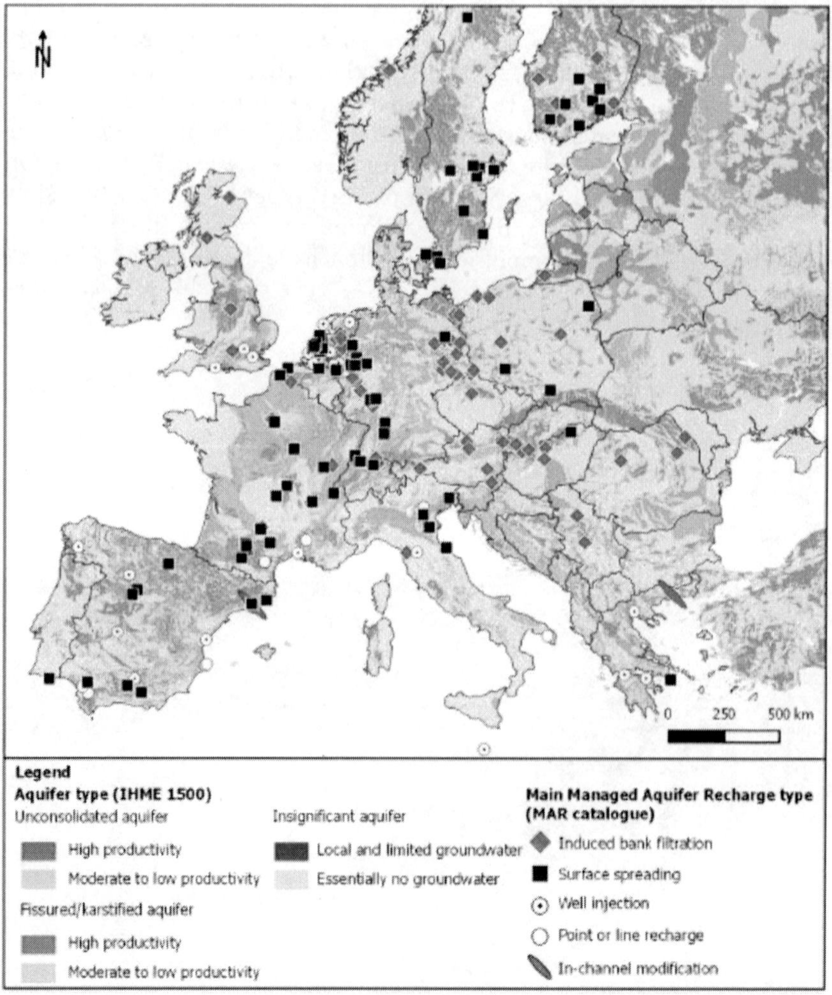

Figure. 3 Overview of MAR sites in Europe and simplified hydrogeological formations (Aquifer types reported in the International Hydrogeological Map of Europe,'IHME 1500', BGR & UNESCO 2014)

The overwhelming majority ($n = 150$) of MAR sites included in the catalogue are situated in unconsolidated geological formations. It is clearly visible that the most common types of MAR are induced bank filtration and surface-spreading methods, located in central and northern countries where large perennial rivers and lakes exist. As shown by IHME 1500, Sweden, Finland, Norway and some parts of Denmark are characterized by local and limited groundwater availability or even essentially no groundwater (Fig. 3). In the Nordic countries, hard rock formations are widespread, while hydraulically conductive formations occur as small unconsolidated glacial deposits (e.g., esker, sander). The occurrence of conductive but small aquifers on one hand and the availability of surface water on the other hand is a main driver for the implementation of MAR (IAEA 2013). In the Nordic countries, surface-spreading methods dominate over induced bank filtration, because surface-spreading methods allow for groundwater replenishment to increase groundwater availability. In the Nordic countries, a special kind of surface spreading has been developed in the last years. Some sprinkling infiltration plants ($n = 3$) have been established, e.g., on forested eskers in Finland (Helmisaari et al. 2005; Lindroos et al. 2002). Sprinkling irrigation (a subtype surface-spreading method) is envisaged to sustain higher dissolved oxygen concentrations in the recharge water in order to avoid anoxic conditions and associated dissolution of geogenic redox-sensitive compounds during subsurface passage. Helmisaari et al. (2005) also highlight that sprinkling irrigation does not cause much disturbance of the environment because the recharge water is "sprinkled directly onto the forest soil from a network of pipes and therefore does not cause as much direct disturbance to the vegetation and soil surface as, e.g., basin recharge". On the other hand, more land is required for sprinkling irrigation as compared to basin recharge, because the recharge areas in the forests are operated in alternating wet/dry cycles. Apart from different types of surface-spreading methods in the Nordic countries, the MAR catalogue depicts a number of IBF sites which are, in contrast to IBF sites, in central Europe, mostly situated at lakes rather than rivers.

In central Europe—i.e., Belgium, the Netherlands, Germany, Poland, Czech Republic, Slovakia, Slovenia, France etc.—MAR sites are usually found in productive unconsolidated aquifers, e.g., in the North European plain in North Germany and large parts of Poland, and along the major rivers. Clusters of MAR sites can be seen in the Netherlands, in Germany along the rivers Rhine and Elbe, in Berlin, and along the Danube River in Austria, Slovakia and Hungary. Most of the IBF sites from the catalogue are found in central Europe.

In the Mediterranean region—i.e., Spain, Italy and some parts of France—IBF sites are only marginally found, but some IBF sites are currently under development, e.g., in the Toscana region near Lucca, Italy (Rossetto et al. 2015). In the Mediterranean countries, mostly surface-spreading sites are found, but also in-channel modifications and point/line recharge schemes. Surface-spreading sites in this region are often designed without point of recovery and mainly aim to replenish the target aquifer, which is often used for agricultural purposes. At a site in Portugal (Campina de Faro aquifer system), river water was recharged through infiltration ponds in order to improve native groundwater quality (Ferreira et al. 2007).

Source water type and end-use

Potential source water types for MAR are river and lake water, stormwater, reclaimed water (treated effluent), desalinated water, and even groundwater from other aquifers or drinking water. The end-use describes the intended final usage of the water from the MAR scheme and includes agricultural, domestic (drinking water), environmental and industrial end-uses. The number of MAR sites related to water end-use are shown in Fig. 4.

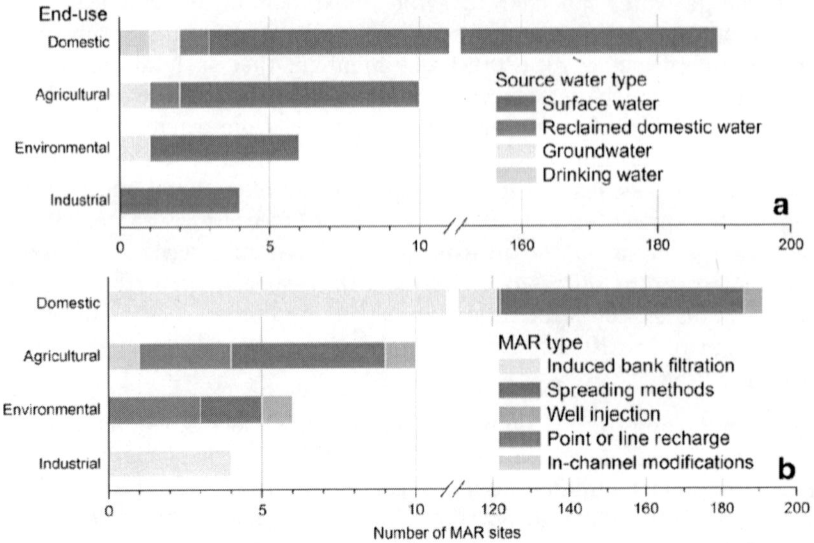

Figure. 4 a Source water types and **b** MAR types, in relation to end-uses for MAR sites in the catalogue (only MAR sites active in 2013 are included)

Figure 4a shows that most MAR sites producing domestic water rely on surface water as their source water, whereas Fig. 4b shows which MAR type is producing for which end-use type. It lies in the nature of MAR that multiple end-uses may exist and a single end-use was sometimes difficult to define. Some sites, for example, do not recover the recharged water and the end-use is then attributed to the dominant usage of the target aquifer—e.g., if the aquifer is mainly used for drinking water abstraction, the end-use is then drinking water production. An environmental end-use is realized when the site is mainly for achieving environmental goals, e.g., sustaining groundwater dependent eco-systems or counteracting salinity ingress.

River and lake water are the most frequent source water types utilized for domestic drinking water supply to a large extent (Fig. 4a). The most frequent MAR type is induced bank filtration and has obviously two primary water sources: river and lake water (Fig. 4b). However, groundwater, which is in many cases of bank filtration also a source, is not shown here, despite the fact that native groundwater may contribute significantly to the abstracted water. Reclaimed water as source water is found at two active MAR sites. One MAR site uses this source water type for agricultural purposes in Spain (Ayuso-Gabella et al. 2011), and the other site produces domestic drinking water with reclaimed water in Belgium (van Houtte et al. 2012). In Torreele/St-Andre (Belgium), treated wastewater is infiltrated in a dune area. The MAR system, in combination with advanced technical pre-treatment, produces potable water in the range of $2.5 \times 10^6\ m^3/a$ (van Houtte and Verbauwhede 2008). Another example for MAR using reclaimed water is found at the Llobregat aquifer in Barcelona (Spain), where reclaimed water was injected via injection wells to counteract seawater intrusion (Ortuno et al. 2012). This hydraulic barrier was in stand-by from 2011 (not shown in Fig. 4) due to financial constraints. The main reason for stopping the injection was the high costs of the (pre-)treatment of reclaimed water (a 15,000 m^3/day plant of ultrafiltration and reverse osmosis was constructed to improve reclaimed water quality before injection). Apart from this clearly communicated usage of reclaimed water by MAR, several other MAR sites exist which use treated wastewater or a blend of fresh and treated effluent water as source water. This de-facto use of reclaimed water is often found at MAR sites situated downstream of a sewage treatment plant, e.g., bank filtration at Berlin-Tegel (Germany) or at the infiltration pond in Sant Vicenç dels Horts (Spain).

MAR sites with a clear environmental end-use are rare and account only for 1% ($n = 3$) of all active sites. In Germany, at an open-pit lignite mine (Garzweiler near Cologne), different surface-spreading techniques (e.g., infiltration shafts, wells) are operated by the mining company to stabilize the water table in order to preserve natural wetlands and swamps (RWE Power 2006); the majority of the source water is groundwater from the active mining area and it is then transported through a pipe network (ca. 125 km lengths) to the adjacent recharge areas to sustain the groundwater dependent eco-systems.

Not all potential source water types have been found to be utilized by European MAR sites. Stormwater run-off is not found as source water for MARs in Europe, and industrial end-use is found at three sites only (in Duisburg and Cologne in Germany, where IBF schemes are operated by industrial companies).

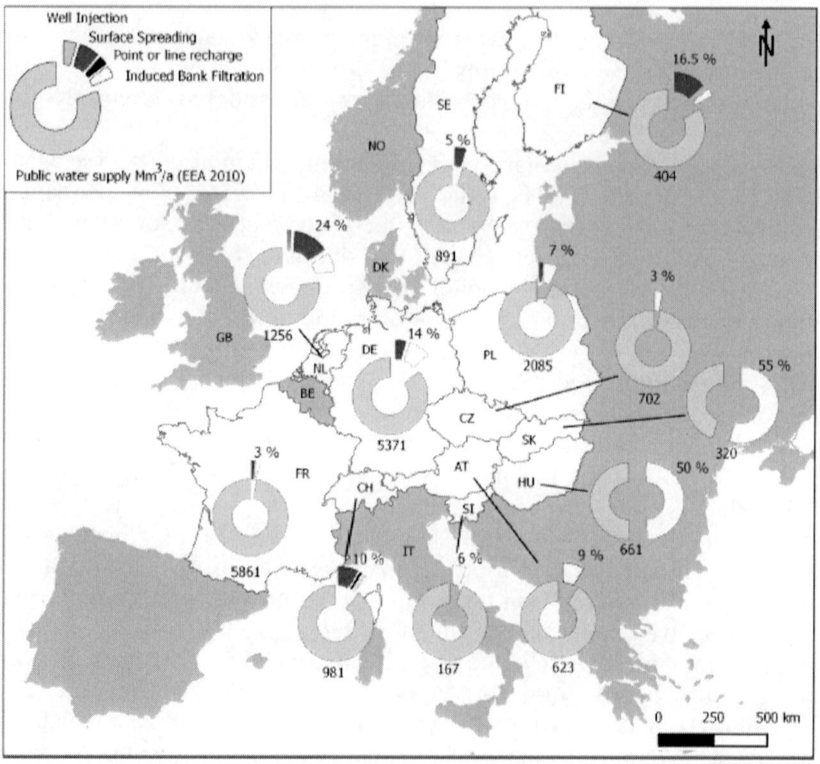

Figure. 5 Percentage contribution of MAR-derived drinking water (calculated from the MAR catalogue) to public water supply (taken from EEA 2010) for European countries. Countries with a MAR contribution <1% are not shown. *NO* Norway, *SE* Sweden, *FI* Finland, *DK* Denmark, *NL* The Netherlands, *BE* Belgium, *DE* Germany, *PL* Poland, *CZ* Czech Republic, *SK* Slovakia, *AT* Austria, *SI* Slovenia, *HU* Hungary, *CH* Switzerland, *FR* France, *IT* Italy, *GB* Great Britain

Contribution of MAR to drinking water supply

The volumetric contribution of MAR-derived water to the drinking water supply for European countries according to the operational scale of MAR sites is shown in Fig. 5. The operational scale gives insight into the total water quantity produced by the MAR scheme. Currently about 190 MAR sites in Europe produce drinking water (see Fig. 4) and are operated by water utilities (mostly public bodies). The percentage contribution of MAR-derived drinking water to drinking water supply is calculated with data from the European Environmental Agency for the year 2007 (EEA 2010).

The contribution of MAR-derived water to drinking water production varies greatly from country to country. In some countries, e.g., Hungary or Slovakia, MAR water may contribute ≥ 50% to the drinking water supply, while other countries, e.g., France, yield only 3% of their drinking water supply from MAR. Countries with a share of MAR-derived drinking water < 1% (e.g., Belgium or United Kingdom) are not shown in Fig. 5.

Some of the largest MAR sites exist on islands in the Danube River, upstream and downstream (IBF sites in Csepel and Szentendre) of Budapest in Hungary. The installed well capacity (indicating the operational scale) of these sites are reported to be 146 and 219 × 10^6 m^3/a, respectively (Grischek et al. 2002). Along with all other MAR sites in Hungary included in the catalogue, the total drinking water volume derived from MAR is about 327 × 10^6 m^3/a, making up ~50% of the public water (total public water supply 661 × 10^6 m^3/a, EEA 2010). Laszlo and Literathy (2002) estimated the share of riverbank filtrated water to the drinking water supply to be around 40% (in total ~470 × 10^6 m^3/a), but the source of these figures remain unclear. Also the Slovakian public water supply relies on MAR to a large extent. The sum of operational scale for all Slovakian MAR sites (entirely IBF) makes up approx. 55% of total public water supply (175 × 10^6 m^3/a from a total 319 × 10^6 m^3/a).

Especially in the Netherlands, different MAR types are used for drinking water production to a large extent. The sum of operational scale for all MAR sites producing drinking water in the Netherlands is about 295 × 10^6 m^3/a, contributing about 24% to the public water supply (1,256 × 10^6 m^3/a from EEA 2010). According to the MAR catalogue, The Netherlands water supply relies on induced bank filtration (7.7%), well injection (0.8%) and spreading methods (15%); Stuyfzand (1989) estimated that approx. 7% of drinking water was produced by IBF systems (valid for the year 1981), which is quite similar to calculations from the MAR catalogue.

The catalogue includes 14 MAR sites from Finland with a total operational scale of 667×10^6 m^3/a. Total annual public water supply in 2007 was 404×10^6 m^3 (EEA 2010), which results in a contribution of 17% MAR water to public water supply. In the literature, estimations for the contribution of MAR water to the public water supply in Finland are in the range of 13–15% from IBF and surface-spreading sites in 2003 (Tapio et al. 2006). Helmisaari et al. (2005) estimated that MAR-derived water accounted for about 15% of the water distributed by Finnish water works in 2005 and is likely to increase to 20% by the year 2010. Figures from the MAR catalogue come to quite similar estimations with 3% from IBF and 13.5% from surface-spreading sites.

For Germany, the catalogue includes 59 active MAR sites producing domestic water (42 sites with dominant induced bank filtration and 17 sites with dominant surface spreading). It should be noted that some water works, e.g., WW Flehe, WW Staad and WW Holthausen in Düsseldorf, are combined to a single site; hence, the real count of MAR sites is likely higher. However, the sum of operational scale from MAR sites in Germany producing drinking water makes up 746×10^6 m^3/a, which is about 14% of the total public water supply—public water supply $5{,}371 \times 10^6$ m^3/a from EEA (2010). Schmidt et al. (2003) estimated that 16% of drinking water in Germany is produced by IBF and surface-spreading sites. Other calculations are found in StatBund (2013)—IBF water makes up 7.8% and surface-spreading sites 9.2%, with a total contribution of 862×10^6 m^3/a; both of these published figures are slightly higher compared to calculations from the MAR catalogue, and may indicate some missing German MAR sites or underestimated operational scale. On the other hand, the operational scale for MAR sites is often based on estimations rather than exact measurements, and is therefore subject to variations. However, by looking on the city scale, e.g., in Berlin, the MAR catalogue includes eight active MAR sites producing about 135×10^6 m^3/a of water, contributing 67% to the total water supply—water supply in Berlin is 202×10^6 m^3/a in 2006 taken from Möller and Burgschweiger (2008). This proportion of MAR water for Berlin water supply is comparable to estimations from Schulze (1997), Hiscock and Grischek (2002) and Massmann et al. (2008b), given as 70–75%.

The MAR catalogue includes eight sites producing domestic water in France. The sum of operational scale from these sites makes up approx. 3% of the public water supply in France—total public water supply $5{,}861 \times 10^6$ m^3/a from EEA (2010). Other figures in the literature, such as 50% of bank filtered water in France (Doussan et al. (1997) and references therein), seem to exaggerate the share of MAR water for public water supply. This large contrast may be explained by different definitions of bank filtration among the countries. In France, production wells which are situated in alluvial strata were considered as surface water influenced and therefore categorized as riverbank filtration wells. This rough simplification may have led to the high share of RBF water, but the actual figures are likely lower.

The MAR catalogue lists nine sites in Switzerland producing domestic water with a total operational scale of approx. 100×10^6 m^3/a. Public water supply in 2007 was 981×10^6 m^3/a (EEA 2010) yielding 10% contribution of MAR water. Diem et al. (2013) estimated that about 25–30% of drinking water originates from induced bank filtration alone in Switzerland. This catalogue entry comes to much lower proportions for all MAR schemes together and it remains unclear how this large difference may be explained.

Aquifer properties and operational parameters

Aquifer properties such as hydraulic conductivity and thickness of the target aquifer are highly important for MAR. Apart from this hydraulic data, the catalogue classifies target aquifers in consolidated and unconsolidated aquifers and aquifer confinement (confined, semi-confined, and unconfined). The target aquifers can further be differentiated by their specific aquifer properties describing the geological genesis of the aquifer (e.g., glacial, fluvial deposits) and the predominant pore type (unconsolidated, fractured, and karstified). Operational parameters are those parameters which can be, at least within the hydrogeological boundaries, controlled by the operator. Operational parameters include the lengths of the horizontal aquifer passage and achieved recovery rates. The horizontal aquifer passage is the modal distance between the point of recharge (e.g., river banks during induced bank filtration or the injection well during ASTR) and the point of recovery (the abstraction well). The recovery rate describes the volumetric ratio between the recharged and the recovered water. Representative properties, expressed as the 10th and 90th percentile, of target aquifer properties for surface spreading, induced bank filtration and well injection are shown in Table 2.

As shown already in Fig. 3, the overwhelming majority of MAR sites are situated in unconsolidated geological strata. MAR sites situated in consolidated geological media are found near London (UK) close to the Thames River where a fissured chalk aquifer (limestone) is hydraulically connected to the overlying riverbed deposits and used for IBF (Schijven et al. 2002). Other examples of MAR in consolidated media can be found in the Salento region in Italy (Ayuso-Gabella et al. 2011), some sites in Spain (Diaz Murillo et al. 2002) or the large ASTR scheme in north London (Harris et al. 2005). Evidence of MAR systems in karstified aquifers has not been found in the literature. Based on the data included in the catalogue, it was observed that MAR in consolidated aquifers is the exception in Europe. It can be concluded that the complex flow conditions and higher degree of heterogeneities (e.g., flow conduits, secondary porosity) require additional field investigations—an example of a tracer test at an ASR site in fissured chalk aquifer is given by Williams (2000). Moreover, the limited purification capacity in consolidated aquifers compared to unconsolidated aquifers complicates the realization of MAR systems. Flow conduits may act as preferential flow paths which transport water with elevated flow velocities, thereby decreasing the residence time of infiltrated water in the subsurface. Residence time is known to be a crucial factor for many attenuation processes during MAR, e.g., microbial transport (Schijven et al. 2002) or pharmaceutical residues (Massmann et al. 2008a); hence, the overwhelming majority of MAR sites are situated in unconsolidated aquifers. Geological formations such as fluviatile and glacial sediments as well as aeolian deposits (e.g., in the Netherlands, Belgium) are commonly utilized. MAR sites situated in consolidated geological media are very rare and no MAR site in karstified strata was found.

Table 2 Representative properties of target aquifers and operational parameters for induced bank filtration, surface spreading and well injection techniques

Property	MAR methodology		
	Induced bank filtration	Surface spreading	Well injection
(Hydro)geological properties[a]	Unconsolidated, unconfined, fluvio-glacial-detrital deposits	Unconsolidated, unconfined, fluvio-glacial-detrital deposits	Unconsolidated, confined, fluvio-glacial-detrital deposits
Hydraulic conductivity (m/s)[b]	5.5×10^{-4}–5.5×10^{-3} ($n = 67$)	3.1×10^{-4}–5.5×10^{-3} ($n = 28$)	2.7×10^{-5}–3.3×10^{-4} ($n = 3$)
Target aquifer thickness (m)[b]	10–48 ($n = 69$)	10–75 ($n = 28$)	28–165 ($n = 5$)
Horizontal aquifer passage (m)[b]	50–1,270 ($n = 78$)	40–682 ($n = 29$)	NA
Residence time during subsurface passage (d)[b]	27–300 ($n = 19$)	15–150 ($n = 10$)	NA
Recovery rate/share of bank filtrate (%)[b]	21–100 ($n = 36$)	40–96 ($n = 4$)	NA

[a] Majority of sites for the respective MAR type

[b] 10th and 90th percentile;

NA not applicable

Out of 127 IBF sites, 60 have been reported to be under unconfined conditions and 9 under semi-confined conditions, and out of 77 surface-spreading sites, 29 have been found to be under unconfined conditions and 4 semi-confined (for the remaining no information was found). Both MAR types require unconfined conditions and it must be assumed that sites where this specific information was not found are also under unconfined conditions. Well injection sites were often reported to be under confined conditions—for example, the ASR site in Barcelona-Cornellà is located between the phreatic aquifer (upstream) and the confined aquifer (downstream).

Induced bank filtration and surface-spreading techniques show a similar range of hydraulic conductivities. Well injection techniques indicate hydraulic conductivities approximately one order of magnitude lower, but the number of records is relatively low. Target aquifer thicknesses are also similar between IBF and surface-spreading sites. This finding is not surprising as many IBF and surface-spreading sites were developed conjunctively and are situated in the same aquifers.

The horizontal aquifer passage roughly relates to the residence time of the infiltrated source water in the subsurface, but is not equal to the flow path. The horizontal aquifer passage may also give insight on the share of source water abstracted by the recovery well(s). As a rule of thumb, a short horizontal aquifer passage within a thin aquifer implies short residence times and high shares of source water in the abstraction well. However, both parameters also depend on aquifer properties, well design and operational parameter for a particular site, which makes it challenging to assess residence times or share of source water based on horizontal aquifer passage alone. The data from the MAR catalogue show that IBF sites have a wide range of horizontal aquifer passages, from a few tens of meters—e.g., 20 m at the Eura IBF site in Finland (Kivimäki 2001)—to even a few kilometers such as the 3.5 km at Aalst IBF site in the Netherlands (Stuyfzand and Doomen 2004). Exceptionally long horizontal aquifer passages are found in the Netherlands where well fields, consisting of some 15–20 wells, are situated in a row perpendicular to the river bank. The horizontal distances reported by Stuyfzand et al. (2006a, 2006b) were measured between the central part of the well field and the riverbank during low flow conditions in summer. At other IBF sites, the well fields are situated in a row parallel to the river or lake bank—e.g., IBF in Berlin as described in Massmann et al. (2004). IBF sites with short aquifer passage are usually characterised by high shares of bank filtrate—e.g., Remmerden in the Netherlands with 82% (Stuyfzand and Doomen 2004). However, also IBF sites with long horizontal aquifer passages (≥3000 m, e.g., Aalst, Kolff and Druten) may abstract substantial shares of bank filtrate (29–68%) in their recovery well(s) (Stuyfzand et al. (2006a, 2006b). Analogous to IBF, the surface-spreading sites also show a wide range of horizontal aquifer passages, from 30 m at an infiltration pond in Poland (Blazejewski 1983) to 1,450 m at a sprinkler irrigation site in Finland (Lindroos et al. 2002). Only a little information about the horizontal aquifer passage and achieved recovery rates for well injection schemes was found and therefore not included in Table 2.

The residence time during MAR is a critical parameter to ensure sufficient attenuation of hygiene-related parameters and other undesired substances. From a hygiene perspective, a subsurface travel time of around 50–60 days (in the UK 400 days are defined as the outer source protection zone) are often demanded by European directives (Chave et al. 2006; DVGW 2006). Representative values from the MAR catalogue shows that residence time for some IBF sites can be substantially shorter than 50–60 days. It has to be taken into account that travel-time estimations are subject to large variations depending on the assessment method applied. Travel time estimations based on only hydraulic data must be considered less accurate than estimations based on tracer breakthrough curves. However, median values for IBF and surface-spreading sites were calculated to be 70 and 55 days, respectively.

Recovery rates for surface-spreading sites were often not reported in the literature and found only at four sites. Whenever possible, this value was calculated based on information from the literature source, e.g., at the Solleveld site (infiltration ponds in dune areas in the Netherlands), the average annual abstracted water volume is lower than the average annual infiltrated volume and the overall recovery rate was approximated to be 90%. For IBF systems, the share of bank filtrate describes the volumetric ratio between bank filtrate and native groundwater in the recovery well(s). Information about the share of bank filtrate was found at 36 IBF sites and ranges from minimum 10% to maximum 100%, resulting in representative values of 21–100% (10th and 90th percentile, respectively).

PERSPECTIVE OF MAR IN EUROPE AND CONCLUSIONS

For over a century, various forms of managed aquifer recharge have been used in Europe. This development has occurred autonomously, with "trial-and-error" within the full range of climatically and hydrogeologically diverse conditions of the European countries. Although, over the years, the use of MAR has grown and spread independently throughout Europe in support of the EU strategy for resource and energy efficient water production and management (EEA 2014), the benefits of MAR could develop towards it becoming a "first-option" in securing water availability for the future. In the face of numerous stresses on the availability of water such as climate change, increased weather variability, salinization, as well as increased urbanization of coastal zones and emerging substances, MAR has the potential to facilitate optimal (re)use and storage of available water resources and to take advantage of the natural purification and low energy requirements during MAR operations. Particularly with respect to the re-use of wastewater treatment-plant effluent and stormwater, which is currently underdeveloped, the use of MAR can support the public acceptance of such water-resource efficient schemes.

In developing water availability strategies in the face of numerous challenges, Europe is not alone; worldwide, MAR is being considered as an integral and essential technique to meet objectives and demands for the future—e.g., Sheng and Zhao (2014); Megdal et al. (2014); Dillon et al. (2010). As globally, the pressure on freshwater supplies increases by growing water demand, intensified by continued urbanization, increased agricultural needs for food production and the desire to preserve ecosystem integrity, MAR is expected to be increasingly relied on.

The most densely populated and (economically) productive regions of the world are in the coastal zones, particularly in Europe along the Mediterranean Sea and the North Sea. It was estimated that about half of the world's population lives within 200 km of a coastline (UN 2010). While these areas produce many economic benefits, the associated high water demand puts tremendous pressure on freshwater resources and coastal ecosystems, leading to problems like seasonal water shortage, overexploitation of groundwater resources, saltwater intrusion, and disappearance of wetlands. Further economic growth, population increase, and climate change will aggravate these problems, ultimately blocking the sustainable development of coastal zones in industrialized, emerging, and developing countries (EC 2012); therefore, in 2015, water crises were identified as the main global risk (WEF 2015).

Traditionally, aboveground solutions such as construction of reservoirs or saltwater desalination are often sought to solve freshwater problems. However, the subsurface may provide more robust, effective, sustainable, and cost-efficient freshwater management solutions—for instance, artificial recharge of aquifers with temporary freshwater surpluses is increasingly applied worldwide for water storage and treatment (Dillon et al. 2010). Typically, artificial recharge in unconfined aquifers is used to increase volumetric water availability by filling part of the overlying unsaturated zone. In (semi)confined aquifers, however, artificial recharge can also increase water availability by displacing native groundwater that is not suitable or less suitable for use than the water source available for storage. In fact, ASR and ASTR have been in confined/semi-confined aquifers, in Europe and beyond. Depending on the intended use, a wide range of water quality parameters—e.g., the presence of manganese, Antoniou et al. (2014)—can make native groundwater less suitable; however, in recent years, particular attention has been on the artificial recharge of freshwater in water-stressed areas with brackish or saline native groundwater (Ward et al. 2007, 2009). These studies have highlighted the impact of the density difference between the injected fresh and native saline water on recovery efficiencies, and recent advances using either horizontal or multiple partially penetrating wells enable the improved recovery in both confined and unconfined saline aquifers (Zuurbier et al. 2014a, b, 2016).

During data compilation for the MAR catalogue, it was realized that clear economic case studies of European MAR sites are often not available. Economic feasibility can be evaluated using cost-benefit analysis, but site specifics and multiple benefits of MAR are often challenging to monetize. The lack of data for economic feasibility of the various types of MAR due to the wide range of benefits is considered to be a major barrier for implementation; however, recent non-European studies about MAR economics related to water re-use (Vanderzalm et al. 2015), and conceptual frameworks for MAR economics including methods for monetizing typical MAR benefits (Maliva 2014), are readily applicable to European sites.

Aquifer storage and recovery (ASR) for instance can be a successful technique for storage and recovery of both potable and irrigation water (Maliva and Missimer 2010); however, as in other areas in the world, it has not reached its full potential in Europe. The advantages of ASR consist of the limited space requirement above ground, the lack of losses by evaporation, the protection from atmospheric, biologic and anthropogenic contamination, and the protection from earthquake damage—for example, in the Netherlands, this increased ASR reliability is very much welcomed with the prospects of longer periods of drought, despite an increase in yearly gross precipitation (KNMI 2014) induced by climate change, while water is increasingly recognized as a scarce resource (WEF 2015). The need for water harvesting and storage is therefore expected to increase; furthermore, an increase in extreme rainfall events is expected in the Netherlands (KNMI 2014; Royal Netherlands Meteorological Institute 2014), which will require better exploitation of aboveground water reservoirs for retention of intense rainfall. ASR provides the means to lower the levels of these reservoirs by early infiltration once potential extreme rainfall events are predicted, and to provide retention, without having to discharge (and loose) the water to the sea. Therefore, the water remains available for later times of demand, while the unrecoverable part can counteract the ongoing salinization. Another European example from the coastal zone is Spain (e.g., the Llobregat aquifer in Barcelona) where MAR is considered as a promising technique to be implemented at a large scale using reclaimed water to improve groundwater availability without compromising surface-water availability for direct users.

ACKNOWLEDGEMENTS

This study has received funding from the 7th Framework Programme of the European Union under the grant agreement number 308339 within the DEMEAU project.

REFERENCES

1. Antoniou EA, Hartog N, van Breukelen BM, Stuyfzand PJ (2014) Aquifer pre-oxidation using permanganate to mitigate water quality deterioration during aquifer storage and recovery. Appl Geochem 50:25–36
2. Ayuso-Gabella N, Page D, Masciopinto C, Aharoni A, Salgot M, Wintgens T (2011) Quantifying the effect of managed aquifer recharge on the microbiological human health risks of irrigating crops with recycled water. Agric Water Manag 99:93–102
3. Azcon A, Dolz J (1978) Estudio de la recarga artificial mediante pozos en Cornellà [Study of artificial recharge by wells in Cornellà]. XII Curso Internacional Hidrología Subterránea [XII International Course Subsurface Hydrology], FCIHS, Barcelona
4. BGR & UNESCO (eds.) (2014): International Hydrogeological Map of Europe 1:1,500,000 (IHME1500). Digital map data v1.1. Hannover/Paris.
5. Blazejewski M (1983) Perspektywy Wykorzystania Sztucznej Infiltracji W Uzadatnianiu Zanieczyszonych Wod Powierzchniowych W Polsce W Swietle Dotychczasowych Doswiadczen [Perspectives of artificial recharge to treat contaminated groundwater in Poland]. Sztuczna infiltracja w uzdatnianiu WOD powierzchniowych [Experiences with artificial recharge]. Instytut Ksztaltowania Srodowiska (Institute of Environmental Science) – Poznan, Poland, pp 7–12
6. BMI (1985) Künstliche Grundwasseranreicherung: Stand der Technik und des Wissens in der Bundesrepublik Deutschland. Bundesministerium des Inneren [Artificial groundwater replenishment: state of the technology and knowledge in the Federal Republic Germany, Ministry of the Interior]. Schmidt, Berlin
7. Chave P, Howard G, Schijven J, Appleyard S, Fladerer F, Schimon W (2006) Groundwater protection zones. In: Schmoll O, Howard G, Chilton J, Chorus I (eds) Protecting groundwater for health: managing the quality of drinking-water sources. IWA, London
8. Diaz Murillo JM, de la Orden G, Lopez Geta JA, Rubio Campos A, Gonzalez Ramon Martin Machuca A (2002) Lessons from groundwater recharge projects in Spain: management of aquifer recharge for sustainability. In: Dillon P (ed) International Symposium on Artificial Recharge of Groundwater (ISMAR), Adelaide, September 2002, pp 393–398
9. Diem S, Schrirmer M, von Rohr MR, Kohler H-P, Hering JG, von Gunten U (2013) Qualität des Uferfiltrats: Einfluss der Klimabestimmten Variablen Temperatur und Abfluss [Quality of bank filtrate: influence of the climate induced variables temperature and discharge]. Aqua & Gas 11, Eawag, Dübendorf, Switzerland, 21 pp
10. Dillon P (2005) Future management of aquifer recharge. Hydrogeol J 13:313–316
11. Dillon P, Toze S, Page D, Vanderzalm J, Bekele E, Sidhu J, Rinck-Pfeiffer S (2010) Managed aquifer recharge: rediscovering nature as a leading edge technology. Water Sci Technol 62:2338–2345
12. Doussan C, Poitevin G, Ledoux E, Detay M (1997) River bank filtration: modelling of the changes in water chemistry with emphasis on nitrogen species. J Contam Hydrol 25:129–156

13. DVGW (2006) Richtlinien für Trinkwasserschutzgebiete; Teil 1: Schutzgebiete für Grundwasser - Arbeitsblatt W 101. Deutscher Verein des Gas- und Wasserfaches [Guidelines for drinking water protection zones, part 1: protection zones for groundwater. Issue W101. German Association for Gas and Water]. DVGW, Bonn, Germany
14. EC (2012) A blueprint to safeguard Europe's water resources. European Commission, Brussels, Belgium
15. EEA (2010) European Environmental Agency Core Set Indicator CSI 18, based on data from Eurostat data table: annual water abstraction by source and by sector. European Environmental Agency, Copenhagen
16. EEA (2014) Performance of water utilities beyond compliance: sharing knowledge bases to support environmental and resource-efficiency policies and technical improvements. European Environmental Agency, Copenhagen
17. Ferreira JPL, Oliveira M, Diamantino C, João Moinante M, Medeiros A (2007) DELIVERABLE 31: inventory of alternative water sources for each test site. In: GABARDINE Project: groundwater artificial recharge based on alternative sources of water—advanced integrated technologies and management. National Laboratory for Civil Eng., Lisbon
18. Grischek T, Schoenheinz D, Worch E, Hiscock K (2002) Bank-filtration in Europe: an overview of aquifer conditions and hydraulic controls. In: Dillon P (ed) Management of aquifer recharge for sustainability: proceedings of the 4th International Symposium on Artificial Recharge of Groundwater, Adelaide, September 2002. CRC, Boca Raton, FL, pp 485–488
19. Hannappel S, Scheibler F, Huber A, Sprenger C (2014) M11.1 Characterization of European Managed Aquifer Recharge (MAR) sites: analysis. DEMEAU. http://demeau-fp7.eu/. Accessed February 2017
20. Harris SJ, Adams MJ, Jones MA (2005) NLARS: evolution of an artificial recharge scheme. ISMAR5, Berlin, June 2005
21. Helmisaari HS, Derome J, Hatva T (2005) Artificial recharge in Finland through basin and sprinkling infiltration: soil process, retention time and water quality. International Symposium on Management of Aquifer Recharge, Berlin
22. Hernández M, Camprovín P, Bernat X, Massana J, Castelló J (2015) ASR en Barcelona: Nuevo régimen de operación para hacer frente a nuevos escenarios [ASR in Barcelona: New operating regime to deal with new scenarios]. IV Jornadas de Ingeniería del Agua La precipitación y los procesos erosivos, Cordoba, October 2015
23. Hiscock KM, Grischek T (2002) Attenuation of groundwater pollution by bank filtration. J Hydrol 266:139–144
24. Homonnay Z (2002) Use of bank filtration in Hungary. In: Ray C (ed) Riverbank filtration: understanding contaminant biogeochemistry and pathogen removal. Kluwer, Dordrecht, Germany, pp 221–228
25. IAEA (2013) Using isotopes for design and monitoring of artificial recharge systems. International Atomic Energy Agency, Vienna
26. IGRAC (2016) Global MAR Inventory, International Groundwater Resources Assessment Centre. https://www.un-igrac.org/special-project/global-mar-inventory. Accessed 1 Nov 2016
27. Kivimäki AL (2001) Site description Jokioinen, Forssa and Eura, Finland, Artificial recharge of groundwater. EC project ENV4-CT95-0071, EC, Brussels, pp 23–25
28. KNMI (2014) Climate Change scenarios for the 21st Century: a Netherlands perspective. In: van den Hurk B, Siegmund P, Klein Tank A (eds) Royal Netherlands Meteorological Institute, De Bilt, The Netherlands
29. Laszlo F, Literathy P (2002) Laboratory and field studies of pollutant removal. In: Ray C (ed) Riverbank filtration: understanding contaminant biogeochemistry and pathogen removal. Kluwer, Dordrecht, The Netherlands, pp 229–234
30. Lindroos A-J, Kitunen V, Derome J, Helmisaari H-S (2002) Changes in dissolved organic carbon during artificial recharge of groundwater in a forested esker in southern Finland. Water Res 36:4951–4958

31. Maliva RG (2014) Economics of managed aquifer recharge. Water 6(5):1257–1279
32. Maliva RG, Missimer TM (2010) Aquifer storage and recovery and managed aquifer recharge using wells: planning, hydrogeology, design and operation. Methods in Water Resources Evolution. Schlumberger, Houston, TX
33. Massmann G, Knappe A, Richter R, Pekdeger A (2004) Investigating the influence of treated sewage on groundwater and surface water using wastewater indicators in Berlin, Germany. Acta Hydrochim Hydrobiol 32:336–350
34. Massmann G, Dünnbier U, Heberer T, Taute T (2008a) Behaviour and redox sensitivity of pharmaceutical residues during bank filtration: investigation of residues of phenazone-type analgesics. Chemosphere 71:1476–1485
35. Massmann G, Sültenfuß J, Dünnbier U, Knappe A, Taute T, Pekdeger A (2008b) Investigation of groundwater residence times during bank filtration in Berlin: a multi-tracer approach. Hydrol Process 22:788–801
36. Megdal S, Dillon P, Seasholes K (2014) Water banks: using managed aquifer recharge to meet water policy objectives. Water 6:1500
37. Möller K, Burgschweiger J (2008) Wasserversorgungskonzept für Berlin und für das von den BWB versorgte Umland (Entwicklung bis 2040) [Water supply concept for Berlin and the surrounding area (development until 2040) supplied by BWB]. BWB, Berlin
38. Ortuno F, Molinero J, Custodio E, Juárez I, Garrido T, Fraile J (2012) Seawater intrusion barrier in the deltaic Llobregat aquifer (Barcelona, Spain): performance and pilot phase results. SWIM21 - 21st Salt Water Intrusion Meeting 21, Azores, Portugal, June 2010, pp 135–138
39. RWE Power (2006) Ökologie im Nordrevier: Wasser für die Feuchtgebiete [Ecology in the Nordrevier: water for the wetlands]. RWE Power, Essen, Germany
40. Pyne D (2005) Aquifer storage recovery: a guide to groundwater recharge through wells. ASR Systems, Gainesville, FL
41. Ray C, Schubert J, Linsky RB, Melin G (2002) Introduction. In: Ray C (ed) Riverbank filtration: improving source-water quality. Kluwer, Dordrecht, The Netherlands
42. Richert JG (1900) On artificial underground water. C. E. Fritze's Royal Book-Store, Stockholm
43. Rojanschi V, Mlenajek L, Stanciulescu M (2002) Riverbank filtration in water supply: old solutions, new problems. In: Ray C (ed) Riverbank filtration: understanding contaminant biogeochemistry and pathogen removal. Kluwer, Dordrecht, The Netherlands, pp 235–245
44. Rossetto R, Barbagli A, Borsi I, Mazzanti G, Vienken T, Bonarim E (2015) Site investigation and design of the monitoring system at the Sant'Alessio Induced RiverBank Filtration plant (Lucca, Italy). Rend Online Soc Geol Ital 35:248–251
45. Royal Netherlands Meteorological Institute (2014) KNMI'14 climate scenarios for the Netherlands: guideline for professionals in climate adaptation. KNMI, De Bilt, The Netherlands, 34 pp
46. Schijven J, Berger P, Miettinen I (2002) Removal of pathogens, surrogates, indicators, and toxins using riverbank filtration. In: Ray C, Melin G, Linsky RB (eds) Riverbank filtration: improving source-water quality. Springer, Heidelberg, Germany, pp 73–116
47. Schmidt CK, Lange FK, Brauch H-J, Kühn W (2003) Experiences with riverbank filtration and infiltration in Germany. DVGW-Water Technology Center (TZW), Karlsruhe, Germany, 17 pp
48. Schulze D (1997) Wassermengenwirtschaft im Ballungsraum Berlin [Water quantity economy in the conurbation area Berlin]. Arch Nat Lands 35:289–300
49. Sheng Z, Zhao X (2014) Special issue on managed aquifer recharge: powerful management tool for meeting water resources challenges. J Hydrol Eng. doi:10.1061/(ASCE)HE. 1943-5584.0001139
50. StatBund (2013) Öffentliche Wasserversorgung und öffentliche Abwasserentsorgung: öffentliche Wasserversorgung [Public water supply and

public sanitation: public water supply]. Statistisches Bundesamt, Wiesbaden, Germany
51. Stuyfzand PJ (1989) Hydrology and water quality aspects of Rhine bank groundwater in the Netherlands. J Hydrol 106:341–363
52. Stuyfzand PJ, Doomen A (2004) The Dutch experience with MARS (Managed Aquifer Recharge and Storage): a review of facilities, techniques and tools. KIWA NV, Rijswijk, The Netherlands
53. Stuyfzand P, Juhàsz-Holterman MA, de Lange W (2006a) Riverbank filtration in the Netherlands: well fields, clogging and geochemical reactions. In: Hubbs SA (ed) Riverbank filtration hydrology. Springer, Dordrecht, The Netherlands, pp 119–153
54. Stuyfzand PJ, Juhasz-Holterman MHA, de Lange WJ (2006b) Riverbank filtration hydrology. In: Hubbs SA (ed) Riverbank filtration hydrology. Springer, Dordrecht, The Netherlands, pp 119–153
55. Stuyfzand PJ, Nienhuis P, Antoniuou A, Zuurbier K (2012) Haalbaarheid van ondergrondse berging via A(S/T)R in Hollands kustduinen [Feasibility of underground storage through A (S/T) R in Dutch coastal dunes]. KWR, Nieuwegein, The Netherlands
56. Tapio SK, Annukka Lipponen M, Rönkö EKT (2006) Groundwater use and policy in community water supply in Finland. Hydrogeol J 14:69–78
57. UN (2010) United Nations atlas of the oceans. UN, New York
58. van Houtte E, Verbauwhede J (2008) Operational experience with indirect potable reuse at the Flemish Coast. Desalination 218:198–207
59. van Houtte E, Cauwenberghs J, Weemaes M, Thoeye C (2012) Indirect potable reuse via managed aquifer recharge in the Torreele/St-Andre project. In: Kazner C, Wintgens T, Dillon P (eds) Water reclamation technologies for safe managed aquifer recharge. IWA, London, pp 33–33
60. Vanderzalm JL, Dillon PJ, Tapsuwan S, Pickering P, Arold N, Bekele EB, Barry KE, Donn MJ, McFarlane D (2015) Economics and experiences of managed aquifer recharge (MAR) with recycled water in Australia. Australian Water Recycling Centre of Excellence, Brisbane
61. Ward JD, Simmons CT, Dillon PJ (2007) A theoretical analysis of mixed convection in aquifer storage and recovery: how important are density effects? J Hydrol 343:169 – 186. doi:10.1016/j.jhydrol.2007.06.011
62. Ward JD, Simmons CT, Dillon PJ, Pavelic P (2009) Integrated assessment of lateral flow, density effects, and dispersion in aquifer storage and recovery. J Hydrol 370:83– 99. doi:10.1016/ j.hydrol.2009.02.055
63. WEF (2015) Global risks 2015. World Economic Forum, Davos, Switzerland
64. Williams A (2000) Using an aquifer storage and recovery (ASR) trial as a large-scale tracer test. TraM'2000, Liege, Belgium
65. Zuurbier K, Kooiman J, Groen M, Maas B, Stuyfzand P (2014a) Enabling successful aquifer storage and recovery of freshwater using horizontal directional drilled wells in coastal aquifers. J Hydrol Eng 20:B4014003
66. Zuurbier KG, Zaadnoordijk WJ, Stuyfzand PJ (2014b) How multiple partially penetrating wells improve the freshwater recovery of coastal aquifer storage and recovery (ASR) systems: a field and modeling study. J Hydrol 509:430–441
67. Zuurbier KG, Hartog N, Stuyfzand PJ (2016) Reactive transport impacts on recovered freshwater quality during multiple partially penetrating wells (MPPW-)ASR in a brackish heterogeneous aquifer. Appl Geochem 71:35–47

CHAPTER 11

Spatio-temporal Prediction of Root Zone Soil Moisture Using Multivariate Relevance Vector Machines

Bushra Zaman, Mac McKee

Utah Water Research Laboratory, Utah State University, Logan, USA

ABSTRACT

Root zone soil moisture at one and two meter depths are forecasted four days into the future. In this article, we propose a new multivariate output prediction approach to root zone soil moisture assessment using learning machine models. These models are known for their robustness, efficiency, and sparseness; they provide a statistically sound approach to solving the inverse problem and thus to building statistical models. The multivariate relevance vector machine (MVRVM) is used to build a model that forecasts soil moisture states based upon current soil moisture and soil temperature conditions. The methodology combines the data at different depths from 5 cm to 50 cm, the largest of which corresponds to the depth at which the soil moisture sensors are generally operational, to produce soil moisture predictions at larger depths. The MVRVM test results for soil moisture predictions at 1 m and 2 m depth on the 4th day are excellent with RMSE = 0.0131 m^3/m^3 for 1 m; and RMSE = 0.0015 m^3/m^3 for 2 m forecasted values. The statistics of predictions for 4th day (CoE = 0.87 for 1 m and CoE = 0.96 for 2 m) indicate good model generalization capability and computations show good agreement with actual measurements with R^2 = 0.88 and R^2 = 0.97 for 1 m and 2 m depths, respectively. The MVRVM produces good results for all four days. Bootstrapping is used to check over/under-fitting and uncertainty in model estimates.

Keywords: relevance vector machines, statistics, predictions, soils, soil moisture, data management

1. INTRODUCTION

Root zone soil moisture is regarded as key factor governing surface water and energy balances and plays a vital role in hydroclimatic and environmental predictions. Soil moisture content (SMC) measurements are important for irrigation scheduling and crop yield forecast modeling, understanding rainfall/runoff generation processes. Information on soil moisture helps in explaining processes related to crop growth, forest dynamics and other vadose zone processes which play a vital role in water resources planning and management.

Soil moisture varies both in space and time because of spatial and temporal variations in precipitation, soil, topographic features, and vegetation characteristics [1]. The spatio-temporal prediction of SMC is difficult though capturing these variations and having an accurate estimation of soil moisture is necessary for soil and land survey [2] [3] , soil and land evaluation [4] , hydrologic modeling and watershed management [5] [6]. Also, there is a need to develop methods for estimating SMC which make the best possible use of ancillary information, particularly that which is relatively cheap to obtain [7] - [9]. Much work has been done in the past where soil moisture at larger depths was retrieved using surface soil moisture estimates [10] - [14]. The surface soil moisture content is physically related to root-zone soil moisture through diffusion processes, and both surface and root-zone soil layers are commonly simulated by land surface models (LSMs) [15] [16] demonstrated the relationship between the soil moisture profile and surface soil moisture and fluxes. It was found that soil moisture can be predicted using low-level atmospheric and meteorological inputs [17] [18].

SMC retrieval using different techniques has been the subject of research for almost four decades. In general, soil moisture measurements are made as point measurements, mainly using gravimetric, nuclear, electromagnetic, tensiometric, or hygrometric techniques [19] , or by measuring SMC with imbedded sensors, such as time- and frequency-domain reflectometers (TDRs and FDRs). Physically based models for soil moisture estimation include the Soil-Plant-Atmosphere-Water (SPAW) model of [20] [21] - [23] , the US Department of Agriculture Hydrograph Laboratory (USDAHL) model [24] [25] , and the Sacramento Soil Moisture Accounting (SAC-SMA) Model [26] [27] used by the National Weather Service River Forecast System (NWSRFS) [28] , soil vegetation atmosphere transfer (SVAT) models, among others. However, the difficulty associated with measurement of the physical parameters required by these models serves as an impediment.

This has furthered the interest of researchers to look for data-driven modeling tools such as artificial neural networks (ANNs) [29] - [31] , higher order neural networks [32] , support vector machines (SVMs) [33] [34]. [18] used soil moisture and meteorological data to generate SVM predictions for four and seven days ahead. The RVM and SVM models were used for forecasting soil moisture five days in the future by [35]. In the present study, we are applying a relatively new data-driven tool, the multivariate relevance vector machine (MVRVM) for soil moisture estimation. The purpose of this research was to develop a new model which forecasts soil moisture at different root zone depths, so both spatial and temporal predictions are done simultaneously. With this goal in mind, past measurements of soil moisture at shallower depths, soil temperature and precipitation were used as inputs to a MVRVM model that forecasts soil moisture at large depths and for several days in the future. This model used available data acquired by the data collection station for previous days. The procedure employs point analyses, and no horizontal spatial patterns are taken into consideration. This learning machine tool automatically learns to recognize complex patterns that reside in data and that can be exploited to model input-output relationships. This technique, which has generated promising results, has never been tried before.

2. MULTI-VARIATE RELEVANCE VECTOR MACHINE (MVRVM)

"Sparse Bayesian Learning" is used to describe the application of Bayesian automatic relevance determination (ARD) concepts to models that are linear in their parameters. The motivation behind the approach is that one can infer a regression model that is both accurate and sparse in that it makes its predictions using only a small number of relevant basis functions that are optimally selected from a potentially large initial set. A special case of this concept is the RVM which is applied to linear kernel models. The RVM was originally introduced by [36] [37] proposed an extension of the sparse Bayesian regression model developed by [38] and this extension enables a single relevance vector machine (RVM) to handle multiple output dimensions. The multivariate regression code developed by [37] is an open source code. This code was used as a base to build the MVRVM model which was particular to this application.

The data set is in the form of input-output pairs, $\{x_n, t_r\}_{n=1,r=1}^{N,P}$, where x is the input matrix, t is the target

vector, P is the number of output dimensions and N is the number of observations. The major goal is to learn a model of dependency of the outputs on the inputs with the objective of making accurate predictions for previously unseen values of x [39]. Each output vector (t_r) is written as $t_r = (t_1, \cdots, t_N)^T$ and is expressed as the sum of an approximation vector $y = (y(x_1), \cdots, y(x_N))^T$ and an "error" vector, the elements of which are considered as independent samples from some noise process: $\epsilon = (\epsilon_1, \cdots, \epsilon_N)^T$:

$$t_r = y_r + \epsilon_r = \Phi w_r + \epsilon_r, \tag{1}$$

where, w_r = weight vector for the rth component of the output vector t_r and $\Phi = [\Phi(x_1) \ \cdots \ \Phi(x_M)]$, the N ´ M "design" matrix whose columns comprise the complete set of M "basis vectors". According to the sparse Bayesian approach [39] , "the errors are conventionally assumed to be zero-mean Gaussian, with variance σ_r^2. The parameter σ^2 is estimated from the data and the error model implies a multivariate Gaussian likelihood for the target vector t_r:

$$p(t_r \mid w_r, \sigma_2^r) = (2\pi)^{-N/2} \sigma_r^{-N} \exp\left[\left\{\|t_r - y_r\|/2\sigma_r^2\right\}^2\right] \tag{2}$$

"There are as many parameters in the model as training examples, therefore we would expect maximum likelihood estimation of w_r and σ^2 from (2) to lead to severe over-fitting" [39]. A prior constraint over w_r is imposed by adding a complexity penalty to the likelihood to avoid overfitting. The "hyperparameters" are used to constrain an explicit zero-mean Gaussian prior probability distribution over the weights, w_r:

$$p(w_r \mid \alpha) = (2\pi)^{-M/2} \prod_{m=1}^{M} \alpha_m^{1/2} \exp\left(-\left(\alpha_m w_{rm}^2/2\right)\right) \tag{3}$$

[38] introduced "M" independent hyperparameters, $\alpha = (\alpha_1, \mathrm{L}, \alpha_M)^T$, where each element α is called a hyperparameter that determines the relevance of the associated basis function φ_j, which is one component of the basis vector

Φ. Each hyperparameter individually controls the strength of the prior over its associated weight. It is this form of prior that ultimately makes the model sparse. Given α, the posterior parameter distribution conditioned on the data is given by combining the likelihood and prior within Bayes' rule:

$$p\left(w_r \mid t_r, \alpha, \sigma_r^2\right) = p\left(t_r \mid w_r, \sigma_r^2\right) p\left(w_r \mid \alpha\right) / p\left(t_r \mid \alpha_r, \sigma_r^2\right) \quad (4)$$

and is Gaussian N(μ_r,Σ_r) with, $\Sigma_r = \left(A + \sigma_r^{-2}\Phi^T\Phi\right)^{-1}$ and $\mu = \sigma_r^{-2}\Sigma_r\Phi^T t_r$,

where A is defined as $\mathrm{diag}(\alpha_1, \mathrm{L}\ , \alpha_M)$.

The algorithm proposed by [37] for training an RVM with multivariate outputs by finding the optimal hyper- parameters is as follows:

(1) Initialization of the noise variance σ_r and the hyperparameter α: σ_r = variance of $t_r \times 0.1,\ r \in 1, \cdots, P;\ \alpha =$

α = infinity (∞); P = number of output dimensions

(2) Iterate: (a) Compute $\{\mu_r, \Sigma_r\}_{r=1}^{P}$ using∈

$$\Sigma_r = \left(\sigma_r^{-2}\Phi^T\Phi + A\right)^{-1};\quad \mu_r = \sigma_r^{-2}\Sigma_r\Phi^T t_r \quad ; \quad (5)$$

(b) Compute $\left\{s_{ri}, q_{ri}\right\}_{r=1,i=1}^{P,M}$, where M is the number of basis functions, using,

$S_{ri} = \alpha_i S_{ri} / \left(\alpha_i - S_{ri}\right)$ and;

$$S_{ri} = \sigma_r^{-2}\Phi_i^T\Phi_i - \sigma_r^{-4}\Phi_i^T\Phi\Sigma_r\Phi^T\Phi_i;\ Q_{ri} = \sigma_r^{-2}\Phi_i^T t_i - \sigma_r^{-4}\Phi_i^T\Phi\Sigma_r\Phi^T t_i \quad (6)$$

(c) Find the basis function, Φ_m, and the corresponding optimal hyperparameter α_m^{opt} that minimize the function,

L(α) using the following Equations:

$$\alpha_i^{opt} = \arg\min_{\alpha_i} l\left(\alpha_i\right);\ L(\alpha) = L\left(\alpha_{-i}\right) + \sum_{r=1}^{M}\left\{\log\alpha_i - \log\left(\alpha_i + s_{ri}\right) + q_{ri}^2 / \left(\alpha_i + s_{ri}\right)\right\};\ m = \arg\min_i l\left(\alpha_i^{opt}\right) \quad (7)$$

If $\alpha_m^{old} = \infty$ and $\alpha_m^{opt} < \infty$, then add Φ_m to the model with $\alpha_m = \alpha_m^{opt}$

If $\alpha_m^{old} < \infty$ and $\alpha_m^{opt} = \infty$, then remove the Φ_m from the model with $\alpha_m = \infty$

If $\alpha_m^{old} < \infty$ and $\alpha_m^{opt} < \infty$, then update α_m with α_m^{opt}

(d) Re-estimate the noise parameters using,

$$\sigma_r^2 = \|t_r - \phi\mu_r\|^2 \Big/ \left(M - \sum_i^M \gamma_i\right), \quad r \in 1, \cdots, P \tag{8}$$

The optimal hyperparameters and the noise parameters are then used to obtain the optimal weight matrix:

$$A^{opt} = \mathrm{diag}\left(\alpha_1^{opt}, \cdots, \alpha_M^{opt}\right); \Sigma_r^{opt} = \left(\left(\sigma_r^{opt}\right)^{-2} \boldsymbol{\Phi}^{\mathrm{T}}\boldsymbol{\Phi} + A\right)^{-1}; \mu_r^{opt} = \left(\sigma_r^{opt}\right)^{-2} \Sigma_r^{opt}\boldsymbol{\Phi}^{\mathrm{T}}\boldsymbol{t}_r \tag{9}$$

Readers interested in detailed descriptions of the model are referred to [37].

3. DATA DESCRIPTION

The data used for this study were taken from the Soil Climate Analysis Network (SCAN) site at Rees Center, Texas, USA. There are about 86 SCAN stations across the United States where daily and hourly measurements for meteorological and soil moisture data are made using various sensors and instruments. Of these 86 stations, most collect soil moisture data up to a depth of 40 inch (around 100 cm), and there a few which collect soil moisture data up to a depth of 80 inch (around 200 cm). Rees Center, Texas is one such SCAN station which collects soil moisture data up to a depth of 200 cm. In this particular application, meteorological inputs (precipitation and soil temperature) and soil moisture data were used. The location of the data collection station at SCAN Site, Rees Center, Texas is 33°37'N and 102°02'W, at an elevation of 3333 feet (1015.9 meter (m)) (Figure 1). This Sensor is reporting since March 10, 2005 and it collects hourly and daily time series data. The sensor recordings are instantaneous and it reports a single observation, taken at the beginning of an interval, which is midnight for a daily time series data. The instrument used for measuring soil moisture and soil temperature is a Hydra Probe Soil Sensor (2.5 volt) with an accuracy of +/− 0.03 wfv ($m^3 \cdot m^{-3}$) and +/− 0.6 Degrees Celsius (From −10°C to 36°C) respectively. Detailed information about the data is available on the website http://www.wcc.nrcs.usda.gov/scan.

4. METHODOLOGY

Our goal was to forecast root zone soil moisture at 1 and 2 m depths. This was done by assimilating soil moisture (m^3/m^3) at shallower depths (5, 10, 20, 30 and 50 cm), soil temperature (Celsius) and precipitation (mm), and predicting soil moisture at depths of 1 and 2 m. The time series soil moisture data for 12 months were downloaded from the Natural Resource Conservation service (NRCS) website. A stratified sampling of the 12 months of data was carried out and the training and testing data were extracted from this stratified sample. The test data were kept aside for validating the performance of the machine. The stratified sampling was done to train the MVRVM model for different values of soil moisture in different seasons. It was observed that normalization of the data between −1 and 1 produced better results as compared to the case where raw data were used. Hence the data were normalized. The MVRVM model was trained with 227 days of soil moisture data spread over a year and corresponding soil temperature and meteorological data. The inputs to the model were precipitation, soil temperature, and soil moisture data on days "d−4", "d−3", "d−2", "d−1". The output of the model was forecasted soil moisture values at "d", "d+1", "d+2" and "d+3". Time steps were in days. The performance of the model was tested with 100 days of input data spread over a year. Figure 2 shows the flow diagram of the model approach in the training phase. Three analyses were done with different inputs. For the first analysis, the MVRVM model was trained using soil moisture at depths of 5, 10, 20, 30 and 50 cm below ground surface, soil temperature (Celsius), and precipitation (mm) as inputs. The second analysis was similar to the first one but it only used soil moisture at 5 and 10 cm, soil temperature (Celsius), and precipitation (mm) as inputs to train the MVRVM model. The third analysis used soil moisture at 30 and 50 cm, soil temperature (Celsius), and precipitation (mm) as inputs to train the MVRVM model. The output in all three cases was soil moisture at 1 and 2 m depths on d, d+1, d+2 and d+3, i.e. four days into the future. The latter two analyses were carried out to observe the variation in the model output when moisture in the topsoil (5, 10 cm) and then at larger depths (30, 50 cm) were used to train the learning machine. Figure 2 shows the MVRVM model approach. The model inputs were: $x = [XP_{d-4}, XS_{d-4}, XT_{d-4}, XP_{d-3}, XS_{d-3}, XT_{d-3}, XP_{d-2}, XS_{d-2}, XT_{d-2}, XP_{d-1}, XT_{d-1}, XS_{d-1}]$, where, d = time (day) and the model outputs were: $y = [Y1_{d}, Y2_{d}, Y1_{d+1}, Y2_{d+1}, Y1_{d+2}, Y2_{d+2}, Y1_{d+3} Y2_{d+3}]$.

To test the degree of association between the observed and estimated data, goodness-of-fit evaluation measures were used. The mean absolute error (MAE), a linear measure and root mean square error (RMSE), a quadratic scoring rule, were used to measure average magnitude of error. The index of agreement (IoA) and coefficient of efficiency (CoE) were also used to check model performance. These statistics are calculated as:

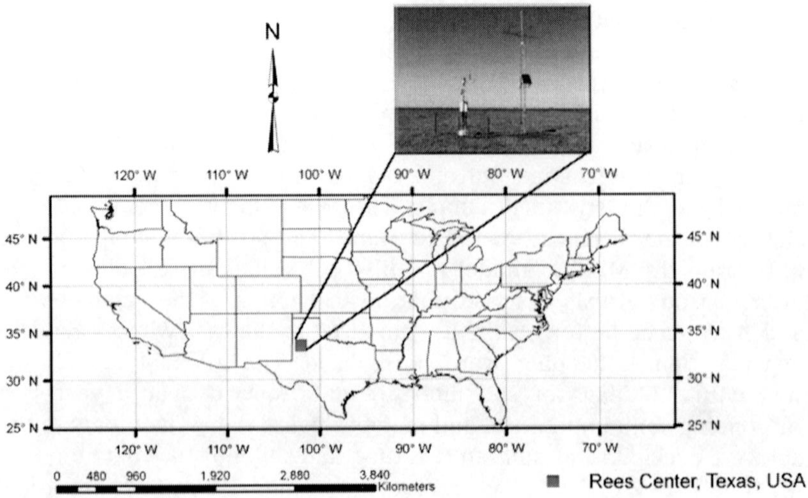

Figure 1. Location of data collection station.

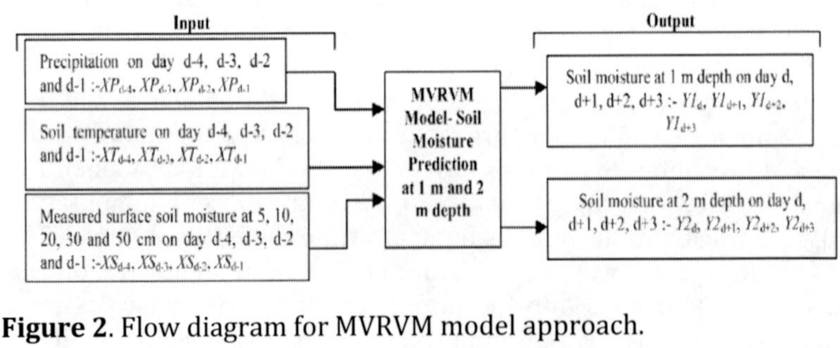

Figure 2. Flow diagram for MVRVM model approach.

$$\text{RMSE} = \sqrt{\sum_{i=0}^{N}\left(y_i - f(x)_i\right)^2 / N} \quad ;;$$

$$\text{IoA} = 1.0 - \left[\sum_{i=1}^{N}\left(y_i - f(x)_i\right)^2 \Big/ \sum_{i=1}^{N}\left(\left|f(x)_i - \bar{y}\right| + \left|y_i - \bar{y}\right|\right)^2\right] = 1.0 - N[\text{MSE/PE}] \quad ,$$

where N is the number of testing samples. In this formulation, y and f(x) are the measured and modeled values, respectively. The RMSE and MAE measure the error between the actual data and modeled values. Large values of RMSE or MAE mean that the difference between the actual measurements and the modeled values is large; hence the model is not performing well enough. Both the MAE and RMSE can range from zero to infinity. Lower values are better. The RMSE has the same dimensionality as the data and therefore it is easy to interpret. The IoA is calculated by comparing an observed group variance with an expected random variance. It varies from zero (inferior model) to one (excellent model). Potential Error (PE) is defined as the sum of the squared absolute values of the distances from $f(x)_i$ to $\bar{y}$ and from y_i to $\bar{y}$ and represents the largest value that it can attain for each actual observation/simulated value pair [40]. CoE ranges from $-\infty$ (inferior model) to 1 (excellent model) [41]. Thus, a value of zero for the CoE indicates that the observed mean, $\bar{y}$, is as good an estimator as the model, while negative values indicate that the observed mean is a better estimator than the model [42].

5. RESULTS AND DISCUSSION

The goal of this research was to forecast spatio-temporal estimates of root zone soil moisture at depths of 1 and 2 m below ground surface. A MVRVM model was used to build the model. This section discusses the selection of model parameters, MVRVM model performance, and the bootstrap analyses. Evaluation of RVM performance was based on selection of optimal kernel width and optimized iterations. Several trials were performed for obtaining the optimal values of these parameters. For the MVRVM, the parameters α and σ^2 were automatically estimated by the learning procedure. The optimal kernel width for the MVRVM model was obtained through a trial and error procedure, and the optimal number of iterations was obtained by plotting parameter beta against the number of iterations. The number at which the value of the parameter beta became almost constant was considered to be optimal. Figure 3 shows the plot of the parameter beta with respect to the number of iterations.

The MVRVM model exhibits good performance. Figure 4 and Figure 5 show predicted outputs versus original data and the confidence bounds for the test phase. The results produced by the MVRVM model in the test phase for SMC estimate at 1 m depth for four consecutive days are shown in Figures 4(a)-(d). The MVRVM model showed good results with the forecasted values of soil moisture closely following the pattern of the time series. Table 1 shows the goodness-of-fit test results for the test data. The average maximum value of soil moisture at 1 m depth is about 30%, and the minimum is about 15%. The correlation result for the MVRVM model on the fourth day at 1 m depth (see Table 1) demonstrated good performance ($R^2 = 0.877$ and RMSE = 1.31%), with an IoA of 0.96 and the CoE of 0.87. This indicated that the observed data and modeled values were close. The bias is very small indicating that the estimator is robust. The average MAE for 1 m depth was 0.5 which suggested that model estimates differed from the data on an average of 0.50%. The results produced by the MVRVM model in the test phase for SMC estimates at a depth of 2 m are shown in Figures 5(a)-(d). Again, the MVRVM model showed excellent results with forecasted values closely following the pattern of the time series. The average maximum value of soil moisture at 1 m depth is about 19%, and the minimum is about 15%. The correlation result for the MVRVM model on the fourth day at a depth of 2 m (see Table 1) again demonstrated good performance ($R^2 = 0.968$ and RMSE = 0.15%), with an IoA of 0.99 and a CoE value of 0.97. This indicated that observed data and modeled values were very close. The average MAE for 1 m depth was 0.08 which suggested that model estimates differed from the data by an average of only 0.08%. The number of relevance vectors (RVs) used in the MVRVM model was 81 out of 227 training points.

Better results were obtained for soil moisture forecast at 2 m depth in comparison to those obtained at 1 m depth. The variation in moisture at larger depths is smaller than at shallower depths. Due to subdued soil moisture dynamics at 2 m depth, the machine has a simpler pattern to learn and forecasting soil moisture at 2 m depth becomes comparatively easier.

Figures 5(a)-(d) show that the soil moisture pattern is followed very accurately by the MVRVM model. The machine was able to capture the spatio-temporal variation of soil moisture at the root zone depths during peak agricultural seasons. Figure 4 and Figure 5 show that the forecast results have fairly narrow confidence bounds (at 95% confidence interval), which indicates that there is low variance in predictions. Most of the measured data points lie inside the confidence bounds, indicating that the model is robust.

The full MVRVM model used soil moisture values at four different depths as inputs: 5, 10, 30, and 50 cm for soil moisture prediction at deeper depths. Two additional analyses were done to reveal the effect of only using the surface data (5 cm and 10 cm) for prediction of soil moisture at 1 m and 2 m, and then using data at 30 cm and 50 cm for the same prediction. The SMC predictions obtained by using input data at 30 cm and 50 cm were closer to the actual soil moisture measurements (see Table 2) and this model produced better results compared to the results generated by the MVRVM model which used data at 5 cm and 10 cm (Table 3). The results for both the analyses were good but not as good as were obtained from the full MVRVM model. However, depending upon the availability of data, the MVRVM model can be applied for soil moisture prediction at larger depths. This article brings into light the capability of the MVRVM model to learn the pattern of soil moisture variation and predict acceptable estimates of soil moisture.

Bootstrapping was performed for the MVRVM model to check for over-fitting and evaluate model generalization capability. Figure 6 shows bootstrap results for RMSE, as estimated from 1000 bootstrap samples. Conforming to the nonparametric approach, no assumption was made about the form of the data, and repeated samples were drawn from the population with replacement. The basic idea is that if the sample is a good approximation of the population, the bootstrap method will provide a good approximation of the sampling distribution of the statistic, in this case, the RMSE. Although beyond the scope of this article, our goal here was to ensure good generalization of the inductive learning algorithm. A narrow confidence interval indicated that the available training dataset was adequate to determine the machine parameters. From Figure 6(a) and Figure 6(b), one could deduce rough confidence bounds that are more revealing of model performance than single values [43]. In Figure 6, we see that the RMSE values for all the three MVRVM models are centered around one maximum value with highest frequency. Also the MVRVM model shows a fairly narrow confidence bound in all the cases, which implies that the model was robust and its parameter values were well determined.

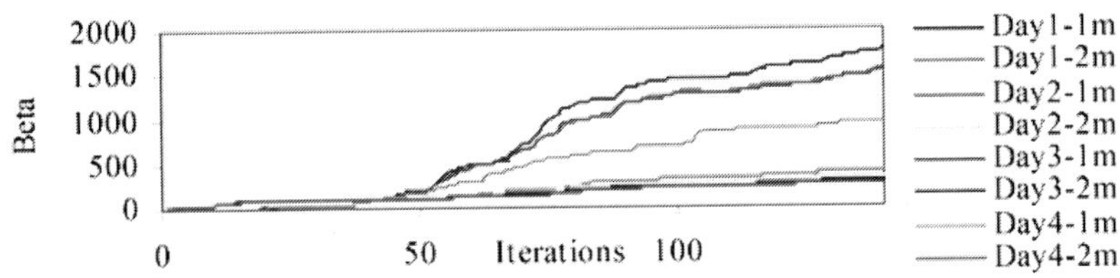

Figure 3. Variation of parameter beta with number of iterations for different outputs.

Table 1. MVRVM model results (Kernel Width, r = 3, Iterations = 140).

Statistics	Multivariate Relevance Vector Machine Model							
	Day d		Day d+1		Day d+2		Day d+3	
	1 m depth	2 m depth	1 m depth	2 m depth	1 m depth	2 m depth	1 m depth	2 m depth
RMSE, %	1.14	0.13	0.99	0.12	1.27	0.13	1.31	0.15
R^2	0.904	0.974	0.926	0.977	0.873	0.972	0.877	0.968
CoE	0.898	0.972	0.92	0.974	0.870	0.970	0.869	0.965
IoA	0.971	0.993	0.977	0.993	0.962	0.992	0.96	0.991
Bias	−0.1548	0.0168	−0.1168	0.0178	0.0251	0.0107	0.0151	0.0193
MAE	0.51	0.08	0.44	0.08	0.53	0.084	0.55	0.087

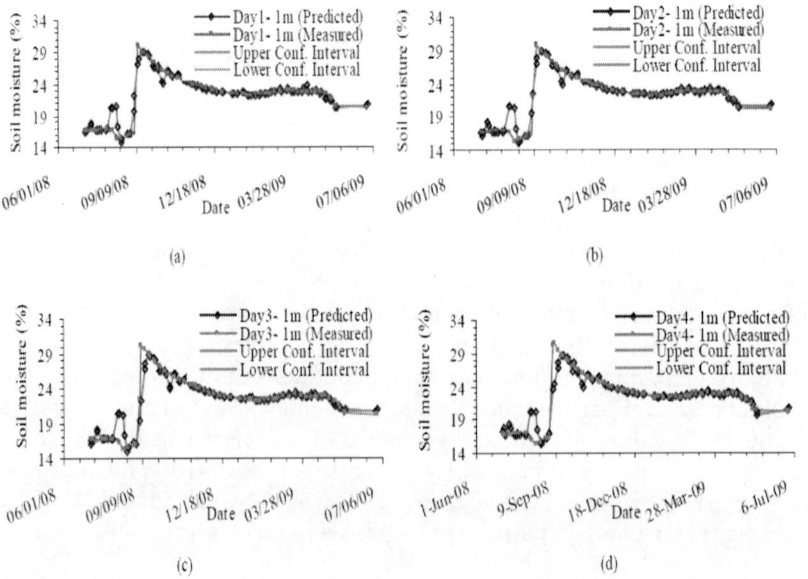

Figure 4. Root zone soil moisture prediction at 1 meter depth on Day: (a) d; (b) d+1; (c) d+2; (d) d+3. (a) Day 1: Root zone SMC at 1 m depth; (b) Day 2: Root zone SMC at 1 m depth; (c) Day 3: Root zone SMC at 1 m depth; (d) Day 4: Root zone SMC at 1 m depth.

Table 2. MVRVM model results when SMC at a depth of 30 cm and 50 cm are used as inputs.

Statistics	Multivariate Relevance Vector Machine Model results at 1 m and 2 m depths							
	Day d		Day d+1		Day d+2		Day d+3	
	1 m	2 m	1 m	2 m	1 m	2 m	1 m	2 m
RMSE, %	1.84	0.16	1.78	0.15	1.95	0.17	2	0.17
CoE	0.74	0.95	0.74	0.96	0.69	0.95	0.7	0.95
IoA	0.94	0.98	0.94	0.99	0.93	0.98	0.9	0.98
Bias	−0.09	0.04	−0.02	0.03	0.14	0.03	0.1	0.03

Table 3. MVRVM model results when only surface SMC at a depth of 5 cm and 10 cm are used as inputs.

Statistics	Multivariate Relevance Vector Machine Model results at 1 m and 2 m depths							
	Day d		Day d+1		Day d+2		Day d+3	
	1 m	2 m	1 m	2 m	1 m	2 m	1 m	2 m
RMSE, %	1.96	0.31	1.83	0.3	1.81	0.32	1.9	0.32
CoE	0.7	0.83	0.72	0.85	0.73	0.83	0.7	0.83
IoA	0.91	0.95	0.92	0.96	0.92	0.95	0.9	0.95
Bias	−0.1	0.12	−0.06	0.08	0.07	0.08	0.1	0.12

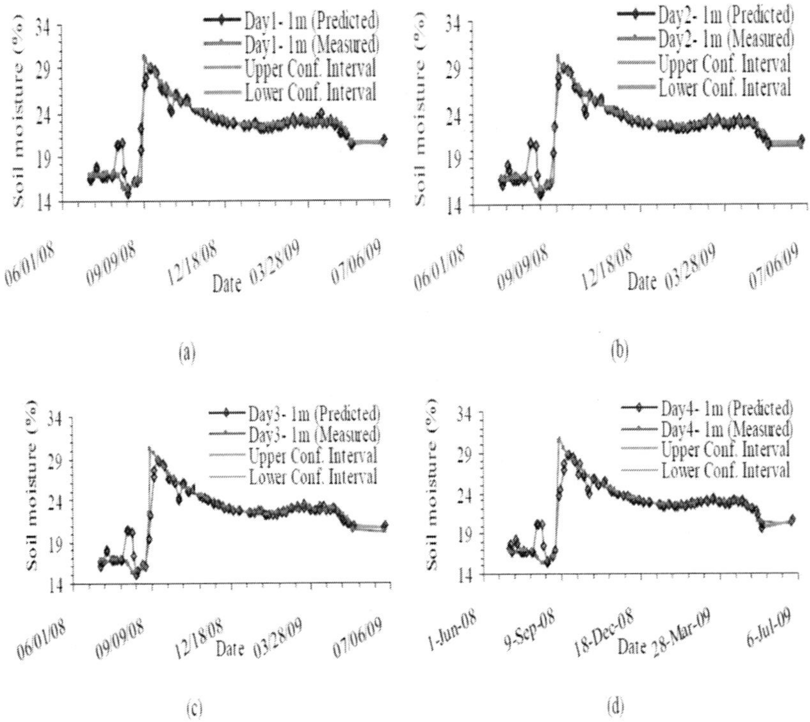

Figure 5. Root zone soil moisture prediction at 2 meter depth on Day: (a) d; (b) d+1; (c) d+2; (d) d+3. (a) Day 1: Root zone SMC at 2 m depth; (b) Day 2: Root zone SMC at 2 m depth; (c) Day 3: Root zone SMC at 2 m depth; (d) Day 4: Root zone SMC at 2 m depth.

6. SUMMARY AND CONCLUSION

This article presents a first attempt to forecast spatial and temporal variation of soil moisture simultaneously using machine learning techniques. This model is based on a sparse Bayesian learning machine approach wherein the machine learns the input output pattern with high accuracy. A MVRVM model is built for developing the prediction functions that forecast soil moisture at 1 m and 2 m depth four days into the future. Three different analyses were done using input data at different depths. The best results were obtained for the full MVRVM model where the input data at 5, 10, 30, and 50 cm depths were used. The results showed excellent performance by the machine for all four days. It was observed that the SMC predictions at 2 m depth were more accurate than those at 1 m depth. The second analysis where the SMC were predicted at 1 m and 2 m depths using data at 5 cm and 10 cm suggested that it is possible to estimate soil moisture in the root zone using surface data by applying the MVRVM model. The inputs for the third analysis were chosen keeping in mind that the soil moisture sensors are generally operational at these depths, i.e. 30 cm and 50 cm. The MVRVM model performance for this third analysis was also very good, leading to the conclusion that soil moisture conditions at larger depths can be predicted using the MVRVM model if soil moisture data from the sensors are available at 30 cm and 50 cm.

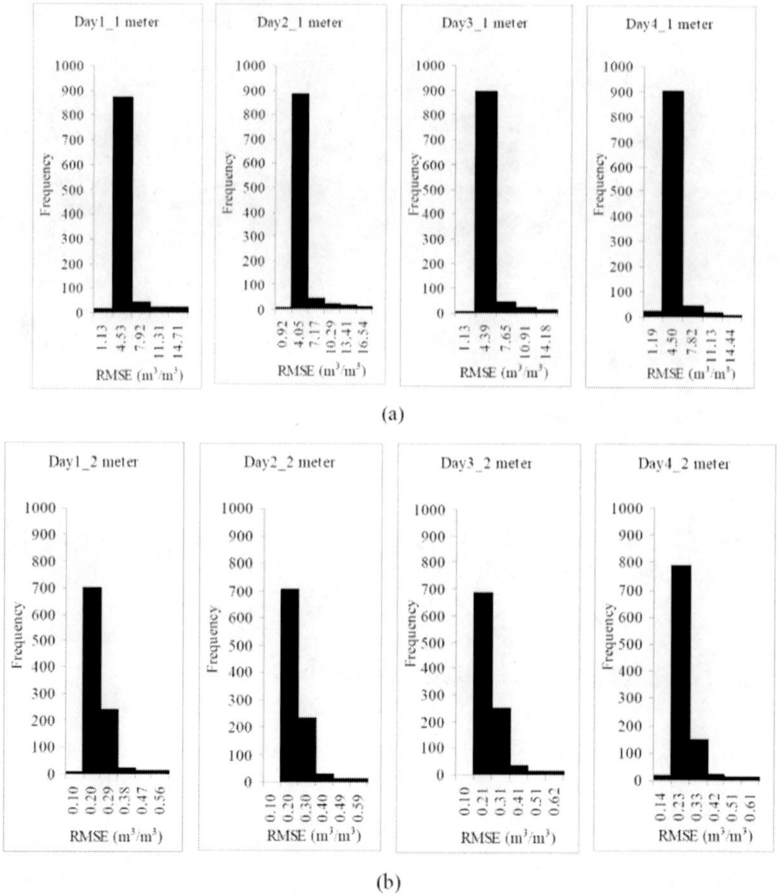

Figure 6. Bootstrap analysis results for uncertainty in the RMSE of the MVRVM model with 1000 bootstrap samples for the test phase: (a) Prediction of Soil Moisture at 1 meter depth; (b) Prediction of Soil Moisture at 2 meter depth.

Computation of statistics of interest in conjunction with bootstrapping analyses accomplished a broad operational evaluation of the full MVRVM soil moisture model. These analyses allow us to conclude that the model can predict spatial and temporal variation of soil moisture at large depths with a high degree of accuracy. The model also had good generalization capabilities providing robustness, as demonstrated by the bootstrap analyses results. The MVRVM scheme discussed in this article can be employed to obtain soil moisture estimates from the model in real time and is a potentially useful approach for obtaining short term forecasts in situations where new data can be rapidly exploited as they become available. The results are encouraging and confirm the relevance of the proposed methodology which can benefit soil moisture monitoring and can be extended to other fields of hydrologic science.

ACKNOWLEDGEMENTS

Bushra Zaman, Mac McKee (UWRL) and the Utah Centre for Water Resources Research (UCWRR) for their support of this research. The authors would like to thank Garry L. Schaefer of National Water and Climate Center (NRCS) for providing help regarding data details when needed. The authors are grateful to the reviewers as their feedback helped in improving the quality of this paper.

REFERENCES

1. Das, N.N. and Mohanty, B.P. (2006) Root Zone Soil Moisture Assessment Using Remote Sensing and Vadose Zone Modeling. Vadose Zone Journal, 5, 296-307. http://dx.doi.org/10.2136/vzj2005.0033.
2. Webster, R. and Butler, B. (1976) Soil Classification and Survey Studies at Ginninderra. Soil Research, 14, 1-24. http://dx.doi.org/10.1071/SR9760001.
3. McKenzie, N.J. and Austin, M.P. (1993) A Quantitative Australian Approach to Medium and Small Scale Surveys Based on Soil Stratigraphy and Environmental Correlation. Geoderma, 57, 329-355. http://dx.doi.org/10.1016/0016-7061(93)90049-Q.
4. Fu, B. and Gulinck, H. (1994) Land Evaluation in an Area of Severe Erosion: The Loess Plateau of China. Land Degradation & Development, 5, 33-40. http://dx.doi.org/10.1002/ldr.3400050105.
5. Western, A.W., Grayson, R.B., Blöschl, G., Willgoose, G.R. and McMahon, T.A. (1999) Observed Spatial Organization of Soil Moisture and Its Relation to Terrain Indices. Water Resources Research, 35, 797-810. http://dx.doi.org/ 10.1029/ 1998WR900065.
6. Qiu, Y., Fu, B., Wang, J. and Chen, L. (2003) Spatiotemporal Prediction of Soil Moisture Content Using Multiple-Li- near Regression in a Small Catchment of the Loess Plateau, China. CATENA, 54, 173-195. http://dx.doi.org/10.1016/S0341-8162(03)00064-X.

7. Moore, I.D., Gessler, P.E., Nielsen, G.A. and Peterson, G.A. (1993) Soil Attribute Prediction Using Terrain Analysis. Soil Science Society of American Journal, 57, 443-452. http://dx.doi.org/10.2136/sssaj1993.03615995005700020026x.
8. Lark, R.M. (1999) Soil-Landform Relationships at Within-Field Scales: An Investigation Using Continuous Classification. Geoderma, 92, 141-165. http://dx.doi.org/10.1016/S0016-7061(99)00028-2
9. Qiu, Y., Fu, B., Wang, J. and Chen, L. (2001) Soil Moisture Variation in Relation to Topography and Land Use in a Hillslope Catchment of the Loess Plateau, China. Journal of Hydrology, 240, 243-263. http://dx.doi.org/10.1016/S0022-1694(00)00362-0.
10. Camillo, P. and Schmugge, T.J. (1983) Estimating Soil Moisture Storage in the Root Zone from Surface Measurements. Soil Science, 135, 245-264. http://dx.doi.org/10.1097/00010694-198304000-00006.
11. Entekhabi, D., Nakamura, H. and Njoku, E.G. (1994) Solving the Inverse Problem for Soil Moisture and Temperature Profiles by Sequential Assimilation of Multifrequency Remotely Sensed Observations. IEEE Transactions on Geoscience and Remote Sensing, 32, 438-448.
12. Calvet, J.C. and Noilhan, J. (2000) From Near-Surface to Root-Zone Soil Moisture Using Year-Round Data. Journal of Hydrometeorology, 1, 393-411. http://dx.doi.org/10.1175/1525-7541(2000)001<0393:FNSTRZ>2.0.CO;2.
13. Albergel, C., Rudiger, C., Pellarin, T., Calvet, J.C., Fritz, N., Froissard, F., et al. (2008) From Near-Surface to Root- Zone Soil Moisture Using an Exponential Filter: An Assessment of the Method Based on In-Situ Observations and Model Simulations. Hydrology and Earth System Sciences, 12, 1323-1337. http://dx.doi.org/10.5194/hess-12-1323-2008.
14. Calvet, J., Noilhan, J. and Bessemoulin, P. (1998) Retrieving the Root-Zone Soil Moisture from Surface Soil Moisture or Temperature Estimates: A Feasibility Study Based on Field Measurements. Journal of Applied Meteorology, 37, 371-386. http://dx.doi.org/10.1175/1520-0450(1998)037<0371:RTRZSM>2.0.CO;2.
15. Sabater, J., Jarlan, L., Calvet, J., Bouyssel, F. and De Rosnay, P. (2007) From Near-Surface to Root-Zone Soil Moisture Using Different Assimilation Techniques. Journal of Hydrometeorology, 8, 194-206. http://dx.doi.org/10.1175/JHM571.1.
16. Li, J. and Islam, S. (2002) Estimation of Root Zone Soil Moisture and Surface FluxesPartitioning Using Near Surface Soil Moisture Measurements. Journal of Hydrology, 259, 1-14. http://dx.doi.org/10.1016/S0022-1694(01)00589-3.
17. Mahfouf, J.F. (1991) Analysis of Soil Moisture from Near-Surface Parameters: A Feasibility Study. Journal of Applied Meteorology, 30, 1534-1547. http://dx.doi.org/10.1175/1520-0450(1991)030<1534:AOSMFN>2.0.CO;2.
18. Gill, M.K., Asefa, T., Kemblowski, M.W. and McKee, M. (2006) Soil Moisture Prediction Using Support Vector Machines. JAWRA Journal of the American Water Resources Association, 42, 1033-1046. http://dx.doi.org/10.1111/j.1752-1688.2006.tb04512.x
19. Song, J.L., Wang, D.H., Liu, N.J., Cheng, L., Du, L. and Zhang, K. (2008) Soil Moisture Prediction with Feature Selection Using a Neural Network. Digital Image Computing: Techniques and Applications (DICTA), Canberra, 1-3 December 2008, 130-136.
20. Saxton, K.E., Johnson, H.P. and Shaw, R.H. (1974) Modeling Evapotranspiration and Soil Moisture. Transactions of the ASABE, 17, 673-677.
21. Rao, A.S. and Saxton, K.E. (1995) Analysis of Soil Water and Water Stress for Pearl Millet in an Indian Arid Region Using the SPAW Model. Journal of Arid Environments, 29, 155-167. http://dx.doi.org/10.1016/S0140-1963(05)80086-2.
22. Arora, V.K., Singh, C. and Singh, K. (1997) Comparative Assessment of Soil Water Balance under Wheat in a Subtropical Environment with Simplified Models. The Journal of Agricultural Science, 128, 461-468. http://dx.doi.org/10.1017/S0021859697004358.

23. Hill, A. and Neary, V. (2009) Hydrologic Response of a Forested Sinkhole Wetland to Different Land Management Scenarios. Journal of Environmental Hydrology, 17, 1-18.
24. Holtan, H., Stiltner, G., Henson, W. and Lopez, N. (1975) USDAHL-74 Revised Model of Watershed Hydrology. Technical Bulletin, No. 1518.
25. Comer, G.H. and Henson, W.H. (1976) An Optimization Technique Adapted to Usdahl-74 Revised Model of Water- shed Hydrology. JAWRA Journal of the American Water Resources Association, 12, 139-146. http://dx.doi.org/10.1111/j.1752-1688.1976.tb02644.x.
26. Peck, E. (1976) Catchment Modeling and Initial Parameter Estimation for the National Weather Service River Forecast System. NOAA Technical Memorandum NWS Hydro-31, NWS, NOAA, DOC, Washington DC, June 1976.
27. Sorooshian, S., Duan, Q.Y. and Gupta, V.K. (1993) Calibration of Rainfall-Runoff Models: Application of Global Op- timization to the Sacramento Soil Moisture Accounting Model. Water Resources Research, 29, 1185-1194. http://dx.doi.org/10.1029/92WR02617.
28. Burnash, R. (1995) The NWS River Forecast System-Catchment Modeling. In: Singh, V., Ed., Computer Models of Watershed Hydrology, Water Resources Publication, Colorado, 311-366.
29. Atluri, V., Hung, C.C. and Coleman, T.L. (1999) An Artificial Neural Network for Classifying and Predicting Soil Moisture and Temperature Using Levenberg-Marquardt Algorithm. IEEE Proceedings Southeastcon'99, Lexington, 25-28 March 1999, 10-13.
30. Chang, D. and Islam, S. (2000) Estimation of Soil Physical Properties Using Remote Sensing and Artificial Neural Network. Remote Sensing of Environment, 74, 534-544. http://dx.doi.org/10.1016/S0034-4257(00)00144-9.
31. Jiang, H. and Cotton, W.R. (2004) Soil Moisture Estimation Using an Artificial Neural Network: A Feasibility Study. Canadian Journal of Remote Sensing, 30, 827-839. http://dx.doi.org/10.5589/m04-041.
32. Elshorbagy, A. and Parasuraman, K. (2008) On the Relevance of Using Artificial Neural Networks for Estimating Soil Moisture Content. Journal of Hydrology, 362, 1-18. http://dx.doi.org/10.1016/j.jhydrol.2008.08.012.
33. Gill, M., Kemblowski, M. and McKee, M. (2007) Soil Moisture Data Assimilation Using Support Vector Machines and Ensemble Kalman Filter. Journal of the American Water Resources Association, 43, 1004-1015. http://dx.doi.org/10.1111/j.1752-1688.2007.00082.x
34. Yang, S. and Huang, Y. (2009) Application of Support Vector Machine Based on Time Series for Soil Moisture and Nitratenitrogen Content Prediction. Computer and Computing Technologies in Agriculture II, Beijing, 18-20 October 2008, 2037-2045.
35. Khalil, A., Gill, M. and McKee, M. (2005) New Applications for Information Fusion and Soil Moisture Forecasting. 8th International Conference on Information Fusion, Philadelphia, 25-28 July 2005. http://dx.doi.org/10.1109/ICIF.2005.1592050.
36. Tipping, M. (2000) The Relevance Vector Machine. Advances in Neural Information Processing Systems, 12, 652-658.
37. Thayananthan, A., Navaratnam, R., Stenger, B., Philip, H., Torr, S. and Roberto, C. (2006) Multivariate Relevance Vector Machines for Tracking. In: Computer Vision-ECCV 2006, Graz, 7-13 May 2006, 124-128.
38. Tipping, M.E. and Faul, A.C. (2003) Fast Marginal Likelihood Maximisation for Sparse Bayesian Models. Proceedings of the 9th International Workshop on Artificial Intelligence and Statistics, Florida, 3-6 January 2003, 3-6.
39. Tipping, M. (2001) Sparse Bayesian Learning and the Relevance Vector Machine. Journal of Machine Learning Research, 1, 211-244.
40. Legates, D.R. and McCabe, G.J. (1999) Evaluating the Use of "Goodness-of-Fit" Measures in Hydrologic and Hydroclimatic Model Validation. Water Resources Research, 35, 233-241. http://dx.doi.org/10.1029/1998WR900018

41. Nash, J.E. and Sutcliffe, J.V. (1970) River Flow Forecasting through Conceptual Models Part I—A Discussion of Principles. Journal of Hydrology, 10, 282-290. http://dx.doi.org/10.1016/0022-1694(70)90255-6.
42. Wilcox, B.P., Rawls, W.J., Brakensiek, D.L. and Wight, J.R. (1990) Predicting Runoff from Rangeland Catchments: A Comparison of Two Models. Water Resources Research, 26, 2401-2410. http://dx.doi.org/10.1029/WR026i010p02401.
43. Willmott, C.J. (1984) On the Evaluation of Model Performance in Physical Geography. In: Spatial Statistics and Models, Springer, Berlin, 443-460.

CHAPTER 12

Catchment Hydrology during Winter and Spring and the Link to Soil Erosion: A Case Study in Norway

Torsten Starkloff[1], Rudi Hessel[2], Jannes Stolte[1] and Coen Ritsema[2]

[1] *Norwegian Institute of Bioeconomy Research (NIBIO), Fredrik A Dahlsvei 20, N-1431 Ås, Norway;*

Jannes.Stolte@nibio.no

[2] *Wageningen Environmental Research (Alterra), P.O. Box 47, 6700 AA Wageningen, The Netherlands;*

ABSTRACT

In the Nordic countries, soil erosion rates in winter and early spring can exceed those at other times of the year. In particular, snowmelt, combined with rain and soil frost, leads to severe soil erosion, even, e.g., in low risk areas in Norway. In southern Norway, previous attempts to predict soil erosion during winter and spring have not been very accurate owing to a lack of catchment-based data, resulting in a poor understanding of hydrological processes during winter. Therefore, a field study was carried out over three consecutive winters (2013, 2014 and 2015) to gather relevant data. In parallel, the development of the snow cover, soil temperature and ice content during these three winters was simulated with the Simultaneous Heat and Water (SHAW) model for two different soils (sand, clay). The field observations carried out in winter revealed high complexity and diversity in the hydrological processes occurring in the catchment. Major soil erosion was caused by a small rain event on frozen ground before snow cover was established, while snowmelt played no significant role in terms of soil erosion in the study period. Four factors that determine the extent of runoff and erosion were of particular importance: (1) soil water content at freezing; (2) whether soil is frozen or unfrozen at a particular moment; (3) the state of the snow pack; and (4) tillage practices prior to winter. SHAW performed well in this application and proved that it is a valuable tool for investigating and simulating snow cover development, soil temperature and extent of freezing in soil profiles.

Keywords: SHAW; soil freezing; snow; infiltration; modelling; soil erosion

1. INTRODUCTION

In the Nordic countries, soil erosion rates in winter and early spring can exceed those occurring during other seasons of the year. A factor of particular importance is the incidence of frozen soil, which modifies surface runoff generation and also the erosivity of the soil material [1]. In addition, water infiltration into frozen soils is more complicated than water infiltration into unfrozen soils, because it involves coupling water and heat transport (temperature of the infiltrating water) with phase change (from liquid to ice and vice versa) [2].

A large number of laboratory studies has investigated different processes occurring in soils during freezing and thawing. Using a rain simulator, Edwards and Burney [3] showed that soil freezing and thawing can significantly increase soil erosivity. They concluded that only plant cover is effective in reducing soil losses due to rain and overland flow on frozen ground. Other more recent studies, e.g., Ban et al. [4], have shown that water flows much faster over a frozen slope than over a thawed slope. Watanabe et al. [5] found that the speed of snowmelt and/or rain infiltration into frozen soils is largely dependent on initial water content, frost depth and temperature of the soil. In addition, Yami et al. [6] showed that increasing soil moisture and finer soil structure advance the speed and depth of the freezing front.

Al-Houri et al. [7] added to knowledge about water transport in frozen soils by showing that the amount of time available for soil water redistribution before freezing affects the infiltration capacity of frozen soil, with more time resulting in better infiltration capacity under frozen conditions. In addition to laboratory studies, a great number of studies has investigated infiltration processes under field conditions at the plot or point scale. In a study examining nine different plots in North Dakota, Willis et al. [8] showed that soils that were dry in autumn freeze faster and deeper than wet soils and that a dry profile thaws upward, while a wet soil thaws both upward and downward. They also recorded less runoff from dry soils.

Stähli et al. [9] and Nyberg et al. [10] concluded that frost has little effect on runoff from forest soils, in contrast to reported effects on agricultural soils [11], and that forests probably do not increase runoff episodes in winter and spring. Furthermore, they predicted that critical initial conditions, such as high water content and early frost penetration combined with heavy rain on still frozen soil, could have a decisive effect on the amount of runoff. Iwata et al. [12,13] showed that a frozen soil layer can significantly impede snowmelt infiltration and thus increase runoff of spring snowmelt water. Zhao et al. [13] demonstrated that soil freezing can reduce hydraulic conductivity by blocking pores and retaining water in the profile, thereby reducing the infiltration capacity during snowmelt.

In a study at five locations in Finland, Sutinen et al. [14] found that a snow pack with a thickness exceeding 30 cm can reduce or even prevent soil freezing. In addition, Zhao et al. [15] showed that snow pack less than 20 cm deep can cause deeper soil freezing than no snow cover, due to an increase in ground albedo caused by the snow. The effects of different tillage practices on soil freezing was investigated by Parkin et al. [16] in a soil profile over 10 years. They found that conventional autumn tillage resulted in lower soil temperatures than no-till.

In a recent five-year field study in Canada, He et al. [17] presented results that confirmed many of the above-mentioned effects and demonstrated the complexity of the interaction between the topsoil and snow cover, especially during snowmelt. They also pointed out that only a limited amount of field studies to date has taken all of these processes into account in a series of measurements that covers several winter periods. The complexity of the different processes occurring in soil during winter is amplified when all of these interacting processes have to be monitored and interpreted at the catchment scale, where different soil types, terrain and water flow at the surface and in the soils interact [18,19]. In addition, detailed observations have to be made over several years to identify processes that can only occur during certain conditions or are masked by confounding factors in a catchment [20].

In Norway, the incidence of soil erosion from agricultural land is greatest during spring [21,22], and the severity of the erosion is often amplified by preceding winter conditions. Snowmelt, combined with rain and soil frost, can lead to severe gully and rill erosion, even in low risk areas in

Norway [23]. In southern Norway, previous attempts to predict soil erosion during winter and spring have not been very accurate [24], probably owing to a lack of catchment-based studies covering several winters, resulting in a lack of knowledge about the interacting processes described above. In the present study, field measurements covering three winter periods (2013, 2014 and 2015) were carried out, with the aim of improving overall understanding of how soil hydraulic properties behave during winter and affect surface runoff caused by snowmelt and rain and how these processes are linked to soil erosion. Furthermore, the data collected were used to calibrate and validate a hydrodynamic model (Simultaneous Heat and Water (SHAW) model; [25], in order to acquire better insights into the complex interactions between freezing, thawing, snowpack and runoff and erosion dynamics.

The focus of this study lies in Norway; however, severe soil erosion on agricultural areas during winter and spring is a problem in many other countries around the world (e.g., USA [26], Belgium [27], the U.K. [28], Germany [29], Russia [30]). In these areas, like in Norway, soil erosion during winter and spring depletes the irreplaceable nutrient-rich top layer of agricultural soils and results in a major part of the annual input of phosphorous and nitrogen from agricultural catchments to fresh water bodies [31]. It is therefore hoped that this study will also contribute to the understanding of winter processes outside of Norway.

2. METHODOLOGY

2.1. Study Area

The study area is located in the Skuterud catchment (4.5 km^2) in Ås and Ski municipalities, approximately 30 km south of Oslo, Norway. For the field investigations and sampling, a sub-catchment, Gryteland (0.29 km^2), in the southeastern part of the main catchment (Figure 1), was chosen. This area has been used for different hydrological studies in the past, and it can easily be reached under all weather conditions. A monitoring station was installed at the outlet of the sub-catchment in 2008. This station measures precipitation, air temperature, surface runoff and drainage discharge. In addition, five stations (one at the outlet) were installed along a transect in the catchment [32] (Figure 1), in order to measure soil moisture and soil temperature at three depths (5, 10 and 20 cm).

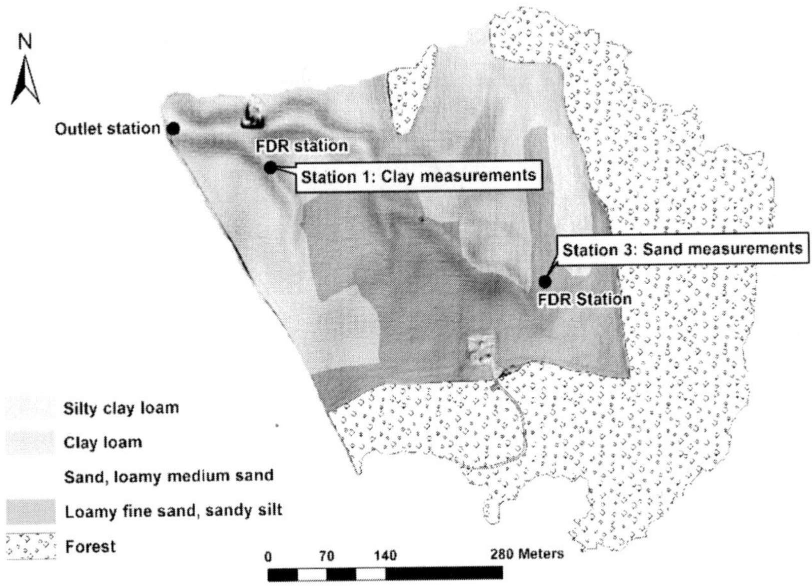

Figure 1. Soil and hill shade map of the Gryteland catchment in southern Norway. FDR, Frequency Domain Reflectometry.

The sub-catchment is characterized by an undulating landscape (elevation 106–141 m, slope 2%–10%) covered by approximately 60% arable land and 40% coniferous forest. Soil types for the arable land are a levelled clay loam (Stagnosol) and a silty clay loam (Albeluvisol) (Group 1), as well as a sandy silt on clay (Umbrisol) and a sand to loamy medium sand (Histic Gleysol) (Group 2). The two soil groups are often not clearly distinguishable in the field. Within the groups, the soils have similar physical properties. Hereafter, Groups 1 and 2 are referred to as clay and sand, respectively (Figure 1).

Mean annual temperature in the study area is 5.3 °C, with an average minimum of −4.8 °C in January/February and an average maximum of 16.1 °C in July. Mean annual precipitation is 785 mm, with a minimum monthly amount of 35 mm in February and a maximum of 100 mm in October [33]. Winter is usually relatively unstable, with alternating periods of freezing and thawing and several snowmelt events [34].

There was no tillage (no-till system) after harvest in 2013, leaving the fields covered in stubble. In 2014 and 2015, secondary tillage was performed after harvest with a cultivator on the slopes, leaving the depressions still covered with stubble.

2.2. Weather Data

A weather station was installed in the catchment outlet at the end of 2013, providing hourly data on net solar radiation, air temperature, wind speed and wind direction for the winters (December–March) of 2014 and 2015. For winter 2013 (January–April), data from a station 6 km away from the catchment were used ([33].

2.3. Soil Temperature and Soil Moisture Measurements

To obtain more detailed measurements of soil water content and temperature during winter, the measuring Stations 1 (clay measurements) and 3 (sand measurements) were upgraded to measure soil water content and soil temperature at four depths, 5, 20, 30 and 40 cm, using Decagon 5 TM temperature and Frequency Domain Reflectometry (FDR) sensors. Measurements from these two stations and the outlet station were used in the present study. However, it should be noted that the soil water content, calculated from the dielectricity of the soil, measured with the FDR probes, represents only the liquid soil water content, not water in the form of ice, and therefore, only the liquid soil water content

2.4. Discharge Measurements

To estimate how the catchment reacted to precipitation and to analyze the infiltration capacity of the soils in the catchment, data on discharge measured at the outlet were analyzed. Besides measuring how much discharge was produced during the winter periods, the runoff coefficient was calculated as:

$$D_{ro} = 100 \times D_M / P_A$$

where D_{ro} = runoff coefficient (%); D_M = discharge (m^3); P_A = precipitation on area (m^3).

2.5. Snow Cover Properties

Snow has a significant influence on changes in soil temperature and soil water content [14]. Therefore, snow properties (depth and density) were monitored in the catchment during the three winters. The measurements taken at the outlet, Stations 1 and 3, are presented in this study. Snow Water Equivalent (SWE) was sampled after weather changes expected to result in changes in SWE [35]. The measured snow depth data were used to validate snow depth values simulated with the SHAW model.

2.6. Erosion Mapping

In addition to the other measurements carried out in the field, soil erosion features were documented. Minor erosion damage was recorded by taking pictures. The extent of any large features observed was mapped using a differential GPS, and the depth and width were measured at several points using a ruler.

2.7. SHAW Model Setup and Calibration

The SHAW model, which was originally developed to simulate soil freezing and thawing [36], simulates heat, water and solute transfer within a one-dimensional profile extending downwards from the vegetation canopy to a specified depth within the soil. A layered system is established through the plant canopy, snow, residue and soil, and each layer is represented by an individual node [25]. Infiltration is calculated using a Green–Ampt approach for a multi-layered soil. Water flow in frozen soil is assumed to be similar to flow in unsaturated soil. Therefore, the relationships for matric potential and hydraulic conductivity of unsaturated soils are assumed to be valid for frozen soils. However, hydraulic conductivity is reduced linearly with ice content, assuming zero conductivity at an available porosity of 0.13 [37]. A detailed description of the model can be found in Flerchinger [38].

Input to the SHAW model includes: initial conditions for snow, soil temperature and water content profiles; daily or hourly weather conditions (temperature, wind speed, humidity, precipitation and solar radiation); general site information; and parameters describing the vegetative cover, snow, plant residues and soil. General site information includes slope, aspect, latitude and surface roughness parameters. Input soil parameters are bulk density, saturated conductivity, albedo and coefficient for the soil water potential-water content relationship [38].

To obtain the necessary soil input data (Table 1), undisturbed samples were taken in April 2014 at three different depths (0, 25 and 35 cm) at Stations 1 and 3. These depths corresponded to the depth between FDR probes, avoiding the disturbed area around the probes. For determination of saturated soil hydraulic conductivity (K_{sat}) and saturated water content (θ_S), two samples (volume of sample ring 250 cm^3) were taken. Two additional samples (volume of sample ring 98 cm^3) were used for determination of bulk density and soil organic matter. In total, 12 samples were taken at each of the two stations.

Table 1. Input parameters for the SHAW model. Only the three layers marked with an asterisk (*) were used for the 2013 simulation, and the depth was reduced from 15 down to 10 cm and from 25 cm down to 20 cm.

Soil Type	Clay				Sand			
Location	59°40′ N, North Facing (22.5°), Slope 12°, Elevation ASL 100 m				59°40′ N, Northwest Facing (330.5°), Slope 0°, Elevation ASL 140 m			
Surface	Albedo of dry soil: 0.15 Wind profile surface roughness: 0.1 cm							
Depth	5 cm *	15 cm *	25 cm *	35 cm	5 cm	15 cm	25 cm	35 cm
Campbell's b	20	20	20	20	3	1	1	1
Air entry potential (hPa)	−31	−31	−34	−35	−31	−31	−34	−35
K_{sat} ($cm \cdot h^{-1}$)	2.60	1.86	1.00	0.60	16.80	18.00	22.00	24.00
Bulk density ($kg \cdot m^{-3}$)	1331	1400	1535	1537	1190	1346	1346	1347
θ_S	0.40	0.40	0.40	0.35	0.43	0.40	0.40	0.31
Sand (%)	13				70			
Silt (%)	58				13			
Clay (%)	29				7			
Organic matter content (%)	4.5	4.0	3.7	3.5	3.4	3.4	2.7	2.0

Saturated soil hydraulic conductivity was determined using the constant head method [39]. The two soil profiles defined for SHAW are presented in Table 1, with the corresponding depths of the simulation nodes (same as the installation depth of the FDR/temperature probes). For the simulations of winter 2013, only the clay was included, as a three-layered soil profile with the location of the nodes at 5, 10 and 20 cm, due to missing data for the sand and only three FDR/temperature probes in the clay.

The model was calibrated to fit the measured snow depth and soil temperature by adjusting the site-specific parameters maximum temperature at which precipitation is snow ($T_{\max}$) (only for winter 2015) and the wind profile roughness for momentum transfer of the snow cover (z_m) (for all three winters).

3. RESULTS

3.1. Weather Measurements

Measured air temperature and precipitation are presented in Figure 2. Periods with continuous snow cover, indicated with blue bands in the diagram, were of differing duration in the different years.

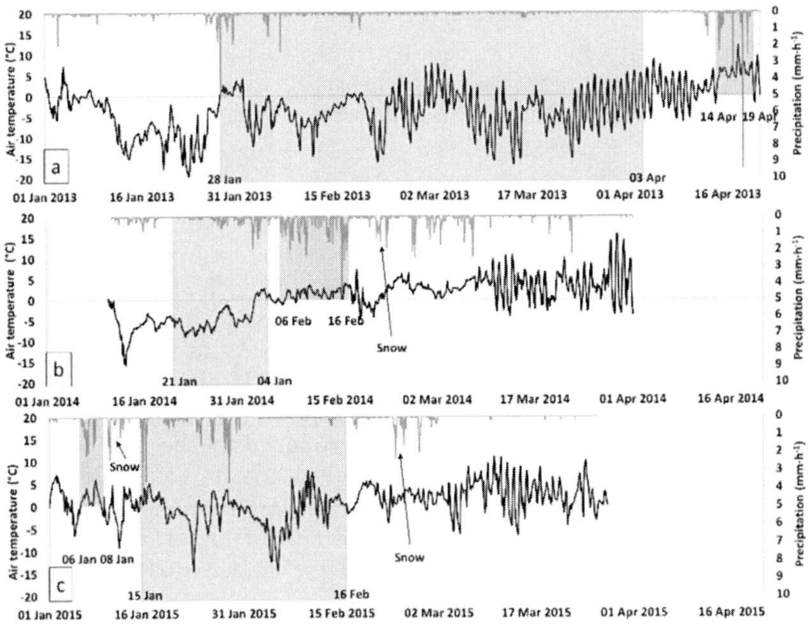

Figure 2. Measured air temperature and precipitation for the three winter periods: 2013 (**a**); 2014 (**b**); and 2015 (**c**). Rain events of interest for this study with start and end dates are marked with grey bands. The duration of unbroken snow cover with start and end dates is indicated with blue bands. Snow events outside the unbroken snow cover period are marked as 'snow'.

The number, duration and intensity of rain events during the three winter periods also differed considerably. For each winter period, rain events of interest were selected (grey bands in Figure 2) for detailed analysis. A rain event was classified as 'interesting' for this study when the precipitation fell as rain on completely or partially frozen ground. The total amount of rain and the duration of the 'main event' when the highest measured intensities occurred are shown in Table 2.

Table 2. Start and end dates of the rain events of interest, with measured total amount of precipitation, highest measured intensity and the duration and amount of precipitation of the main event within the whole event.

Date	Total (mm)	Main Event	Highest Intensity ($mm \cdot h^{-1}$)
14–19 April 2013	42.1	15.8 mm in 2 h	9.3
6–16 February 2014	105	10 mm in 4 h	4.6
6–8 January 2015	28	7 mm in 4 h	2.0

In terms of intensity, the 2013 event listed in Table 2 would be classified as heavy rain and the other events as moderate rain. The rain event in 2014 occurred on top of ongoing snowmelt after 4 February.

3.2. Surface Discharge Measurements

Measured surface discharge for the selected rain events is presented in Table 3, together with the estimated discharge coefficients. The main discharge events in the three winter periods occurred during these rain events, rather than as a result of snowmelt. While the rain event in 2013 was the largest of the three events, the rain event in 2015 produced the highest amount of discharge and had the highest discharge coefficient. During winter 2014, little discharge was produced by the rain event of interest (1008 m^3) compared with the other two years (2594–3096 m^3) (Table 3). However, it should be noted that some tunneling below the flume was observed, resulting in the by-pass of water and too low discharge values. This was repaired and did not happen in 2015.

Table 3. Start and end date of the rain events of interest with measured surface discharge, precipitation per area and percentage of precipitation water, which reached the outlet (discharge coefficient) for these rain events. Values in parenthesis show the discharge coefficient for the agricultural area only (no forest). Values marked with an asterisk (*) are incorrect measurements due to the by-pass of water below the flume.

Date	Discharge (m^3)	Precipitation (m^3)	Discharge Coefficient (%)
14–19 April 2013	3096	12,180	25 (50)
6–16 February 2014	1008 *	30,450	4 *
6–8 January 2015	2594	8120	32 (63)

By comparing the discharge measured at the Gryteland sub-catchment outlet with discharge measured at the outlet of the main Skuterud catchment, it was determined that a surface discharge coefficient of about 12%, rather than 4%, was more realistic for the event in 2014.

3.3. Liquid Soil Water Content and Soil Temperature

A malfunction in Stations 1 and 3 created a gap in the data for winter 2013. The two stations were repaired just before the extreme event at the end of the winter period, so that soil moisture and soil temperature measurements for the clay and sand were available for this event (insert diagram in Figure 3). For the clay, measured data were taken close by the outlet station, where undisturbed soil moisture and soil temperature data were obtained throughout winter 2013 (Figure 3).

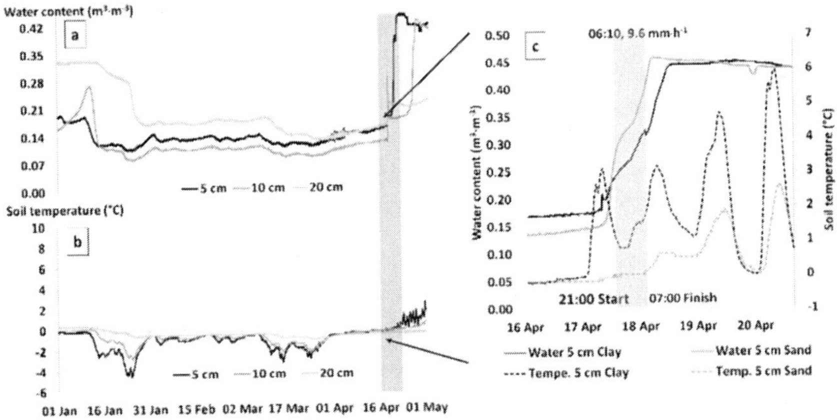

Figure 3. Measured soil water content (**a**) and soil temperature (**b**) in the clay during winter 2013 with the insert diagram (**c**) showing measured soil water content and soil temperature at 5 cm depth in the clay and sand during the rain event of interest in 2013 (duration marked with grey band in (**a**,**b**)). The period with the highest precipitation intensities is marked with a green band in (**c**).

During winter 2013, the soil temperature and liquid soil water content measurements were characterized by a long period of soil frost at all soil depths (Figure 3). Shortly after 2 December, the soil started to freeze at 5 and 10 cm and stayed frozen until the highest precipitation intensities (9.6 mm·h^{-1}) occurred on 18 March (Figure 3). Frozen soil was indicated by a low liquid soil water content (~0.12 m^3·m^{-3}) compared with pre-freezing (~0.38 m^3·m^{-3} in November 2012) and soil temperature below 0

°C. As can be seen in Figure 3, the soil was still frozen after the first rain event on 14 April. The liquid soil water content was still low in both soils (~0.17 $m^3 \cdot m^{-3}$ in the clay and 0.13 $m^3 \cdot m^{-3}$ in the sand), and the measured soil temperature was below 0 °C in both soils. On 17 April, the soil temperature in the top 5 cm in the clay soil started to rise above 0 °C, followed by a continuous increase in liquid soil water content at 5 and 10 cm depth. The soil temperature in the topsoil dropped to about 0.8 °C when the major rain event started in the evening of 17 April. However, the liquid soil water content continued to increase steadily and more rapidly in the sand than in the clay. When the highest precipitation rates occurred, the clay soil had reached a liquid soil water content of about 0.25 $m^3 \cdot m^{-3}$. The temperature in the sand soil rose much more slowly than in the clay soil, but the liquid soil water content increased rapidly in the sand topsoil layer at the same rate as in the clay soil, reaching a liquid soil water content of 0.34 $m^3 \cdot m^{-3}$ when the highest precipitation intensities occurred. During the final period of the high intensity rain, the soil water content continued to rise to about 0.45 $m^3 \cdot m^{-3}$ in both the clay and the sand. No further increase in soil water content occurred during the rest of the rain event, probably because both soils were fully saturated.

Measured soil water content during winter 2014 for the clay and sand is shown in Figure 4, and measured soil temperature is presented in Figure 5.

Measured soil water content and temperature declined significantly in both the clay and sand immediately after the air temperature started to drop below zero (Figure 4 and Figure 5) around 6 January. Measured soil water content did not change much after 13 January, whereas soil temperature followed the changes in air temperature, although this effect decreased with increasing depth for both soil types. After 27 January, the liquid soil water content rose slowly, due to an increase in air temperature (Figure 2), resulting in thawing of ice in the soil. After 3 February, the liquid soil water content increased rapidly due to incoming snowmelt water. While the soil water content did not change much in the sand at 30 and 40 cm depth, a decrease in liquid soil water content was observed at these depths in the clay. Both soils showed similar temperature profiles for all four depths (Figure 5). The soil water content at all depths was much lower in the sand than in the clay, reducing the penetration depth of freezing [6]. Furthermore, at 30 and 40 cm depth in the sand, there was little water left to freeze (30 cm: 0.08 $m^3 \cdot m^{-3}$; 40 cm: 0.06 $m^3 \cdot m^{-3}$).

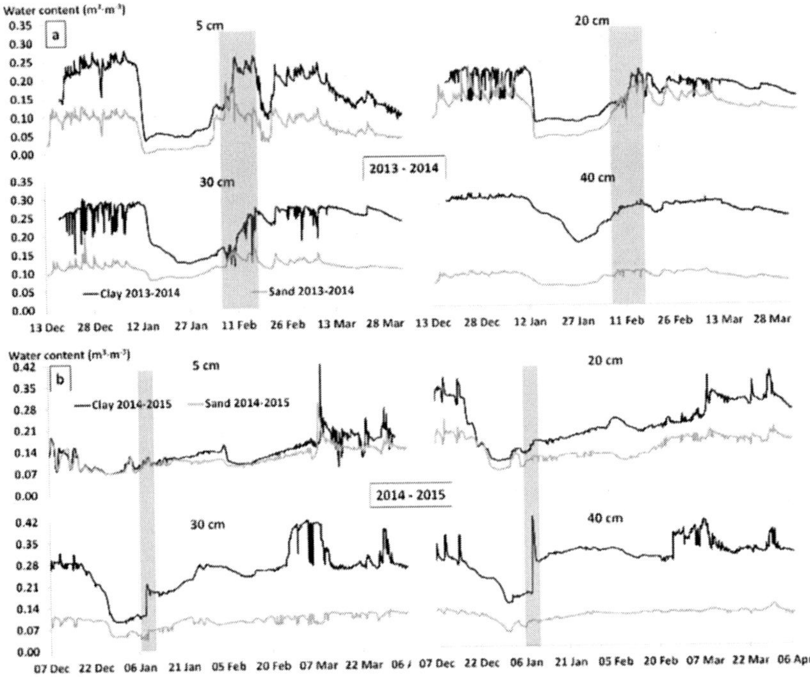

Figure 4. Measured soil water content in the clay and sand for the winter periods 2014 (**a**) and 2015 (**b**) at four depths: 5, 20, 30 and 40 cm. Durations of the rain events of interest (according to Table 2) are marked with grey bands.

When the combined rain and snowmelt event occurred in 2014 (Table 2), both soils had just started to thaw. During the event, similar soil water contents as measured before freezing (~0.25 $m^3 \cdot m^{-3}$) were reached in the clay. In the sand, soil water content at 5 cm depth initially rose rapidly above (~0.18 $m^3 \cdot m^{-3}$) pre-freezing values (~0.10 $m^3 \cdot m^{-3}$), but decreased to pre-freezing values during the event, while at the other three depths studied, the soil water content continued to increase during the whole event. At all four soil depths in both soils, the soil temperature rose significantly to above 0 °C, reaching the highest temperature (3.6 °C) at 5 cm.

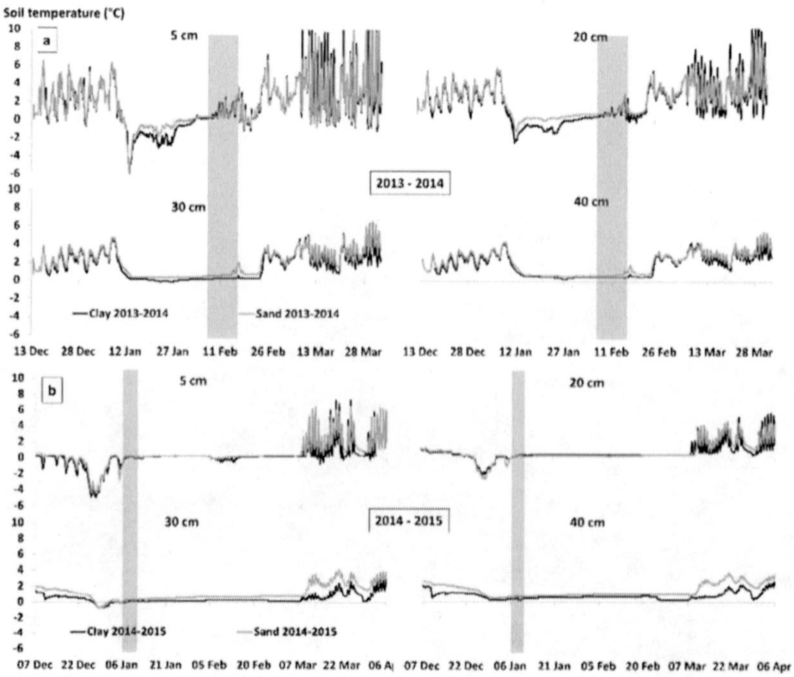

Figure 5. Measured soil temperature in the clay and sand for the winter periods 2014 (**a**) and 2015 (**b**) at four depths: 5, 20, 30 and 40 cm. Durations of the rain events of interest (according to Table 2) are marked with grey bands.

The change in liquid soil water content in winter 2015 (Figure 4) was characterized by low values at 5 cm in both soils (minimum of 0.07 $m^3 \cdot m^{-3}$) due to early freezing in the beginning of December. This was followed by a decrease in liquid soil water content at 20, 30 and 40 cm in the clay on 17 December, reaching a minimum of 0.10 $m^3 \cdot m^{-3}$ at 20 and 30 cm and 0.14 $m^3 \cdot m^{-3}$ at 40 cm. A sudden increase in liquid soil water content occurred on 7 March at all depths in the clay and at 5 cm in the sand, when the air (Figure 2) and soil temperature (Figure 5) rose significantly to above 0 °C. In winter 2015, the rain event (Table 2) was preceded by freezing temperatures at 5–30 cm soil depth and low liquid soil water content at all soil depths in both soils (Figure 5 and Figure 6). During the event, the liquid soil water content rose rapidly at 30 and 40 cm depth in the clay, reaching its highest value (0.42 $m^3 \cdot m^{-3}$) at 40 cm depth. At 5- and 20 cm depth, the change in liquid soil water content was less pronounced. Soil temperature rose in the upper three depths in both soils to about 0 °C, while at 40 cm depth, the soil temperature stayed above 0 °C at all times in both soils.

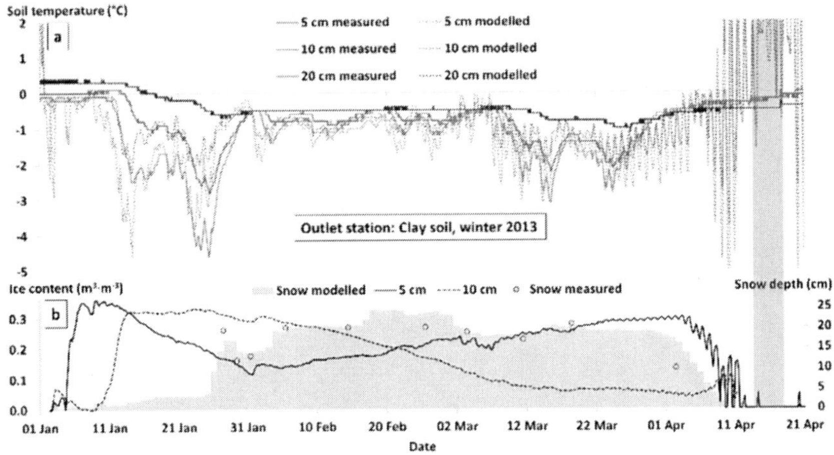

Figure 6. Diagram of (**a**) modelled and measured soil temperature at three depths (5, 10 and 20 cm) in the clay soil (at outlet station) during winter 2013; diagram of (**b**) measured and modelled snow depths at the outlet station and modelled ice content at the three depths in the clay soil (was zero at 20 cm). The duration of the rain event of interest is marked with a grey band.

3.4. SHAW Modelling

The simulated snow depth, soil temperature and ice content values for the clay during winter 2013 are presented in Figure 6, for the clay and sand during winter 2014 in Figure 7 and Figure 8 and for winter 2015 in Figure 9 and Figure 10. As mentioned, due to missing measurements for the sand during winter 2013, the sand was not modelled for that year.

As can be seen from Figure 6, the SHAW model simulated the changes in the snow pack very accurately for 2013. To fit the simulated snow depth to the measured values, the z_m parameter was set to 0.15 cm. The simulated soil temperature at 5 and 10 cm showed some fluctuations around the measured temperature, which increased after the snowpack disappeared in the end of March, when the air temperature fluctuated between −7 and 8 °C on a daily basis (Figure 2). However, it should be noted that the soil temperature was measured below a thick grass layer

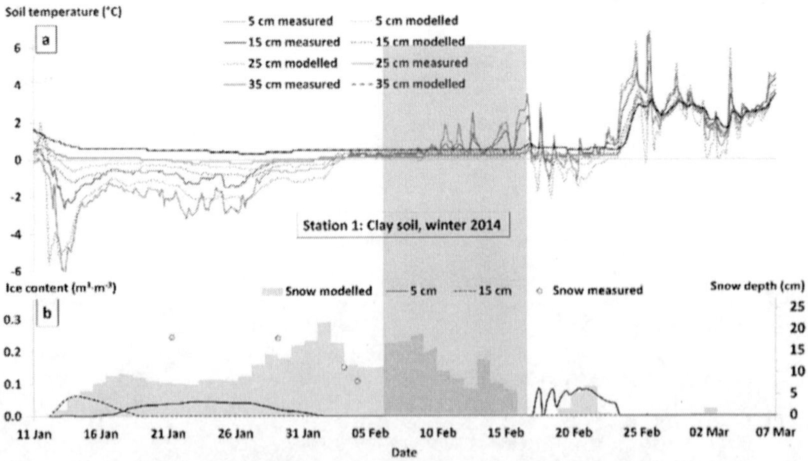

Figure 7. Diagram of (**a**) modelled and measured soil temperature at four depths (5, 10 and 20 cm) in the clay soil (at Station 1) during winter 2014; diagram of (**b**) measured and modelled snow depths at Station 1 and modelled ice content at four depths in the clay soil (was zero at 25 and 35 cm). The duration of the rain event of interest is marked with a grey band.

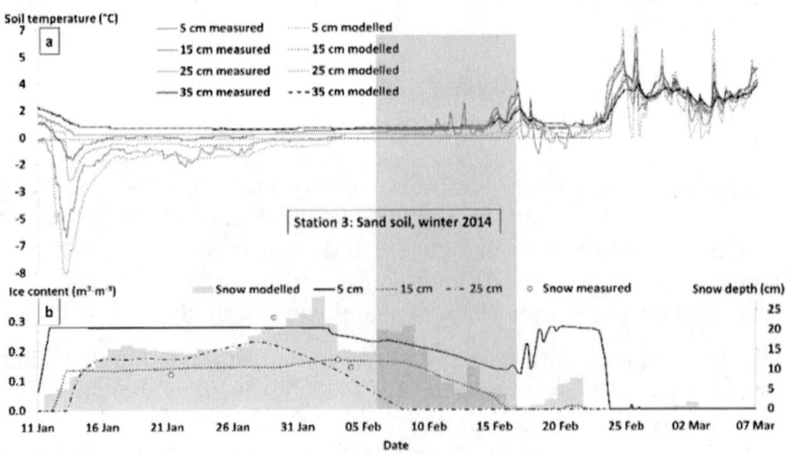

Figure 8. Diagram of (**a**) modelled and measured soil temperature at four depths (5, 15, 25 and 35 cm) in the sand soil (at Station 3) during winter 2014; diagram of (**b**) measured and modelled snow depths at Station 3 and modelled ice content at four depths in the sand soil (was zero at 35 cm). The duration of the rain event of interest is marked with a grey band.

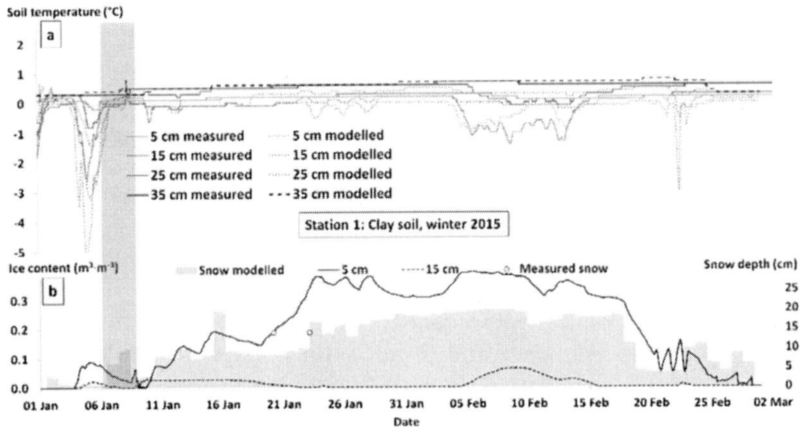

Figure 9. Diagram of (**a**) modelled and measured soil temperature at four depths (5, 15, 25 and 35 cm) in the clay soil (at Station 1) during winter 2015; diagram of (**b**) measured and modelled snow depths at Station 1 and modelled ice content at four depths in the clay soil (was zero at 25 and 35 cm). The duration of the rain event of interest is marked with a grey band.

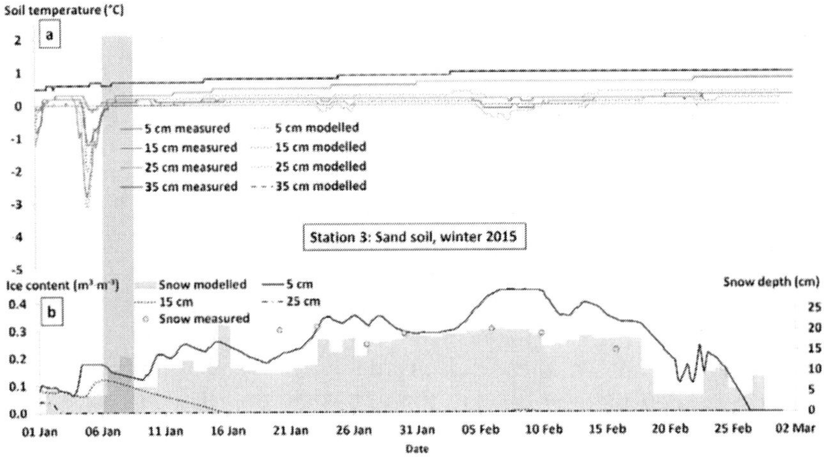

Figure 10. Diagram of (**a**) modelled and measured soil temperature at four depths (5, 15, 25 and 35 cm) in the sand soil (at Station 3) during winter 2015; diagram of (**b**) measured and modelled snow depths at Station 3 and modelled ice content at four depths in the sand soil (was zero at 35 cm). The duration of the rain event of interest is marked with a grey band.

at the outlet station, and this grass layer probably acted as insulation [40], buffering the soil from the fluctuating air temperature. The model results show soil temperatures for bare ground, for comparison with the other winters. Due to the satisfactory simulation of snow depth and soil temperature, it was assumed that SHAW also performed well in simulating ice content in the soil profile. Simulated ice content was high (0.35 $m^3 \cdot m^{-3}$) at 5 and 10 cm depth and zero at 20 cm depth, indicating that almost all of the water at these depths was frozen. The ice at these depths was apparently formed during the low soil temperatures in January, before the snow pack was established. SHAW was able to simulate the rise in soil temperature after the snow pack reached a depth of about 20 cm in February and predicted a decrease in ice content to zero at all depths by 12 April. The SHAW model was also able to simulate the observed decrease in soil temperature in the middle of April.

Because only four snow depth measurements were possible in 2014, it was difficult to compare the simulated and measured snow depth values. However, it can be seen that SHAW underpredicted the snow depth above the clay soil during January and overpredicted it above the sand soil (Figure 7 and Figure 8). The z_m parameter was adjusted to 0.5 cm for the clay and 1.5 cm for the sand.

The simulated soil temperature in the clay soil (Figure 7) followed the measured temperature rather closely during the freezing period in January, but during the start of the snowmelt in the middle of February, the measured temperature showed a higher fluctuation than the simulated temperature. For the sand, SHAW simulated slightly lower temperature (−8 °C at 5 cm) than the measured values (−6 °C at 5 cm), but in general, the simulated temperature followed the trend in the measured data quite well (Figure 8). SHAW predicted low ice content for the clay at 5 and 15 cm (max. 0.06 $m^3 \cdot m^{-3}$ at 15 cm) and no ice at lower depths, while for the sand, it predicted an ice content of 0.28 $m^3 \cdot m^{-3}$ at 5 cm and 0.22 $m^3 \cdot m^{-3}$ at 25 cm depth. At 5 and 15 cm, the model predicted that all the available water was frozen, which can be seen as the constant ice contents between 12 January and 2 February. Moreover, the period with frozen water in the sand profile was considerably longer (11 January–23 February at 5 cm) than that in the clay (12 January–1 February at 5 cm).

Except for a slight underprediction of the snow depth for the period when the first two snow measurements were taken in 2015, SHAW simulated the snow depth well for winter 2015 (Figure 10). To obtain this fit, z_m was set to 0.05 cm for the clay and 0.08 cm for the sand. In addition, T_{max} had to be set to 3.8 °C to fit the modelled snow thicknesses to the measured, which resulted in the simulation of a thin snow cover, which did not occur in the field, during the rain event of interest in 2015.

This adjustment of T_{max} was in accordance with [32], who used a comparable snow model for the study area and had to make a similar adjustment to this parameter.

The simulated soil temperature showed similar low fluctuations as the measured (Figure 9 and Figure 10), and the first negative peak was also simulated, with a good fit in the sand and slightly lower temperature in the clay (−5.8 °C simulated compared with −3 °C measured at 5 cm).

For both the clay and the sand, SHAW predicted high ice content at 5 cm depth, mainly during the period when the snow pack was between 10 and 20 cm thick (Figure 9 and Figure 10). The predicted values were 0.39 $m^3 \cdot m^{-3}$ in the clay and 0.45 $m^3 \cdot m^{-3}$ in the sand. For both soils, SHAW simulated an increase in ice content at 5 cm with growing snow thickness and a decrease in ice content with decreasing snow thickness.

3.5. Erosion Mapping

During winter 2013, no erosion was observed anywhere in the catchment. The combined snowmelt and rain event in 2014 formed several small rills with a maximum width of 20 cm and 5–10 cm deep on two slopes in the catchment (Figure 11). The observed rills were formed mainly in the tracks created by the cultivator.

In winter 2015, a continuous rill system (total length 493 m) was formed in depressions, where surface runoff was concentrated, during the main rain event. The rills varied in width and depth (Figure 11, Image 2,3), with measured maximum width of 40 cm and a depth of 20 cm. Several sedimentation areas were observed where the surface runoff was slowed down, due to reduced slope angle and increased flow width, indicating that not all of the eroded soil reached the outlet. At the other slope where rills were observed in 2014 (Figure 11, Image 1), the rills were connected to the extensive rill system in 2015.

4. DISCUSSION

The winters of 2013, 2014 and 2015 differed significantly from each other in terms of the number of freezing periods, the length of period with a continuous snow cover and the number of rain events (Figure 2). The soils stayed frozen throughout the whole time with a continuous snow cover in the three winter periods, confirming the finding by Zhao et al. [15] that a thin (<25 cm) snow pack can increase soil freezing, e.g., due to an increase in ground albedo.

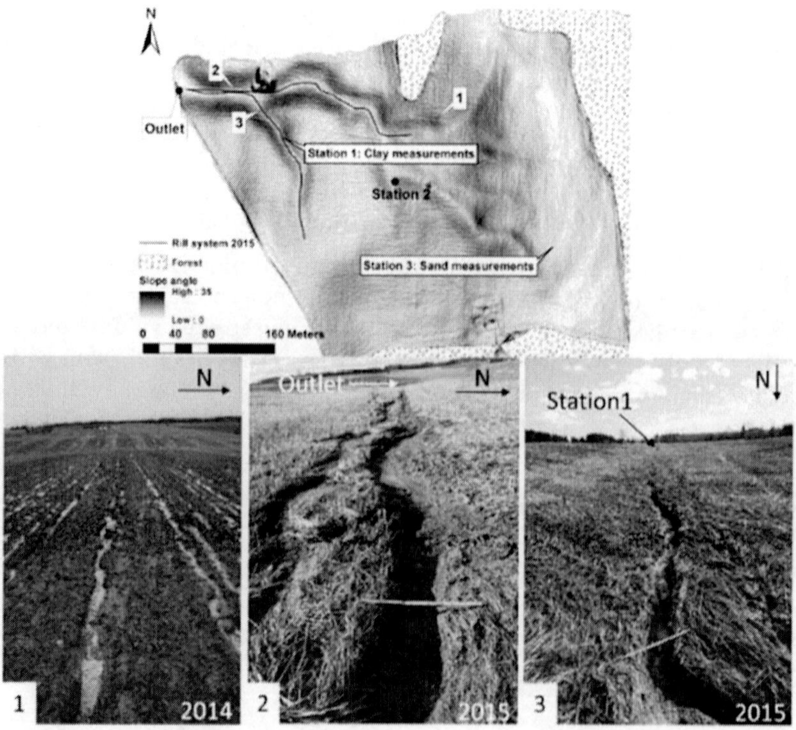

Figure 11. Slope angle map showing; the extent of the rill system, which formed during winter 2015 and the location where images 1–3 were taken. Image 1 shows rills which formed during the combined snowmelt and rain event in 2014 and images 2,3 show rills that formed during winter 2015.

The amount of water that can infiltrate depends on the water permeability of the soil and the speed of surface runoff. In frozen soil, the water permeability depends on pre-freezing conditions [41], i.e., what the soil water content was before freezing started. In saturated soils the macropores are filled with water, which when frozen, clogs the pores. In the sand studied here, the initial soil water content was similar in 2013, 2014 and 2015, at between 0.10 and 0.25 $m^3 \cdot m^{-3}$, which is far below the measured saturation of 0.40–0.43 $m^3 \cdot m^{-3}$. Therefore, it can be assumed that the sand areas in the catchment contributed little to surface runoff during these three winters. This could explain why no erosion occurred in the depression on the sand soil in the catchment (Figure 1).

In the clay, however, the conditions differed between the years. In clays, water transport and infiltration capacity are highly dependent on macropores, particularly in the levelled clays found in the study area [42]. When freezing started in 2014, the clay had a soil water content of 0.25–0.30 $m^3 \cdot m^{-3}$, which according to the soil hydraulic characteristic curve for the clay represents a matric potential of 32 kPa (5 cm depth) to 315 kPa (20 cm depth), at which macropores are filled with air [43]. Furthermore, the rain event of interest in 2014 was less intensive than that in 2013 (Table 2), and the thawed soil allowed more water to infiltrate, resulting in a smaller amount of water reaching the outlet (Table 3). The SHAW results suggested that the clay was completely thawed by the time the rain event occurred (Figure 7).

However, despite a smaller discharge coefficient compared with 2013, the 2014 event caused erosion (Figure 11), in the tracks created by the cultivator.

Similarly to 2013, in winter 2015 the soil water content was 0.40–0.45 $m^3 \cdot m^{-3}$ at all soil depths in the clay when freezing started in November 2014, resulting in macropores filled with ice. Both events had a high discharge coefficient, of ~25% (50% when forest area was excluded) in 2013 and 32% (63%) during the first event in 2015 (Table 3). This, together with the low liquid soil moisture values, suggests that infiltration was restricted due to frozen soil during both events. Therefore, the fact that erosion did not occur in 2013, but did occur in 2015, cannot be explained by the infiltration capacity of the soil during the events. Moreover, the stability of the soils cannot explain the differences between 2013 and 2015, as confirmed by shear strength measurements carried out on the clay immediately before the event in 2013, which revealed low shear strength of 5–10 kPa at the soil surface (vane shear test, Eijkelkamp, The Netherlands).

Another process that could explain the differences between 2013 and 2015 is the speed of surface runoff, which determines the erosivity of surface runoff [44]. In 2013, the whole catchment was covered with stubble, but in 2015, only the depressions had intact stubble, while secondary tillage with a cultivator reduced the amount of plant residues on slopes. This tillage created a loose and smoother surface, probably causing higher speed of surface runoff, in freezing conditions in particular, as previously shown by Ban et al. [4]. This assumption was supported by the occurrence of soil erosion in the form of rills on the tilled slopes in 2014 (Figure 11). These findings were also in agreement with Edwards and Burney [3], who concluded that only a plant cover can significantly reduce soil losses by rain and overland flow on frozen ground, e.g., through reduced runoff speeds and increased soil stability by roots. Our finding that the selected rain events on saturated and frozen soil produced a large amount of surface discharge confirmed the prediction by Stähli et al. [9] and Nyberg et al. [10] that high water saturation and early frost penetration, combined with heavy rain on still frozen soil, cause a marked increase in the amount of runoff.

Contrary to observation made by other studies (e.g., [21,22,23]), in all three winters studied, snowmelt played no significant role in terms of soil erosion. During snowmelt in 2014, the snow layer acted as a buffer for incoming rain. The rain infiltrated into the snow pack and surface runoff was delayed by the snow, reducing the erosive forces of the rain event. In 2013, no erosion occurred, and in 2015, the major soil erosion features had occurred before the first snow fell (Figure 2).

The performance of the SHAW model was satisfactory for all three winter periods. The change in soil temperature and snow pack was well reproduced, and the simulated ice content was in agreement with the measured liquid soil water content. In general, SHAW predicted ice in the soil for the periods when the FDR probes measured low liquid soil water contents. With the adjustment of two snow-related parameters in the model, it was possible to obtain reasonable results for the three different winter periods. Adjustment of T_{max} for winters when snow falls at temperatures above 0 °C, as was the case in 2015, allows the model to partition incoming precipitation into snow and rain based on field observations, rather than using linear interpolation. The performance of SHAW in this study proved that it can be a valuable tool for investigating and predicting: (1) water content at freezing; (2) whether soil is frozen or unfrozen at a particular moment; and (3) the state of the snow cover. These are three important factors that control the amount of runoff during winter and are indispensable for predicting when soil erosion can be expected.

5. CONCLUSIONS AND IMPLICATIONS

Field observations carried out during three winters in a catchment in southern Norway showed how soil hydraulic properties changed due to freezing-thawing, affecting surface runoff caused by snowmelt and rain, and how these processes are linked to soil erosion. The largest amount of soil erosion was caused by a small rain event on frozen ground, before the snow cover was established, while snowmelt played no significant role in terms of soil erosion. Four factors that determine the extent of runoff and erosion were of particular importance: (1) soil water content at freezing; (2) whether soil was frozen or unfrozen at a particular moment; (3) the state of the snow cover; and (4) tillage operations prior to winter. The simulation results showed that the SHAW model, with its accurate snow pack routine, is a useful tool that can help to investigate and identify non-tillage factors (e.g., 1, 2 and 3) influencing erosion.

ACKNOWLEDGMENTS

This research was funded by the Catchy project of NIBIO, Norwegian Institute of Bioeconomy Research.

REFERENCES

1. Ollesch, G.; Sukhanovski, Y.; Kistner, I.; Rode, M.; Meissner, R. Characterization and modelling of the spatial heterogeneity of snowmelt erosion. *Earth Surf. Process. Landf.* **2005**, *30*, 197–211.
2. Iwata, Y.; Nemoto, M.; Hasegawa, S.; Yanai, Y.; Kuwao, K.; Hirota, T. Influence of rain, air temperature, and snow cover on subsequent spring-snowmelt infiltration into thin frozen soil layer in northern Japan. *J. Hydrol.* **2011**, *401*, 165–176.
3. Edwards, L.M.; Burney, J.R. Soil erosion losses under freeze/thaw and winter ground cover using a laboratory rainfall simulator. *Can. Agric. Eng.* **1987**, *29*, 109–115.
4. Ban, Y.; Lei, T.; Liu, Z.; Chen, C. Comparison of rill flow velocity over frozen and thawed slopes with electrolyte tracer method. *J. Hydrol.* **2016**, *534*, 630–637.
5. Watanabe, K.; Kito, T.; Dun, S.; Wu, J.Q.; Greer, R.C.; Flury, M. Water Infiltration into a Frozen Soil with Simultaneous Melting of the Frozen Layer. *Vadose Zone J.* **2013**, *12*.
6. Yami, E.R.; Khalili, A.; Rahimi, H.; Etemad, A. Investigation of moisture on soil temperature regimes and frost depths in a laboratory model. *Int. J. Agric. Sci.* **2012**, *2*, 717–732.
7. Al-Houri, Z.M.; Barber, M.E.; Yonge, D.R.; Ullman, J.L.; Beutel, M.W. Impact of frozen soils on the performance of infiltration treatment facilities. *Cold Reg. Sci. Technol.* **2009**, *59*, 51–57.
8. Willis, W.O.; Carlson, C.W.; Alessi, J.; Haas, H.J. Depth of freezing and spring run-off as related to fall soil-moisture level. *Can. J. Soil Sci.* **1961**, *41*, 115–123.
9. Stähli, M.; Nyberg, L.; Mellander, P.; Jansson, P.; Bishop, K.H. Soil forst effects on soil water and runoff dynamics along a boreal transect: 2. Simulations. *Hydrol. Process.* **2001**, *15*, 927–941.
10. Nyberg, L.; Stähli, M.; Mellander, P.; Bishop, K.H. Soil frost effect on soil water and runoff dynamics along a boreal forest transect: 1. Field investigations. *Hydrol. Process.* **2001**, *15*, 909–926.
11. Lindström, G.; Löfvenius, M.O. *Tjäle Och Avrinning i Svartberget—Studier Med HBV-Modellen*; SMHI hydrologi Nr 84; SMHI: Norrköping, Sweden, 2000.
12. Iwata, Y.; Hayashi, M.; Suzuki, S.; Hirota, T.; Hasegawa, S. Effects of snow cover on soil freezing, water movement, and snowmelt infiltration: A paired plot experiment. *Water Resour. Res.* **2010**, *46*, 1–11.
13. Zhao, Y.; Nishimura, T.; Hill, R.; Miyazaki, T. Determining Hydraulic Conductivity for Air-Filled Porosity in an Unsaturated Frozen Soil by the Multistep Outflow Method. *Vadose Zone J.* **2013**, *12*.
14. Sutinen, R.; Hänninen, P.; Venäläinen, A. Effect of mild winter events on soil water content beneth snowpack. *Cold Reg. Sci. Technol.* **2008**, *51*, 56–67.
15. Zhao, Y.; Huang, M.B.; Horton, R.; Liu, F.; Peth, S.; Horn, R. Influence of winter grazing on water and heat flow in seasonal frozen soil of Inner Mongolia. *Vadose Zone J.* **2013**, *12*.
16. Parkin, G.; von Bertoldi, A.P.; McCoy, A.J. Effect of tillage on Soil Water Content and Temperature under Freeze-Thaw Conditions. *Vadose Zone J.* **2013**, *12*.

17. He, H.; Dyck, M.F.; Si, B.C.; Zhang, T.; Lv, J.; Wang, J. Soil freezing—Thawing characteristics and snowmelt infiltration in Cryalfs of Alberta, Canada. *Geodermal. Reg.* **2015**, *5*, 198–208.
18. Kormos, P.R.; Marks, D.; Williams, C.J.; Marshall, H.P.; Aishlin, P.; Chandler, D.G.; McNamara, J.P. Soil, snow, weather, and sub-surface storage data from a mountain catchment in the rain-snow transition zone. *Earth Syst. Sci. Data* **2014**, *6*, 165–173.
19. Williams, C.J.; Mcnamara, J.P.; Chandler, D.G. Controls on the temporal and spatial variability of soil moisture in a mountainous landscape: The signature of snow and complex terrain. *Hydrol. Earth Syst. Sci.* **2009**, *13*, 1325–1336.
20. Shanley, J.B.; Chalmers, A. The effect of frozen soil on snowmelt runoff at Sleepers River, Vermont. *Hydrol. Process.* **1999**, *13*, 1843–1857.
21. Lundekvam, H.; Skøien, S. Soil erosion in Norway. An overview of measurements from soil loss plots. *Soil Use Manag.* **1998**, *14*, 84–89.
22. Deelstra, J.; Kværnø, S.H.; Granlund, K.; Sileika, A.S.; Gaigalis, K.; Kyllmar, K.; Vagstad, N. Runoff and nutrient losses during winter periods in cold climates—Requirements to nutrient simulation models. *J. Environ. Monit.* **2009**, *11*, 602–609.
23. Øygarden, L. Rill and gully development during an extreme winter runoff event in Norway. *Catena* **2003**, *50*, 217–242.
24. Grønsten, H.A.; Lundekvam, H. Prediction of surface runoff and soil loss in southeastern Norway using the WEPP Hillslope model. *Soil Tillage Res.* **2006**, *85*, 186–199.
25. Flerchinger, G.N.; Xiao, W.; Sauer, T.J.; Yu, Q. Simulation of within-canopy radiation exchange. *NJAS Wagening. J. Life Sci.* **2009**, *57*, 5–15.
26. Hansen, N.C.; Gupta, S.C.; Moncrief, J.F. Snowmelt runoff, sediment, and phosphorous losses under three different tillage systems. *Soil Tillage Res.* **2000**, *57*, 93–100.
27. Govers, G. Rill erosion on arable land in central Belgium: Rates, controls and predictability. *Catena* **1991**, *18*, 133–155.
28. Boardman, J.; Shepheard, M.L.; Walker, E.; Foster, I.D.L. Soil erosion risk-assessment for on- and off-farm impacts: A test case using the Midhurst area, West Sussex, UK. *J. Environ. Manag.* **2009**, *90*, 2578–2588.
29. Weigert, A.; Schmidt, J. Water transport under winter conditions. *Catena* **2005**, *64*, 193–208.
30. Yakutina, O.P.; Nechaeva, T.V.; Smirnova, N.V. Consequences of snowmelt erosion: Soil fertility, productivity and quality of wheat on Greyzemic Phaeozem in the south of West Siberia. *Agric. Ecosyst. Environ.* **2015**, *200*, 88–93.
31. Su, J.J.; van Bochove, E.; Thériault, G.; Novotma, B.; Khaldoune, J.; Denault, J.T.; Zhou, J.; Nolin, M.C.; Hu, C.X.; Bernier, M.; et al. Effects of snowmelt on phosphorus and sediment losses from agricultural watersheds in Eastern Canada. *Agric. Water Manag.* **2011**, *98*, 867–876.
32. Kramer, G.J.; Stolte, J. *Cold-Season Hydrologic Modeling in the Skuterud Catchment. An Energy Balance Snowmelt Model Coupled with a GIS-Based Hydrology Model*; Bioforsk Report 4(126); Bioforsk: Ås, Norway, 2009; Volume 46.
33. Thue-Hansen, V.; Grimenes, A.A. *Meteorologiske Data for Ås*; Norges Miljø og Biovitenskaplige Universitet: Ås, Norway, 2013–2014.
34. Kværnø, S.H.; Øygarden, L. The influence of freeze-thaw cycles and soil moisture on aggreagte stability of three soils in Norway. *Catena* **2006**, *67*, 175–182.
35. Starkloff, T.; Stolte, J.; Hessel, R.; Ritsema, C. Understanding snowpack development at catchment scale, using comprehensive field observations and spatially distributed snow modelling. *Hydrol. Res.* **2017**, in press.
36. Flerchinger, G.N.; Saxton, K.E. Simultaneous heat and water model of a freezing snow-residue-soil system. I. Theory and development. *Trans. Am. Soc. Agric. Eng.* **1989**, *32*, 565–571.

37. Li, R.; Shi, H.; Flerchinger, G.N.; Akae, T.; Wang, C. Simulation of freezing and thawing soils in Inner Mongolia Hetao Irrigation District China. *Geoderma* **2012**, *173*, 28–33.
38. Flerchinger, G.N. *The Simultaneous Heat and Water (SHAW) Model: Technical Documentation*; Technical Report NWRC 2000-09; Northwest Watershed Research Center, USDA Agricultural Research Service: Boise, ID, USA, 2000.
39. Black, C.A.; Evans, D.D.; White, J.L.; Ensminger, L.E.; Clark, F.E. *Methods of Soil Analysis Physical and Mineralogical Properties, Including Statistics of Measurement and Sampling*; American Society of Agronomy: Madison, WI, USA, 1965; Part 1; Volume 9, pp. 1–770.
40. Flerchinger, G.N.; Sauer, T.J.; Aiken, R.A. Effects of crop residue cover and architecture on heat and water transfer at the soil surface. *Geoderma* **2003**, *116*, 217–233.
41. McCauley, C.A.; White, D.M.; Lilly, M.R.; Nyman, D.M. A comparison of hydraulic conductivities, permeabilities and infiltration rates in frozen and unfrozen soils. *Cold Reg. Sci. Technol.* **2002**, *34*, 117–125.
42. Øygarden, L.; Kværner, J.; Jenssen, P.D. Soil erosion via preferential flow to drainage systems in clay soils. *Geoderma* **1997**, *76*, 65–86.
43. Lundberg, A.; Ala-Aho, P.; Eklo, O.; Klöve, B.; Kværner, J.; Stumpp, C. Snow and frost: Implications for spatiotemporal infiltration patterns—A review. *Hydrol. Process.* **2015**, *30*, 1230–1250.
44. Boardman, J.; Poesen, J. *Soil Erosion in Europe*; John Wiley & Sons Inc.: Hoboken, NJ, USA, 2006; p. 855.

CHAPTER 13

Investigating Groundwater Flow Components in an Alpine Relict Rock Glacier (Austria) using a Numerical Model

Marcus Pauritsch[1] & Thomas Wagner[2] & Gerfried Winkler[2] & Steffen Birk[2]

[1]*Tannhofweg 14/8, 8044 Graz, Austria*

[2]*Institute of Earth Sciences, NAWI Graz Geocenter, University of Graz, Heinrichstrasse 26, 8010 Graz, Austria*

ABSTRACT

Relict rock glaciers are complex hydrogeological systems that might act as relevant groundwater storages; therefore, the discharge behavior of these alpine landforms needs to be better understood. Hydrogeological and geophysical investigations at a relict rock glacier in the Niedere Tauern Range (Austria) reveal a slow and fast flow component that appear to be related to the heterogeneous structure of the aquifer. A numerical groundwater flow model was used to indicate the influence of important internal structures such as layering, preferential flow paths and aquifer-base topography. Discharge dynamics can be reproduced reasonably by both introducing layers of strongly different hydraulic conductivities or by a network of highly conductive channels within a low-conductivity zone.

Moreover, the topography of the aquifer base influences the discharge dynamics, which can be observed particularly in simply structured aquifers. Hydraulic conductivity differences of three orders of magnitude are required to account for the observed discharge behavior: a highly conductive layer and/or channel network controlling the fast and flashy spring responses to recharge events, as opposed to less conductive sediment accumulations sustaining the long-term base flow. The results show that the hydraulic behavior of this relict rock glacier and likely that of others can be adequately represented by two aquifer components. However, the attempt to characterize the two components by inverse modeling results in ambiguity of internal structures when solely discharge data are available.

Keywords: Relict rock glacier Austria Groundwater flow Hydraulic properties Numerical modeling

INTRODUCTION

Mountainous and, in particular, alpine groundwater is contributing significantly to the stream flow of rivers in valleys and consequently in the foreland (e.g., Campbell et al. 1995; Clow et al. 2003; Tague and Grant 2009; Muir et al. 2011; Welch et al. 2012). With increasing population and propagation of tourist industries and recreational activities in mountainous areas, knowledge of the hydraulic behavior and storage capacities of mountainous aquifers is getting more important for sustainable water resources management as well as for predicting natural hazards, as for example flash floods and debris flows (e.g. Lauber et al. 2014).

Especially in alpine catchments of crystalline regions, a major portion of the groundwater contributing to the streamflow originates from debris accumulations such as moraines and rock glaciers (e.g., Hood and Hayashi 2015; Wagner et al. 2016). While the hydrogeology of moraines, talus and hillslope aquifers has been the subject of intensive investigations (e.g., Clow et al. 2003; Roy and Hayashi 2009; Muir et al. 2011), knowledge about the hydraulic behavior and storage capacities of rock glaciers is sparse (Krainer and Mostler 2002; Krainer et al. 2007; Millar et al. 2013; Winkler et al. 2016a). Nevertheless, rock glaciers are common landforms all around the globe in mountainous areas and at high latitudes—for instance, in the Austrian Alps, Kellerer-Pirklbauer et al. (2012) and Krainer and Ribis (2012) have identified a total of 4,792 rock glaciers covering an area of about 286 km^2. Bollmann et al. (2012) identified 1,697 rock glaciers in the South Tyrolean Alps in Italy and Schmid et al. (2015) used Google Earth to map 702 rock glaciers in selected areas of the Hindu Kush

Himalayan region; therefore, rock glaciers have a high potential to influence the discharge behavior of rivers downstream of their catchments (Wagner et al. 2016).

Rock glaciers in general are scree masses that are supersaturated with ice and gravitationally move downslope with velocities of a few centimeters up to several meters per year (e.g., Barsch 1996; Haeberli et al. 2006). The size and shape of rock glaciers depends on climatic conditions, cirque geometry and debris supply of the comprising headwalls (Degenhardt 2009). Therefore, the range in size is highly variable (e.g., Kellerer-Pirklbauer et al. 2012; Krainer and Ribis 2012). The thickness of rock glaciers depends among other things on the topography of the underlying bedrock/base and is a priori difficult to estimate; however, previous investigations reported thicknesses on the order of several tens of meters (e.g., Krainer et al. 2015; Monnier and Kinnard 2015; Winkler et al. 2016a). Rock glaciers can be classified into active (currently moving), inactive (no movement, but ice is present) and relict forms (no ice mass and no movement). Active and inactive rock glaciers can also be summarized to intact rock glaciers, as both types contain ice in contrast to relict rock glaciers, where ice has melted and morphological structures such as ridges and furrows might become more obvious.

Investigations of rock glaciers mainly focus on the distribution, movement and the inner structure of intact rock glaciers (e.g., Haeberli et al. 2006; Jansen and Hergarten 2006; Leopold et al. 2011; Monnier et al. 2011). Monnier and Kinnard (2015) investigated the internal structure of an active rock glacier in the Dry Andes with ground penetrating radar. Their results suggest a heterogeneous structure with an upper and lower ice-rich layer and an intermediate zone with a high fraction of liquid water. Additionally, their results show inner structures in the form of upward-dipping reflectors caused by the movement of the rock glacier. Krainer et al. (2015) investigated an active rock glacier in the Italian Alps. They also identified a layered structure at two drill core sites on the rock glacier. Relict rock glaciers are often used as an indicator of previous permafrost conditions for climate studies (e.g., Hughes et al. 2013; Matthews and Wilson 2015), but their inner structure and hydraulic properties are largely unknown. Recently, Winkler et al. (2016a) performed hydrogeological investigations at the relict Schöneben rock glacier located in the Austrian Eastern Alps (hereinafter referred to as SRG), describing its storage and flow components by using data of the spring hydrograph as well as natural and artificial tracers and geophysical surveys. They suggest that the rock glacier is a heterogeneous aquifer with a layered internal structure. Their investigation further suggests that there is a rather thin (approximately 10 m) layer at the base of the relict rock glacier, which consists of silty or fine sand material providing a considerable storage capacity controlling the base flow observed at the rock glacier spring (see Fig. 9 in Winkler et al. 2016a). This layer is supposed to represent glacial sediment deposits (probably morainic sediments), which would be in agreement with the findings at an active rock glacier, where moraine deposits have been encountered in two drill cores at its base (Krainer et al. 2015). The hydraulic properties of the layer at the base of the Schöneben rock glacier were additionally estimated by Pauritsch et al. (2015) using analytical models of sloping aquifers by focusing on winter base flow data. The derived hydraulic conductivity on the order of 10^{-5} m/s is consistent with findings of Winkler et al. (2016a). The upper layers of the relict rock glacier likely consist of coarser material with a higher permeability that may explain the rapid response of the spring to recharge events. They can be activated when the lower layer is saturated or simply due to the differences of hydraulic conductivity (Winkler et al. 2016a). Another explanation might be the existence of preferential flow paths. Preferential flow paths related to cracks in the soil and roots of plants have been observed at hillslopes (e.g., Graham et al. 2010). Similarly, rapid localized flow through solution conduits is well known from karst aquifers (e.g., Worthington 2009). In this report, the term "preferential flow paths" refers to channel flow (and not layer-controlled flow); in rock glaciers, preferential flow might be possible through washed-out channels within an otherwise fine-grained layer. The existence of such channels is suggested by excavations at a relict rock glacier nearby the SRG (Untersweg and Proske 1996).

At present, a large portion of the rock glaciers are relict forms, and considering climate change (global warming), it can be assumed that today's intact rock glaciers will also in the future become relicts. Therefore, knowledge about the hydrogeological properties of relict rock glaciers is essential for predicting future changes in the discharge behavior of rivers in alpine catchments which currently contain intact rock glaciers.

Simple lumped-parameter models are able to satisfyingly simulate the discharge behavior of complex aquifers and even of this particular relict rock glacier (Wagner et al. 2016); however, they cannot be used to investigate and distinguish between different types of heterogeneity related to the internal structure (e.g., layered structure or preferential flow paths). To resolve such a task, a distributed numerical model is preferable. Numerical models have been extensively used during the last decades and represent a helpful tool for estimating aquifer parameters (Carrera et al. 2005) and also help to improve the conceptual understanding of a hydrogeological setting (e.g., Eisenlohr et al. 1997a, b; Mansour et al. 2012). Thus, in this report, the current conceptual understanding regarding the drainage processes within the relict Schöneben rock glacier (Winkler et al. 2016a) is tested using a three-dimensional numerical groundwater flow model. The main purpose of this report is to indicate the major features of the internal structure (layering, draining channels, topography) controlling the discharge behavior of the relict rock glacier and to compare them with the current conceptual understanding. To this end, several model runs with increasing complexity are provided. As there are no boreholes or other information about groundwater heads, only the discharge of the draining spring is used for calibration. Due to these limitations, the attention is turned towards reproducing the discharge dynamics in general rather than the observed discharge exactly.

Field site and data set

The investigation area in this research is the relict Schöneben rock glacier (SRG) located in the Niedere Tauern Range in Austria at E 14°40′25″ and N 47°22′39″ (Fig. 1). The SRG is situated in a northeast-facing alpine cirque which also represents the hydrological catchment of the SRG, covering an area of 0.67 km^2 with a maximum elevation of 2,295 m a.s.l. The elevation of the tongue-shaped SRG ranges from 1,715 to 1,912 m a.s.l., it has a length of about 750 m, a maximum width of about 250 m, and covers an area of 0.11 km^2 (Fig. 1). The surface of the SRG is covered by coarse-grained, blocky material consisting of gneissic rocks ranging from cubic decimeters to a few cubic meters. The characteristic morphology with distinct collapse structures (i.e., caused by the melting of the ice and the resulting loss in volume), the partial vegetation cover (mainly grasses and dwarf pines) and the low slope gradient of the rock glacier front indicate that the SRG can be assumed to be a relict rock glacier, which is supported by an average water temperature above 2.2 °C at the spring emerging at the front of the rock glacier (Kellerer-Pirklbauer et al. 2015; Winkler et al. 2016a).

Steep rock faces surround the rock glacier in the eastern, southern and western area with talus slopes situated between the rock faces and the SRG itself (Fig. 1a). The catchment is drained by one large spring (hereinafter referred to as SEQ) located directly at the front of the SRG (Fig. 1a).

The SEQ is a spring belonging to the official spring network of the Hydrographic Service of Styria (HZB No. 396762). The stage is continuously monitored at a rectangular notch weir situated about 40 m downstream of the spring. Stage data have been available since July 2002 and are recorded in hourly intervals. Individual discharge measurements (using the salt dilution method) are used to relate the monitored stage to the actual discharge. Precipitation is continuously monitored at an automatic weather station located directly at the SRG at 1,822 m a.s.l.

(AWS, see Fig. 1a). Precipitation data have been available since November 2011, recorded in hourly time steps.

Figure. 1 Field site impression of the Schöneben rock glacier (*SRG*) catchment. **a** Overview of the investigation area with the *SRG* highlighted in *blue* and the hydrological catchment delineated with a *red-dashed line*; *AWS* automatic weather station; *SEQ* rock glacier spring; *Geierhaupt* is the highest peak in the surroundings (2,417 m a.s.l.); **b** location of the *SRG* within Austria and Styria; **c** rectangular notch weir situated about 40 m downstream of *SEQ*

Model setup

The groundwater model is implemented in MODFLOW (McDonald and Harbaugh 1988) using the NWT solver (Niswonger et al. 2011; Hunt and Feinstein 2012), which employs a Newton–Raphson solution with improved handling of dry cells. This is an important feature as the topography of the investigation area is showing a high relief, frequently causing the upper cells to fall dry. The application of other solvers, for example the PCG2 solver combined with a rewetting package, may result in an unstable model that often fails to converge.

The model domain covers an area of 225,300 m^2 and is set up with uniform cell sizes of 10 m × 10 m in 66 columns and 80 rows. As there is no actual boundary between the SRG and the talus slopes but rather a smooth transition, the SRG and the talus slopes are assumed to form one merged aquifer system (z1 in Fig. 2). The boundaries of the numerical model are therefore chosen to be beyond the actual boundaries of the SRG to include the adjacent talus slopes to the east and south. The boundary conditions (except for the spring and the top of the active cells) are set to no-flow boundaries as the groundwater flow through the bedrock is assumed to be negligible. Surface elevation data are adopted from a downsampled digital elevation model (DEM) with a resolution of 10 m (based on the chosen MODFLOW cell sizes) based on airborne laser scan data (ALS) with a resolution of 1 m provided by the GIS Service of the Federal State Government of Styria (GIS Steiermark). Elevation data of the underlying bedrock are based on a digital elevation model resulting from geophysical surveys on the SRG (Fig. 2a) comprising seismic refraction investigations along three profiles (Winkler et al. 2016a) and ground penetrating radar (GPR) investigations along eight profiles (two additionally available profiles are not used herein as one of them lies outside the model area and the other is discarded due to poor data quality; see Winkler et al. 2016b). Interestingly, the results of the geophysical surveys show that the thickness of the SRG is increasing towards the southeast at the transition to the talus slopes. As the geophysical investigations did not capture the full extent of the aquifer in this part, the aquifer thickness in this area is subject to a high degree of uncertainty; therefore, two approaches were applied to determine the geometry of the aquifer base. In a first approach, the slopes of the rock faces located above the talus slopes are extrapolated below the surface and intersected with the elevation of the bedrock at the last point of the geophysical investigation profile. Including these extended profiles, the depth of the bedrock is interpolated using kriging with the software Surfer (Fig. 2b). The alternative approach uses a linear extrapolation of the geophysical profiles, i.e. the depth is assumed to decrease linearly from the last point of the profile towards the upper end of the debris slope (Fig. 2c).

It is assumed that there is no significant groundwater flow between the SRG catchment and its adjacent catchments because of the compact gneissic geology and the steep relief; moreover, groundwater storage is assumed to occur only where unconsolidated rocks are present (scree and, especially, in the relict rock glacier itself). Bare rocks and steep cliffs are assumed to have negligible storage capacities; additionally, the potential storage in the bedrock is assumed to be of minor importance as no considerable weathering zone is observable.

The recharge to the rock glacier aquifer is determined based on precipitation data from the automatic weather station located on the SRG (AWS in Fig. 1a) and an estimated evapotranspiration. The actual evapotranspiration is computed using an estimate of potential evapotranspiration based on Thornthwaite (1948) as input for a soil water balance model at a daily time step as described by, e.g., Peters et al. (2005).

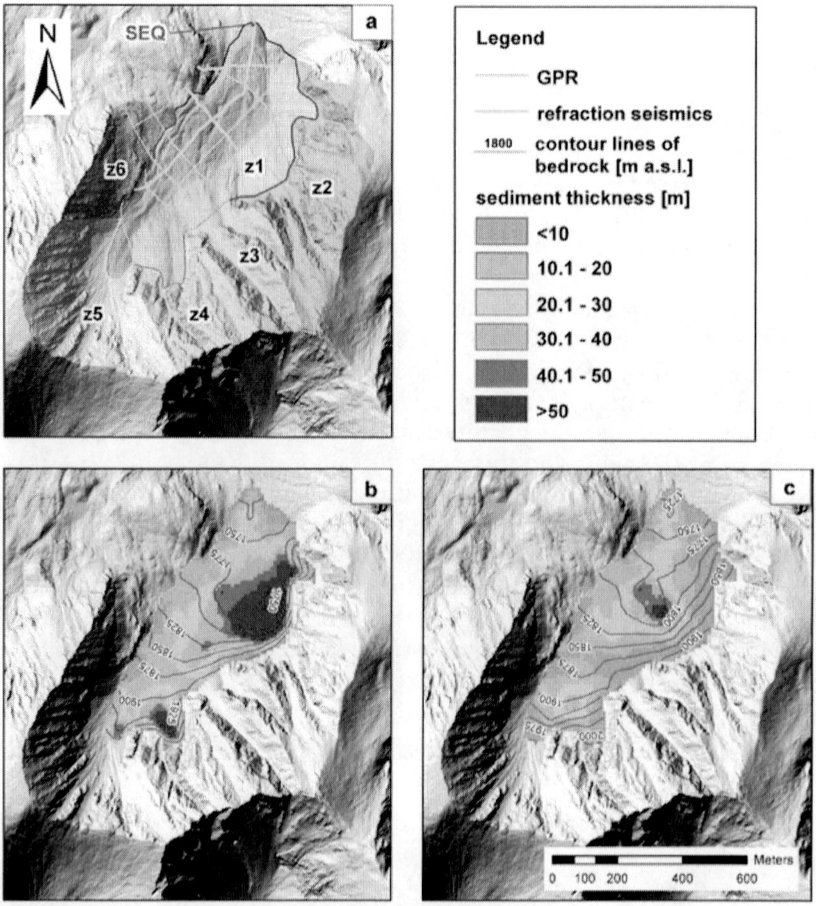

Figure. 2 SRG catchment with model area. **a***Cell grid* and recharge zones (*z1–z6*) with corresponding input cells (*color-coded*); delineation of the actual SRG within the model area highlighted as *blue polygon* (see Fig. 1); profiles of geophysical investigations *blue lines* and constant-head cell at the spring (*SEQ*); **b***Color-coded* sediment thickness; *contour lines* indicate the aquifer base topography based on geophysical data and extrapolated rock faces; **c***Color-coded* sediment thickness; *contour lines* indicate an alternative aquifer base topography based on geophysical data and linearly decreasing depth towards the boundary

A relatively low value of 20 mm is used for the soil moisture storage at field capacity because the surface of the investigation area is mostly covered by coarse rock debris; moreover, the application of a simple lumped-parameter model indicated a minor role of the soil moisture accounting store (Wagner et al. 2016). The actual evapotranspiration is assumed to be equal to potential evapotranspiration if the soil moisture storage amounts at least to 70 % (14 mm) of the field capacity; below this value actual evapotranspiration is linearly decreased towards zero. Recharge occurs if the net inflow (precipitation minus actual evapotranspiration) to the soil moisture storage causes an exceedance of field capacity (time series is shown in Fig. 3).

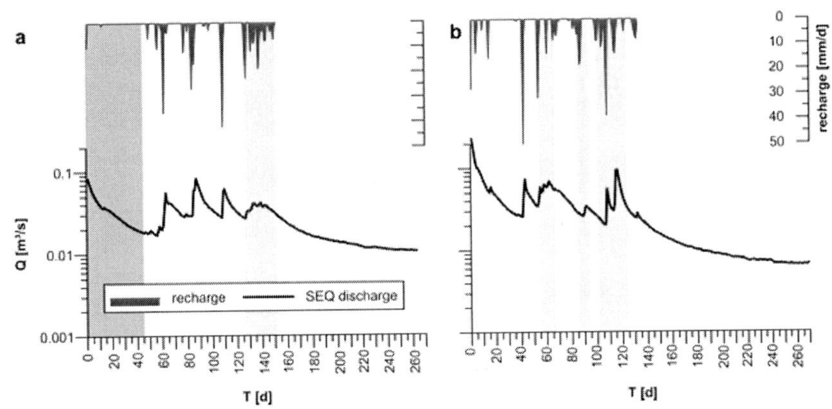

Figure. 3 *Recharge* and *discharge* (*Q*) time series of **a** the calibration period and **b** the verification period. The *grey area* marks the warm-up period. The *light blue areas* mark time periods with snow influence. *T* time

The resulting direct recharge (i.e., precipitation falling directly onto the surface of z1 (Fig. 2a), 31 % of the total catchment area) is distributed homogeneously across the model. The recharge from the surrounding rock faces of the hydrological catchment primarily consists of surface runoff and is added to the recharge of the peripheral cells, proportional to the area percentage of the respective subcatchment (Fig. 2a). The areal percentages are 19, 10, 11, 20 and 9 % for the recharge zones z2, z3, z4, z5, and z6, respectively.

The parameter estimation software PEST (Doherty 2013) is employed for automatic calibration of the groundwater models. The hydraulic conductivity and the specific yield of specified groups of cells are treated as calibration parameters. According to the assumed plausibility of the parameter values and (in the case of the upper limit of hydraulic conductivity) to avoid model instabilities, parameter ranges between 1×10^{-2} and 1×10^{-7} m/s as well as 0.1 and 0.3 were defined for hydraulic conductivity and specific yield, respectively. The specific storage is defined as a constant value of 1×10^{-4} 1/m. As there are no boreholes or other information about groundwater levels, the model has to be calibrated solely to the discharge of the spring (SEQ). During calibration the discharge is weighted inversely proportional to the observed values (Doherty 2013) in order to compensate for the high variability of discharge (Winkler et al. 2016a). The simulation is restricted to time periods when melting snow can be neglected in order to avoid the uncertainties involved in modeling the snow melt; therefore, the model is calibrated using the hydrograph of the SEQ in the time period from 25 June 2013 to 14 March 2014 (263 days; see Fig. 3a). This year has been chosen because it has three distinct recharge events following a long dry period, whereas the hydrograph recession of most other years is interrupted by several smaller recharge events which complicate the interpretation. The transient simulation comprises 151 stress periods and a total of 8,321 time steps with a steady-state water table as initial condition. The stationary boundary conditions are defined as constant recharge (distributed on the six recharge zones z1–z6; Fig. 2). The total recharge rate is thereby specified to be equal to the measured discharge at the beginning of the simulation period. In the time period from day 127 to 150, minor snowfall events occurred according to observations from an automatic digital camera monitoring the area. Because the precipitation was stored temporarily in the snow cover, the recharge occurred delayed (compared to the precipitation records at the automatic weather station) when the snow melted in the following days. As this delay is not considered in the recharge model, the time period is given only 10 % weight in the calibration. Moreover, the first 45 days of the simulation are regarded as a warm-up period (potentially affected by the initial condition) and the discharge data during that time are not used at all for calibration. Special attention is given to the accurate simulation of the winter base flow, as this time period remains unaffected by recharge events, and therefore groundwater flow components other than the base flow can be neglected; thus, the weight in the calibration during this time period is multiplied by a factor of 10. The goodness-of-fit of the model results is analyzed using the root-mean-square error (RMSE), the weighted root-mean-square error (RMSEw; see Table 1) and visual inspection of the simulated versus the observed discharge time series (neglecting the warm-up period). The model is verified by simulating the time period of 21 July 2012 to 15 April 2013 (269 days; see Fig. 3b). This time period also exhibits a winter base flow without intermediate recharge events (see Winkler et al. 2016a), and pictures of the automatic digital camera are available to identify snowfall events (Wagner et al. 2016).

Table 1 Estimated hydraulic parameters of the investigated scenarios and goodness-of-fit measures for the calibration period

Scenario	K_u [m/s]	K_m [m/s]	K_l [m/s]	K_c [m/s]	SY_u [–]	SY_m [–]	SY_l [–]	SY_c [–]	RMSE [m³/s]	RMSEw [m³/s]
i	3.67×10^{-4}	–	–	–	0.30	–	–	–	1.41×10^{-2}	8.70×10^{-3}
ii	5.33×10^{-3}	–	4.93×10^{-5}	–	0.10	–	0.30	–	1.40×10^{-2}	4.01×10^{-3}
iii	1.00×10^{-2}	–	1.22×10^{-5}	–	0.21	–	0.10	–	1.78×10^{-2}	2.46×10^{-3}
iv	8.00×10^{-3}	8.00×10^{-7}	1.70×10^{-4}	–	0.10	0.30	0.30	–	1.38×10^{-2}	3.17×10^{-3}
v	2.72×10^{-5}	–	–	1.00×10^{-2}	0.10	–	–	0.10	9.32×10^{-3}	5.53×10^{-3}
vi	9.67×10^{-5}	–	–	1.00×10^{-2}	0.17	–	–	0.10	8.78×10^{-3}	3.87×10^{-3}
vii	8.61×10^{-3}	–	3.54×10^{-5}	2.13×10^{-3}	0.10	–	0.30	0.19	1.44×10^{-2}	3.08×10^{-3}
viii	2.61×10^{-3}	–	2.60×10^{-6}	2.01×10^{-3}	0.22	–	0.30	0.30	1.14×10^{-2}	3.08×10^{-3}
ii-a	2.02×10^{-3}	–	7.27×10^{-6}	–	0.30	–	0.30	–	1.33×10^{-2}	4.49×10^{-3}
vi-a	7.18×10^{-7}	–	–	1.00×10^{-2}	0.17	–	–	0.10	1.14×10^{-2}	2.66×10^{-3}
viii-a	3.97×10^{-3}	–	7.29×10^{-6}	1.59×10^{-3}	0.30	–	0.30	0.30	1.30×10^{-2}	4.00×10^{-3}

K hydraulic conductivity, *SY* specific yield, subscripts: *u* upper-, *m* middle-, *l* lower layer; *c* channel; *RMSE* unweighted root-mean-square error; *RMSEw* weighted root-mean-square error

In order to investigate the effects of different kinds of internal structures of the SRG and, moreover, to verify the current conceptual understanding of groundwater flow processes in relict rock glaciers, several scenarios (i–viii) with increasing complexity are simulated and compared to each other. For these eight scenarios, the aquifer base topography with extrapolated rock faces (Fig. 2b) is applied. The model setups of the investigated scenarios in this research can be grouped in laterally homogeneous (i–iv) and heterogeneous (v–viii) models which are presented in Fig. 4. The laterally homogeneous scenarios contain models with one (i), two (ii and iii) or three (iv) layers with different hydraulic parameters for each layer. The thickness of the lower layer in ii and iii covers 20 and 80 % of the total aquifer thickness at each cell, respectively. In scenario iv the layered structure of the SRG is further refined by subdividing the lower layer of ii into two layers, each covering 10 % of the total aquifer thickness. A thin lower layer is in accordance with the current conceptual understanding of the relict SRG (Winkler et al. 2016a). Note that the upper two layers presented in Fig. 9 of Winkler et al. (2016a) are not further distinguished in the numerical models as these two layers are, compared to the lower layer, rather similar and highly conductive. Results of seismic refraction surveys showed a maximum total thickness of the rock glacier of more than 50 m (Fig. 2). However, the saturated zone was not detectable because a "blind zone" with a thickness of 19 m is possible with the applied setup of the geophysical survey, meaning that a saturated layer with a thickness of up to 19 m could have been missed in the survey (Winkler et al. 2016a and b).

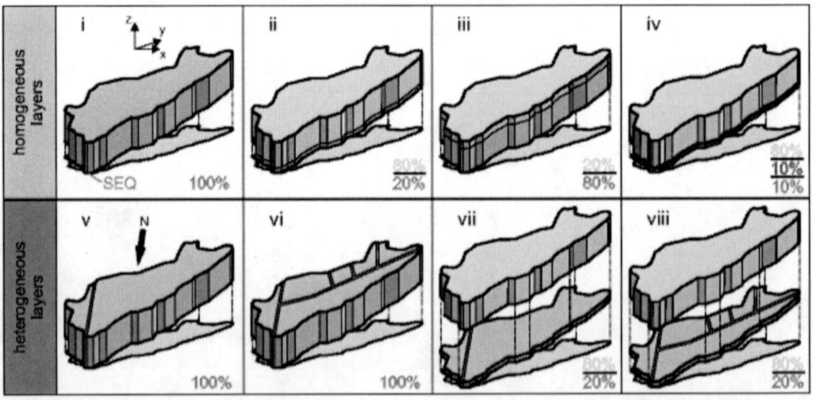

Figure. 4 Three-dimensional sketches of the investigated internal structures of the model. *Colors* indicate zones with uniform hydraulic parameters within one scenario. *Percentages* indicate the thickness of layers. *Red bars* indicate the outflow cell (*SEQ*). Note that the actual surface and aquifer base topography is not represented here for more clarity of the sketch; color codings e.g., *blue colors* in i and v do not necessarily represent similar hydraulic parameters

In the laterally heterogeneous scenarios, a zone with higher hydraulic conductivity is introduced to models with one (v, vi) or two layers (vii, viii). This represents a washed-out channel (according to field observations/excavations from a nearby relict rock glacier; Untersweg and Proske 1996) or channel network which is lacking fine-grained material and is embedded in a less permeable (fine-grained) matrix. The orientation of the channels is defined to follow the topography of the model base and to connect the spring with one (v, vii) and five (vi, viii) dominant gullies from outside the model area (see Fig. 2). In the two-layered scenarios (vii, viii), the channel and channel network are introduced to the lower layer (model runs with channels in both layers did not change the results). During calibration, the channel network is treated as one zone in order to minimize the number of adjustable parameters. A test run of model vi where the hydraulic parameters of the channels were individually calibrated has shown to have a nearly identical discharge behavior as one where channels were calibrated as a single zone due to a narrow range of calibrated hydraulic conductivity in the individual channels (not shown here). Furthermore, three models (ii-a, vi-a and viii-a) are presented with setups equivalent to models ii, vi and viii, but with an alternatively shaped aquifer base topography (Fig. 2c). These models are aimed to investigate to what degree the discharge behavior of the spring is affected by the morphology of the base both without and with a draining channel network.

RESULTS

The comparisons of the simulated to the measured discharge of the SEQ in the calibration and verification periods are presented in Figs. 5 and 6. The estimated aquifer parameters and the goodness-of-fit measures RMSE and RMSEw are presented in Table 1. Some deviation between simulated and observed discharge is expected in the time period from days 127 to 150 that was less weighted in the calibration, as the recharge was affected by snow fall and snow melt processes that are not considered by the model. As a consequence, the observed discharge dynamics are not well reproduced within this period, and therefore are not further discussed; nevertheless, the following winter base flow recession appears to be largely unaffected by these short-term effects. Likewise, several snowfall and melting events occurred during the time period used for verification between days 50 and 125 (Figs. 5 and 6), resulting in delayed responses of SEQ and/or overestimated recharge as precipitation at least was partially stored in the snow cover.

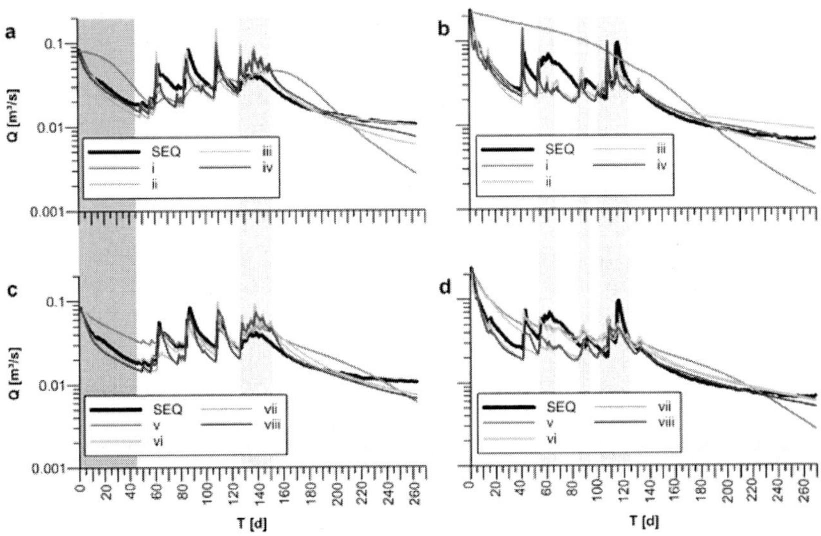

Figure. 5 Hydrographs of the investigated model scenarios compared to the observed data of *SEQ*. The *grey area* marks the warm-up period. The *light blue areas* mark time periods with snow influence. **a–b** Shows the scenarios with homogeneous layers (models i–iv in Fig. 4) for the calibration (25 June 2013 to 14 March 2014) and verification (21 July 2012 to 15 April 2013) time period, respectively. **c–d** Shows the scenarios with heterogeneous layers for the calibration and verification time period, respectively (models v–viii in Fig. 4)

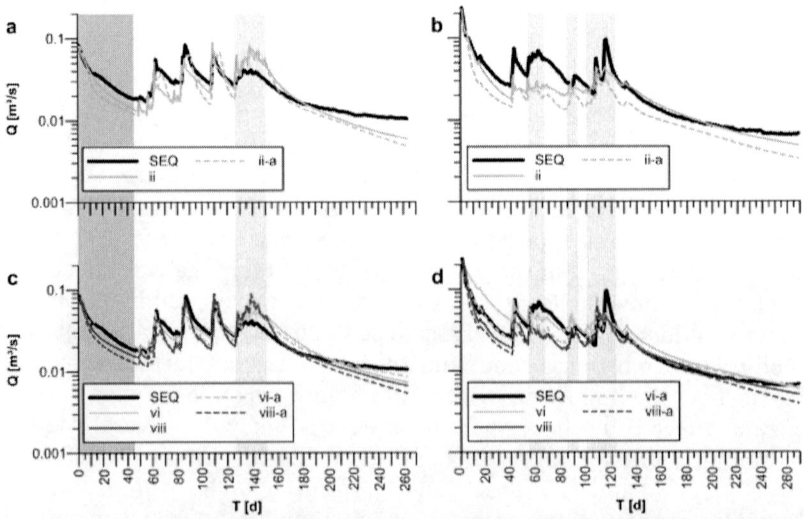

Figure. 6 Hydrographs of the investigated model scenarios compared to the observed data of *SEQ*. The *grey area* marks the warm-up period. The *light blue areas* mark time periods with snow influence. **a–b** Shows a comparison of laterally homogeneous layered scenarios with differently shaped topography of the model base for the calibration (25 June 2013 to 14 March 2014) and verification (21 July 2012 to 15 April 2013) time period, respectively (*solid line* refers extrapolated rock faces; see Fig. 2b); *dashed line* is the linearly decreasing depth towards boundary (see Fig. 2c). **c–d** Shows a comparison of two laterally heterogeneous layered scenarios with differently shaped topography of the model base for the calibration and verification time period, respectively (*solid lines* are extrapolated rock faces; see Fig. 2b); *dashed lines* represent linearly decreasing depth towards boundary (see Fig. 2c)

Using the aquifer base topography of Fig. 2b, the results of the laterally homogeneous scenarios (Fig. 5a,b) show that the single-layered scenario i visually has a very poor fit. This is supported by the goodness-of-fit measures presented in Table 1. This scenario obviously fails to reproduce the discharge behavior of the SEQ, which is characterized by sharp peaks after recharge events and a slowly decreasing base flow. The introduction of a second homogeneous layer in scenarios ii and iii greatly improves the model fit; nevertheless, it can be seen that in ii the short-term recessions after recharge events as well as the winter base flow recessions are too fast and that the absolute value of discharge is too low between the recharge events and during the winter base flow. Scenario iii shows that with a thick lower layer, the absolute value of discharge between the recharge events is closer to the observed values. The thick lower layer also leads to higher hydraulic heads and an improved fit of the winter base flow recession in the calibration (Fig. 5a, RMSEw in Table 1) but overestimates the base flow in the verification time period (Fig. 5b). The introduction of a third layer in iv leads to similar discharge dynamics compared to ii, with similar discharge peaks but a slightly slower base flow recession. In general, the estimated aquifer parameters show a high hydraulic conductivity for the upper layer and a lower hydraulic conductivity in the lower layer (Table 1) with the single-layered scenario i showing intermediate values. Interestingly, calibration of iv results in a very low conductivity middle layer that acts as a barrier between the higher conductivity upper and lower layers.

It can be seen in Fig. 5c,d that the introduction of a draining channel to a single-layered model in v is not sufficient to represent the fast flow component of the SRG (see also Table 1). The discharge recession is generally too slow, resulting in overestimated discharge during base flow. Interestingly, the discharge recession during the winter base flow shows a further decrease after approximately 200 days. At that time the upper part of the aquifer has fallen dry and the drying front has advanced to the area where a relatively flat basin is present in the east (see Fig. 2b). The model fit improves by extending the channel to a channel network in scenario vi; however, the simulated discharge also shows deviations to the measured discharge of the SEQ. While the peaks of the major recharge events in the calibration period are close to the observed data, the discharge response to small recharge events appears too sharp and overestimated; furthermore, similar to most of the aforementioned scenarios, the long-term winter base flow is underestimated, i.e., the simulated discharge recession is too fast. In the verification period, the simulated winter base flow is close to that of the SEQ, but the discharge dynamics of the preceding time period are not well matched. Scenarios vii and viii represent two-layered versions of v and vi, respectively, in which the channel and the channel network is only present in the lower layer (see Fig. 4). The hydrograph of vii is similar to that of ii, indicating that the upper layer is dominating the quick flow component rather than the channel. The channel network of viii obviously has a larger influence than the single channel in vii and results in a quickly responding hydrograph but with broad peaks (resulting in a slightly better RMSE, but identical RMSEw, Table 1). Moreover, the base flow between the recharge events and the winter base-flow recession are similar to the layered scenarios ii, iii and vii.

Examining the results of the scenarios using the alternatively shaped model base (Fig. 2c) shows that the topography of the base has some influence on the discharge behavior of the homogeneous layered model. As can be seen during the calibration (Fig. 6a), ii-a shows double peaks at the major recharge events. In the verification time series (Fig. 6b) the double peaks cannot be observed and the discharge dynamics of ii-a appear more realistic than ii, although the absolute discharge is underestimated. Using the alternatively shaped base (Fig. 2c) results in a winter recession approximately following an exponential decrease (Fig. 6a,b). The heterogeneous layered scenarios vi-a and viii-a visually only slightly differ from vi and viii, respectively (Fig. 6c,d), but yield higher RMSE values (indicating an inferior fit) and different parameter estimates (Table 1). While the matrix of vi-a has a lower hydraulic conductivity compared to vi, the matrix of the lower layer of viii-a has a higher hydraulic conductivity and specific yield compared to viii.

DISCUSSION

The model results indicate that this aquifer has at least two domains with different aquifer properties. This finding is in good agreement with the outcomes of Winkler et al. (2016a), who used spring data as well as natural and artificial tracers to show that the relict Schöneben rock glacier is a heterogeneous aquifer with an assumed layered structure. Importantly, this investigation indicates that both scenarios, one with a layered internal structure and the other with an internal structure with draining channels, can simulate the fast and slow groundwater flow components that were shown to exist in the previous studies (Wagner et al. 2016; Winkler et al. 2016a). Moreover, the numerical modeling exercise supports the existence of a rather thin lower layer with low hydraulic conductivity, which might include channels with a high hydraulic conductivity; however, the findings cannot give a clear preference to one of the investigated internal structures.

If both a layered structure and channels exist, the discharge behavior is dependent on the extent of the channel network. With a single channel (vii), the discharge is dominated by the layered structure, and the channel only plays a subordinate role (Fig. 5c,d). With a network of channels in the lower layer (viii) the influence of the channels increases. Scenarios vii and viii (Fig. 4) illustrate that both types of heterogeneity influence the discharge behavior in a different way. The different extent of the draining channels mainly affects the fit of the discharge peaks and the short-term runoff after recharge events. In contrast, the base flow between the recharge events and the winter base flow is similar in both scenarios and also similar to the corresponding scenario without channels; therefore, the base flow is dominated by the layered structure rather than by the draining channels.

According to the conceptual model of the relict rock glacier of Winkler et al. (2016a) and drill core observations of Krainer et al. (2015), the basal layer might represent morainic remnants. As this layer was not detectable in the geophysical investigations at the SRG, there is no information about the lateral distribution and extension of this layer. Excavations at the neighboring Hochreichart Rock Glacier showed that there is a washed-out zone in the proximity of the spring that is lacking fine-grained material (Untersweg and Proske 1996) and therefore has a high hydraulic conductivity. Thus, it might be reasonable to assume that this layer is not homogeneously distributed across the aquifer. Assuming slow development processes of a talus-derived rock glacier, preexisting water flow paths or newly evolving ones will probably prevent sedimentation of fine-grained sediments or wash them out, therefore leaving a channel network of unknown extent between patchily distributed sediment accumulations of lower hydraulic conductivity. Graham et al. (2010) showed at excavations of a hillslope watershed that a network of preferential flow paths occurred at the soil-bedrock boundary, which was controlled by the bedrock topography. Scenario vi shows that with such a channel network even a single-layered model can reproduce the discharge dynamics of the SEQ reasonably well, and the fit of the simulated hydrograph could be even further optimized by adjusting the number and extent of the channels. Yet, this would suggest a matrix hydraulic conductivity of about 10^{-4}–10^{-5} m/s, which is in discrepancy with the geophysical investigations, suggesting coarse-grained sediments based on seismic velocities. Moreover, fine-grained sediments, for instance from the development, evolution, and movement of the rock glacier itself, are likely to exist as a potential source of low conductive zones (e.g., Zurawek 2002; Hausmann et al. 2012).

As a saturated zone needs to exist within the rock glacier to provide groundwater (especially base flow) for the spring, the thickness of the saturated zone is estimated to be around 10–15 m based on recession analysis but seems to be limited to a maximum thickness of 19 m based on refraction seismic results (see Fig. 9 in Winkler et al. 2016a). However, as can be seen in Fig. 5a, model setups containing thin lower layers lead to an underestimation of the base flow between the major recharge events as well as during the winter base flow recession in the calibration period. Keeping in mind that the specific yield of the thin lower layers is always calibrated to 0.3, which was specified as the upper limit of realistic values (see Table 1), this result indicates that the storage of a thin lower layer is not sufficient to reproduce the observed discharge during base flow conditions, which might be related to the simplified aquifer base topography of the model.

The comparison of the scenarios with differently shaped aquifer bases in Fig. 6 shows that even though a large part of the aquifer is identical (see Fig. 2), the different aquifer base topographies and the resulting differences in aquifer thickness cause changes of the discharge dynamics or the estimated aquifer parameters. Because of the higher sediment thickness, scenario ii shows a slower recession and higher base flow relative to ii-a (Fig. 6a,b). Similarly, the lower thickness of the lower layer of vi-a compared to vi (Fig. 2) results in a slightly faster recession between the major recharge events. However, due to the higher number of adjustable parameters in the more complex scenarios (viii and viii-a), the effect of the different sediment thickness can be compensated by changes of the hydraulic parameters; hence, the role of the base topography appears to be more important in the less complex scenarios.

According to these results, for the SRG the modeled aquifer base topography based on the extrapolated rock faces in the eastern part of the model area seems more likely, because the higher thicknesses result in a higher storage capacity, which is preferable to simulate the observed base flow. The estimated hydraulic conductivity of the SRG (Table 1) is in good agreement with findings of Pauritsch et al. (2015) and Winkler et al. (2016a). Nevertheless, the winter base flow cannot be reproduced correctly in the calibration period with either of the applied aquifer bases. Remarkably, the base flow recession in the verification period is better matched by the scenarios containing the thin base layer, but even here the models tend to underestimate base flow at the late stage. The observed winter base flow clearly deviates from a single exponential recession. In fact, from 10 days onward the recession is found to follow a power law (Winkler et al. 2016a). The simulated recession curves also deviate from an exponential recession (which would be a straight line in the semi-log plots shown in Figs. 5 and 6), but not as much as the observed discharge. Conceptually, a fractal size distribution of storage elements has been proposed to explain the power law recession of spring hydrographs (Hergarten and Birk 2007), which suggests that the assumption of homogeneous layers might be oversimplified and discontinuities should be considered. In particular, the assumption of heterogeneous layers with spatially varying thickness could provide a larger storage compared with the model setups considered here and still be consistent with the results from the geophysical investigations that covered only parts of the aquifer.

Another explanation for the discrepancies of the simulated and observed base flow might be the influence of the vadose zone, which was not considered in the models; however, because of the coarse blocky debris and its low retention capabilities, precipitation can instantaneously infiltrate through the top layer of the rock glacier. Precipitation is therefore assumed to quickly pass through the unsaturated zone to become groundwater recharge. This is supported by the findings of the tracer test conducted in 2012 (Winkler et al. 2016a), where the tracer, injected at a distance of 350 m from the spring at the surface of the rock glacier, was detected after 2–3 h at the spring. In addition, Wagner et al. (2016) showed that parameters of a simple lumped-parameter model indicate a rather fast transfer of seepage water towards the saturated zone (time parameter x_4 in Wagner et al. 2016). Furthermore, it was assumed in the model that the contribution of the mountain block recharge, i.e., groundwater flow out of the bedrock of the contributing catchment into the rock glacier (e.g., Welch et al. 2012), is negligible in such a crystalline catchment with an insignificant weathering zone. Assuming an additional constant inflow from the bedrock (mountain block recharge), a slightly better model fit to the winter base flow is conceivable. However, as this would have little effect on the discharge dynamics and as there is no data or field evidence of such a flow component, such a more complex model was not considered herein.

More plausible explanations for the discrepancies of the simulated and measured discharge might be unrealistic initial conditions and/or underestimated recharge. The influence of the initial condition could be further reduced by extending the simulation to perennial time periods, but this would require the introduction of a snow model, thus leading to more adjustable parameters. An underestimation of the recharge might result from the general uncertainties in the measurement of precipitation in alpine terrains. Wagner et al. (2016) demonstrated that a lumped-parameter model is able to simulate the perennial discharge behavior of the SRG reasonably well, but only when additional water is considered (water exchange coefficient x_2 in Wagner et al. 2016) to compensate for water balance deficits. In the numerical groundwater model, such water balance deficits were not compensated and thus are likely to account at least partly for the deviation of the simulated from the observed discharge. As a consequence, the most complex model structures considered here do not perform much better than the less complex models (apart from the most simple single-layer scenario). A simple lumped-parameter model as employed by Wagner et al. (2016) therefore might be an appropriate choice for predictive hydrological modeling. In this study, however, the modeling is aimed at supporting the aquifer characterization, which requires a process-based groundwater model.

In the absence of spatial information about hydraulic heads, the calibration of the groundwater model solely relies on spring discharge and therefore leads to ambiguous results, in particular with regard to the distribution of fine-grained sediments and preferential flow paths. Nevertheless, the results of the present modeling study are helpful for the design of future investigation and monitoring campaigns. As the installation of piezometers is technically difficult and expensive, the use of additional geophysics seems more beneficial. Based on the outcome of the modeling scenarios, more precise information about the distribution and thickness of the fine-grained sediments, which are assumed to exist as a thin (<19 m) layer at the base of the aquifer, is paramount. For this task, additional refraction seismic investigations with adjusted setups (narrower geophone distance) seem to be particularly promising. In addition, an extension of the previously conducted investigation towards southeast would be of great help, as this area appears to provide the highest thickness of the sediments and, thus, potentially high storage that may account for the observed recession base flow.

There are several other methods that could be envisaged for future investigations, but each of them has some drawbacks that need to be considered as well. Additional ground-penetrating radar measurements (GPR) can be applied considering varying frequencies (especially lower ones) to improve the existing data. Similar results related to challenging GPR data were shown by Merz et al. (2015) who addressed the high level of noise (especially in areas with low ice contents) to interferences caused by the boulders of the surface layer and shallow heterogeneities; however, they have shown that airborne ground-penetrating radar (antenna mounted on a helicopter) greatly improves the quality of the data. Also other airborne geophysical methods such as frequency-domain or time-domain electromagnetics might help to gain further insights; nevertheless, all airborne methods prefer constant elevation above ground (De Barros and Guimaraes 2016), which can be challenging in alpine terrain. Electrical resistivity tomography might be a cost-effective solution, but the coarse blocky material at the surface of relict rock glaciers results in a weak electrode coupling to the ground. Furthermore, the extended, air filled voids, and the lack of mineral soils or ice lead to weak electrical contacts between the individual blocks which makes the application of this method rather challenging (Hilbich et al. 2009; Kellerer-Pirklbauer et al. 2014); however, capacitively coupled resistivity methods, which use an alternating current across a transmitter-earth capacitor, can be used as an alternative to overcome these (Hauck 2013).

Thus, characterizing the hydrogeological properties and functioning of rock glaciers remains a challenging but important task. Rock glaciers represent significant aquifers in alpine catchments and are the source of many important water basins (e.g., Wagner et al. 2016). Furthermore, high heavy metal concentrations within the melt water of an active rock glacier recently reported by Krainer et al. (2015) illustrate the vulnerability of these aquifers to contamination. In addition, rock glaciers are potentially affected by impacts of the future global warming. Assessing groundwater flow and transport processes in a changing environment is particularly challenging. While purely empirical models are considered inadequate for accomplishing this task (e.g., Rehrl and Birk 2010), further progress in the hydrogeological characterization of rock glaciers is needed to reduce the ambiguity of the calibration of process-based distributed models. The afore-mentioned findings provide a guideline towards the further investigation of the rock glacier considered here that can likely be transferred to similar settings elsewhere.

CONCLUSIONS

A number of model scenarios representing the relict SRG have been elaborated using a numerical groundwater flow model. Several scenarios differing in the internal structure and the topography of the aquifer base were considered to indicate the major features that control the discharge behavior of the rock glacier spring. The results of this modeling study converge to a consistent conceptual understanding of the hydrological functioning of the rock glacier. It has been found that two aquifer components are necessary to account for the observed discharge behavior: a highly conductive layer and/or channel network controlling the fast and flashy spring responses to recharge event, as opposed to less conductive sediment accumulations sustaining the long-term base flow. Despite the ambiguity of the specific implementation, the parameter estimates obtained from the model calibration provide order-of-magnitude estimates of the hydraulic properties of these two components. The hydraulic conductivity of the highly conductive component is found to be on the order of 10^{-2}–10^{-3} m/s, whereas the hydraulic conductivity of the low-conductivity elements is about three orders-of-magnitude less. The specific yield of the high-conductivity elements varied between the different model setups. Assuming a low-conductivity layer at the aquifer base covering 20 % of the total thickness, the specific yield of this layer is consistently 0.3, which corresponds to the upper limit assumed to be plausible. As the long-term base flow is underestimated by most of the model scenarios, the specific yield or the thickness of this layer might be even larger than assumed here.

The topography of the aquifer base is found to have an impact on the discharge behavior particularly when a simple internal structure is considered. If more complex aquifer structures with a high number of adjustable parameters are employed, changes in the topography can be compensated by the adjustment of other parameters. This ambiguity can only be resolved if additional data are included as constraint on the aquifer properties and internal structure (especially further geophysical investigations) or as additional calibration target (e.g., hydraulic heads).

The Schöneben Rock Glacier in the Niedere Tauern Range, Austria, is one example of a relict rock glacier and its hydrological functioning might be representative for other relict rock glaciers as well. Thus, the insights obtained herein will have implications for intact rock glaciers in the course of climate warming and future investigations of contaminant transport in rock glaciers. As such, studying the discharge behavior of relict rock glaciers (and other alpine debris accumulations) contributes largely to a better understanding of the complex hydrological processes in alpine regions and of their contribution to the fragile ecological systems in such environments.

ACKNOWLEDGMENTS

Open access funding provided by University of Graz. This study was funded by the European Regional Development Fund (ERDF) and the Federal State Government of Styria. The authors are grateful to the Hydrographic Service of Styria for providing the discharge data. The digital elevation models and the topographic maps were provided by the GIS Service of the Federal State Government of Styria (GIS Steiermark). We acknowledge the constructive suggestions by Majdi Mansour, an anonymous reviewer and the associate editor Alan MacDonald, which greatly improved the report.

REFERENCES

1. Barsch D (1996) Rockglaciers: indicators for the present and former geoecology in high mountain environments. Springer Series in Physical Environment 16, Springer, Heidelberg, Germany
2. Bollmann E, Rieg L, Spross M, Sailer R, Bucher K, Maukisch M, Monreal M, Zischg A, Mair V, Lang K, Stötter, J (2012) Blockgletscherkataster Südtirol: Erstellung und Analyse [Rock glacier registry South Tyrol: creation and analysis]. Innsbrucker Geographische Studien 39, Permafrost in Südtirol, IGS, Innsbruck, Austria, pp 147–171
3. Campbell DH, Clow DW, Ingersoll GP, Mast MA, Spahr NE, Turk JT (1995) Processes controlling the chemistry of two snowmelt-dominated streams in the Rocky Mountains. Water Resour Res 31(11):2811–2821. doi:10.1029/95WR02037
4. Carrera J, Alcolea A, Medina A, Hidalgo J, Slooten LJ (2005) Inverse problem in hydrogeology. Hydrogeol J 13(1):206–222. doi:10.1007/s10040-004-0404-7
5. Clow DW, Schrott L, Webb R, Campbell DH, Torizzo A, Dornblaser M (2003) Ground water occurrence and contributions to streamflow in an Alpine catchment, Colorado Front range. Ground Water 41:937–950. doi:10.1111/j.1745-6584.2003.tb02436.x
6. De Barros CE, Guimaraes SNP (2016) Magnetic airborne survey: geophysical flight. Geosci Instrum Method Data Syst 5:181–192. doi:10.5194/gi-5-181-2016
7. Degenhardt JJ (2009) Development of tongue-shaped and multilobate rock glaciers in alpine environments: interpretations from round penetrating radar surveys. Geomorphology 109(3–4):94–107. doi:10.1016/j.geomorph.2009.02.020
8. Doherty J (2013) PEST: model-independent parameter estimation, user manual. Watermark Numerical Computing, Brisbane, Australia. http://www.pesthomepage.org /downloads.php. Accessed October 2016

9. Eisenlohr L, Bouzelboudjen M, Király L, Rossier Y (1997a) Numerical versus statistical modelling of natural response of a karst hydrogeological system. J Hydrol 202(1–4):244–262. doi:10.1016/S0022-1694(97)00069-3
10. Eisenlohr L, Király L, Bouzelboudjen M, Rossier Y (1997b) Numerical simulation as a tool for checking the interpretation of karst spring hydrographs. J Hydrol 193(1–4):306–315. doi:10.1016/S0022-1694(96)03140-X
11. Graham CB, Woods RA, McDonnell JJ (2010) Hillslope threshold response to rainfall: (1) a field based forensic approach. J Hydrol 393(1–2):65–76. doi:10.1016/j.jhydrol.2009.12.015
12. Haeberli W, Hallet B, Arenson L, Elconin R, Humlum O, Kääb A, Kaufmann V, Ladanyi B, Matsuoka N, Springman S, Mühll DV (2006) Permafrost creep and rock glacier dynamics. Permafr Periglac Process 17:189–214. doi:10.1002/ppp.561
13. Hauck C (2013) New concepts in geophysical surveying and data interpretation for permafrost terrain. Permafr Periglac Process 24:131–137. doi:10.1002/ppp.1774
14. Hausmann H, Krainer K, Brückl E, Ulrich C (2012) Internal structure, ice content and dynamics of Ölgrube and Kaiserberg rock glaciers (Ötztal Alps, Austria) determined from geophysical surveys. Aust J Earth Sci 105(2):12–31
15. Hergarten S, Birk S (2007) A fractal approach to the recession of spring hydrographs. Geophys Res Lett 34, L11401. doi:10.1029/2007GL030097
16. Hilbich C, Marescot L, Hauck C, Loke MH, Mäusbacher R (2009) Applicability of electrical resistivity tomography monitoring to coarse blocky and ice-rich permafrost landforms, Permafr Periglac Proc 20:269–284. doi:10.1002/ppp.652
17. Hood JL, Hayashi M (2015) Characterization of snowmelt flux and groundwater storage in an alpine headwater basin. J Hydrol 521:482–497. doi:10.1016/j.jhydrol.2014.12.041
18. Hughes PD, Gibbard PL, Ehlers J (2013) Timing of glaciation during the last glacial cycle: evaluating the concept of a global 'Last Glacial Maximum' (LGM). Earth Sci Rev 125:171–198. doi:10.1016/j.earscirev.2013.07.003
19. Hunt RJ, Feinstein DT (2012) MODFLOW-NWT: robust handling of dry cells using a Newton formulation of MODFLOW-2005. Ground Water 50:659–663. doi:10.1111/j.1745-6584.2012.00976.x
20. Jansen F, Hergarten S (2006) Rock glacier dynamics: stick–slip motion coupled with hydrology. Geophys Res Lett 33, L10502. doi:10.1029/2006gl026134
21. Kellerer-Pirklbauer A, Lieb KG, Kleinferchner H (2012) A new rock glacier inventory of the Eastern European Alps. Aust J Earth Sci 105(2):78–93
22. Kellerer-Pirklbauer A, Pauritsch M, Morawetz R, Kuehnast B, Schreilechner M, Winkler G (2014) Thickness and internal structure of relict rock glaciers: a challenge for geophysics: examples from two rock glaciers in the Eastern Alps. Geophys Res Abstr 16:EGU2014–EGU12581
23. Kellerer-Pirklbauer A, Pauritsch M, Winkler G (2015) Widespread occurrence of ephemeral funnel hoarfrost and related air ventilation in coarse-grained sediments of a relict rock glacier in the Seckauer Tauern Range, Austria. Geogr Ann A 97(3):453–471. doi:10.1111/geoa.12087
24. Krainer K, Mostler W (2002) Hydrology of active rock glaciers: examples from the Austrian Alps. Arc Antarc Alp Res 34:142–149. doi:10.2307/1552465
25. Krainer K, Ribis M (2012) A rock glacier inventory of the Tyrolean Alps (Austria). Aust J Earth Sci 105(2):32–47
26. Krainer K, Mostler W, Spoetl C (2007) Discharge from active rock glaciers, Austrian Alps: a stable isotope approach. Aust J Earth Sci 100:102–112
27. Krainer K, Bressan D, Dietre B, Haas JN, Hajdas I, Lang K, Mair V, Nickus U, Reidl D, Thies H, Tonidandel D (2015) A 10,300-year-old permafrost core from the active rock glacier Lazaun, southern Ötztal Alps (South Tyrol, northern Italy). Quat Res 83(2):324–335. doi:10.1016/j.yqres.2014.12.005
28. Lauber U, Kotyla P, Morche D, Goldscheider N (2014) Hydrogeology of an Alpine rockfall aquifer system and its role in flood attenuation and maintaining baseflow. Hydrol Earth Syst Sci 18:4437–4452. doi:10.5194/hess-18-4437-2014

29. Leopold M, Williams MW, Caine N, Völkel J, Dethier D (2011) Internal structure of the Green Lake 5 rock glacier, Colorado Front Range, USA. Permafrost Periglac 22:107–119. doi:10.1002/ppp.706
30. Mansour MM, Hughes AG, Robins NS, Ball D, Okoronkwo, C (2012) The role of numerical modelling in understanding groundwater flow in Scottish alluvial aquifers. In: Shepley MG (ed) Groundwater resources modelling: a case study from the UK. Geol Soc London Spec Publ 364:85–98
31. Matthews JA, Wilson P (2015) Improved Schmidt-hammer exposure ages for active and relict pronival ramparts in southern Norway, and their palaeoenvironmental implications. Geomorphology 246:7–21. doi:10.1016/j.geomorph.2015.06.002
32. McDonald MG, Harbaugh AW (1988) A modular three-dimensional finite-difference ground-water flow model. US Geol Surv Tech Water Resour Invest 6-A1, 586 pp
33. Merz K, Maurer H, Buchli T, Horstmeyer H, Green AG, Springman SM (2015) Evaluation of ground-based and helicopter ground-penetrating radar data acquired across an Alpine Rock Glacier. Permafrost Periglac 26:13–27. doi:10.1002/ppp.1836
34. Millar CI, Westfall RD, Delany DL (2013) Thermal and hydrologic attributes of rock glaciers and periglacial talus landforms: Sierra Nevada, California, USA. Quat Int 310:169–180. doi:10.1016/j.quaint.2012.07.019
35. Monnier S, Kinnard C (2015) Internal structure and composition of a Rock Glacier in the Dry Andes, inferred from ground-penetrating radar data and its artefacts. Permafrost Periglac Proc 26(4):335–346. doi:10.1002/ppp.1846
36. Monnier S, Camerlunck C, Rejiba F, Kinnard C, Feuillet T, Dhemaied A (2011) Structure and genesis of the Thabor rock glacier (Northern French Alps) determined from morphological and ground-penetrating radar surveys. Geomorphology 134:269–279. doi:10.1016/j.geomorph.2011.07.004
37. Muir DL, Hayashi M, McClymont AF (2011) Hydrological storage and transmission characteristics of an alpine talus. Hydrol Process 25:2954–2966. doi:10.1002/hyp.8060
38. Niswonger RG, Pandai S, Ibaraki M (2011) MODFLOW-NWT: a Newton formulation for MODFLOW-2005. US Geol Surv Tech Methods 6-A37, 44 pp
39. Pauritsch M, Birk S, Wagner T, Hergarten S, Winkler G (2015) Analytical approximations of discharge recessions for steeply sloping aquifers in alpine catchments. Water Resour Res 51:8729–8740. doi:10.1002/2015WR017749
40. Peters E, Van Lanen HAJ, Torfs PJJF, Bier G (2005) Drought in groundwater: drought distribution and performance indicators. J Hydrol 306(1–4):302–317. doi:10.1016/j.jhydrol.2004.09.014
41. Rehrl C, Birk S (2010) Hydrogeological characterisation and modelling of spring catchments in a changing environment. Aust J Earth Sci 103(2):106–117
42. Roy JW, Hayashi M (2009) Multiple distinct groundwater flow systems of single moraine-talus feature in alpine watershed. J Hydrol 373:139–150. doi:10.1016/j.jhydrol.2009.04.018
43. Schmid MO, Baral P, Gruber S, Shahi S, Shrestha T, Stumm D, Wester P (2015) Assessment of permafrost distribution maps in the Hindu Kush Himalayan region using rock glaciers mapped in Google Earth. Cryosphere 9:2089–2099. doi:10.5194/tc-9-2089-2015
44. Tague C, Grant GE (2009) Groundwater dynamics mediate low-flow response to global warming in snow-dominated alpine regions. Water Resour Res 45, W07421. doi:10.1029/2008wr007179
45. Thornthwaite CW (1948) An approach towards a rational classification of climate. Geogr Rev 38:55–94. doi:10.1097/00010694-194807000-00007
46. Untersweg T, Proske H (1996) Untersuchungen an einem fossilen Blockgletscher im Hochreichhartgebiet (Niedere Tauern, Steiermark) [Investigations at a relict rock glacier in the Hochreichhart area "Niedere Tauern Range, Styria"]. Grazer Schrift Geogr Raumfor 33:201–207

47. Wagner T, Pauritsch M, Winkler G (2016) Impact of relict rock glaciers on spring and stream flow of alpine watersheds: examples of the Niedere Tauern Range, Eastern Alps (Austria). Aust J Earth Sci 109. doi:10.17738/ajes.2016.0006
48. Welch LA, Allen DM, van Meerveld HJ (2012) Deep groundwater contributions to mountain headwater streams and sensitivity to available recharge. Can Water Resour J 37(4):349–371. doi:10.4296/cwrj3703907
49. Winkler G, Wagner T, Pauritsch M, Birk S, Kellerer-Pirkelbauer A, Benischke R, Leis A, Morawetz R, Schreilechner MG, Hergarten S (2016a) Identification and assessment of flow and storage components of the relict Schöneben Rock Glacier, Niedere Tauern Range, Eastern Alps (Austria). Hydrogeol J 24:937–953. doi:10.1007/s10040-015-1348-9
50. Winkler G, Pauritsch M, Wagner T, Kellerer-Pirklbauer A (2016b) Reliktische Blockgletscher als Grundwasserspeicher in alpinen Einzugsgebieten der Niederen Tauern [Relict rock glaciers as groundwater storages in alpine catchment areas in the Niedere Tauern Range]. Berichte zur wasserwirtschaftlichen Planung Steiermark, vol 87. Graz, 134 pp. http://www.wasserwirtschaft.steiermark.at/cms/dokumente/11913323_102332494/6885027d/87.pdf. Accessed 12 September 2016
51. Worthington SRH (2009) Diagnostic hydrogeologic characteristics of a karst aquifer (Kentucky, USA). Hydrogeol J 17:1665–1678. doi:10.1007/s10040-009-0489-0
52. Zurawek R (2002) Internal structure of a relict rock glacier, Ślęża Massif, southwest Poland. Permafr Periglac Process 13:29–42. doi:10.1002/ppp.403

CHAPTER 14

Water Supply Source Evaluation in Unmanaged Aquifer Recharge Zones: The Mezquital Valley (Mexico) Case Study

Antonio Hernández-Espriú [1,*], Claudia Arango-Galván [2], Alfonso Reyes-Pimentel [3], Pedro Martínez-Santos [4], Carlos Pita de la Paz [5], Sergio Macías-Medrano [1], Alberto Arias-Paz [1] and José Agustín Breña-Naranjo[6]

[1] *Hydrogeology Group, Earth Sciences Division, Faculty of Engineering, Universidad Nacional Autónoma de México (UNAM), C.P. 04510 Mexico City, Mexico*

[2] *Department of Geomagnetism and Exploration, Institute of Geophysics, Universidad Nacional Autónoma de México (UNAM), C.P. 04510 Mexico City, Mexico*

[3] *Comisión Nacional de Hidrocarburos, C.P. 03700 Ciudad de México, Mexico*

[4] *Department of Geodynamics, Faculty of Geological Sciences, Universidad Complutense de Madrid, 28040 Madrid, Spain*

[5] *GEOTEM Ingeniería, S.A. de C.V., C.P. 14640 Mexico D.F., Mexico*

[6] *Institute of Engineering, Universidad Nacional Autónoma de México (UNAM), C.P. 04510 Mexico City, Mexico*

ABSTRACT

The Mezquital Valley (MV) hosts the largest unmanaged aquifer recharge scheme in the world. The metropolitan area of Mexico City discharges ~60 m^3/s of raw wastewater into the valley, a substantial share of which infiltrates into the regional aquifer. In this work, we aim to develop a comprehensive approach, adapted from oil and gas reservoir modeling frameworks, to assess water supply sources located downgradient from unmanaged aquifer recharge zones. The methodology is demonstrated through its application to the Mezquital Valley region. Geological, geoelectrical, petrophysical and hydraulic information is combined into a 3D subsurface model and used to evaluate downgradient supply sources. Although hydrogeochemical variables are yet to be assessed, outcomes suggest that the newly-found groundwater sources may provide a long-term solution for water supply. Piezometric analyses based on 25-year records suggest that the MV is close to steady-state conditions. Thus, unmanaged recharge seems to have been regulating the groundwater balance for the last decades. The transition from unmanaged to managed recharge is expected to provide benefits to the MV inhabitants. It will also be likely to generate new uncertainties in relation to aquifer dynamics and downgradient systems.

Keywords: aquifer; groundwater exploration; water supply; well logs; geophysical prospection; TDEM; pumping test; unmanaged recharge; wastewater irrigation; Mezquital Valley

1. INTRODUCTION

Aquifer exploration and water supply evaluation is a key challenge in low- and mid-income countries [1,2,3], where the characterization of potential groundwater sources is still needed and where a spectacular increase in groundwater use has occurred in the past decades [4].

Groundwater exploration is usually most needed in arid regions. This is because scanty rainfall and high evapotranspiration rates tend to make surface sources unreliable. In these areas, aquifers play an essential role in ensuring sustainability, particularly whenever unregulated recharge schemes are involved. This paper focuses on developing a comprehensive and transferable approach to assessing groundwater supply sources. The methodology is illustrated through its application to a semiarid region subject to unmanaged aquifer recharge practices: the Mezquital Valley in central Mexico (Figure 1a). More specifically, we assess a representative sector of the so-called Alfajayucan aquifer (green boundary in Figure 1b,c), which is located downstream to the Mezquital Valley unmanaged region (Figure 1c).

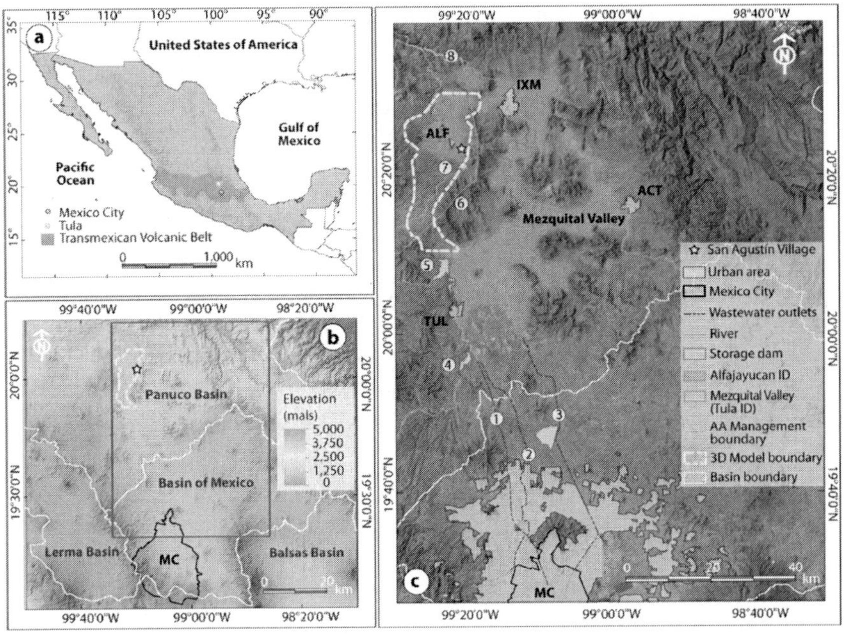

Figure 1. Geographical setting. (**a**) Mexico; (**b**) Mexico City (MC) and hydrologic watersheds; (**c**) study area. For part (**c**), MC: Mexico City, ACT: Actopan, TUL: Tula, IMX: Ixmiquilpan, ALF: Alfajayucan, ID: Irrigation District, AA: Alfajayucan aquifer (the management boundary is shown in green line, parts **b** and **c**), (1) Central Emitter; (2) Western wastewater outlet; (3) Grand Canal; (4) Requena Dam; (5) Endhó Dam; (6) Zinthe Mountain Range; (7) Rojo Gómez Dam; (8) Tula River. Note that the 3D Model boundary (yellow line in parts **b** and **c**) is shown in Section 3.6.

Evaluating groundwater resources in the vicinity of the world's largest unmanaged recharge region poses a major technical challenge. To address it, we develop an inclusive methodology based on geological reservoir modeling techniques. This encompasses all aspects related to stratigraphic, lithological and petrophysical properties of subsurface rocks, and serves the purpose of estimating the spatial distribution of target units and the volume of economic fluids [5]. Available information for integrative modeling combines static and dynamic data at different scales. Static records include geophysical surveys, well logs and geological data [6] whilst dynamic information provides evidence of the interactions between the static model and the fluid over time. This approach delivers an appropriate means to deal with incomplete datasets [7,8] and has been the standard in oil and gas reservoir modeling for at least three decades [9,10].

The aim of reservoir characterization is to understand and identify regional flow units, i.e., hydrofacies, as well as to assess the distribution of relevant properties in the vicinity of a borehole [11]. Static reservoir modeling is frequently coupled with complementary geophysical techniques. This is because drilling data is characterized by high vertical resolution–low horizontal resolution, whereas geophysical soundings present the opposite pattern [8]. In the case of the former, well logs provide a quantitative method to assess near-borehole petrophysical signatures. Within the latter, deep horizons (usually derived from seismic data) deliver the geometric, structural and stratigraphic properties of subsurface units [12,13,14,15,16,17,18]. Moreover, advances in three-dimensional (3D) mapping provide alternative means to developing visual representation of subsurface reservoirs and quantifying some of their key properties [7,19,20,21].

Static modeling frameworks have been successfully used to deal with a variety of practical problems. These include geological CO_2 sequestration and storage [22,23], uncertainty analysis in enhanced oil recovery processes [24], land subsidence related to subsurface gas development [21], correlation between oil field analogues [25] or the understanding of geothermal systems [26], among others.

Mainstream static reservoir modelling techniques, including 3D mapping or well logging, have gained recognition in the field of groundwater exploration in recent years [27,28,29,30,31]. However, integrated approaches (e.g., [32]) such as the one we present are comparatively scarce in the groundwater literature. More specifically, these have seldom been used for the purpose of characterizing groundwater bodies, and their application in the context of unmanaged aquifer recharge is practically non-existent.

A combination of geological, geoelectrical, petrophysical and hydraulic methods is used to gain an understanding of the interaction between lithological factors, hydrogeological properties and the spatial distribution of groundwater. Time domain electromagnetic (TDEM) techniques were used for subsurface modeling. TDEM has been widely used to characterize potential groundwater zones [33,34,35]. The TDEM-resistivity model was correlated with drilling data, and then interpreted jointly with well logs and pumping tests so as to deduce hydrogeological properties and borehole performance.

Within this context, the main contribution of this paper to the literature consists in demonstrating how these techniques may lead to locating and assessing reliable long-term sources of drinking water in unmanaged regions. This research presents additional value by dealing with a rather unique hydrogeological setting in the world.

2. MATERIALS AND METHODS

2.1. Geological and Hydrogeological Setting

The study area corresponds to the Alfajayucan aquifer system (Figure 1b,c), which in turn is located downstream from the Mezquital Valley, central Mexico (Figure 1c). Climate is predominantly semiarid. Rainfall amounts to 450 mm/year, while potential evapotranspiration is 2100 mm/year [36].

Mexico City discharges ~60 m^3/s of raw wastewater without formal treatment into the Mezquital Valley (Figure 1b,c). Wastewater is transported via a complex network that includes the Central Emitter, the Grand Canal and the Western wastewater outlets (Figure 1c). These are all part of a ~90,000 ha irrigation district that straddles across the Tula and Alfajayucan districts [36] (Figure 1c).

Artificial recharge within the Mezquital Valley is controlled by unmanaged wastewater discharge and estimated at 25 m^3/s, that is, over 13 times the natural recharge rate [36]. This has caused the water table to rise since the 1950s. New artesian wells and springs have appeared as a result. Some showcase substantial flowrates. Take for instance the Cerro Colorado spring (~600 L/s) [37].

Overall, the Mezquital Valley is the largest raw wastewater-based agricultural irrigation region in the world [37]. For nearly 110 years, raw wastewater has been used as a source of irrigation water and nutrients for crops. This has allowed local farmers to minimize production costs [38]. Numerous authors deal with the environmental impacts of wastewater irrigation [38], organic micropollutants [39], leaching potential of pharmaceutical components [40], health risk assessments [41] and groundwater pollution [42]. These and other environmental concerns led the Federal Government to build the Atotonilco water treatment plant in 2013 [43]. The plant was designed to treat ~35 m^3/s of wastewater by means of biological and physicochemical processes. Operations are expected to begin in 2016–2017.

The Mezquital Valley provides a unique example of unmanaged aquifer recharge (UAR). UAR is described as the disposal of water that results in unintentional aquifer replenishment [44], in contrast with managed recharge schemes [45,46]. There have been examples of longstanding unmanaged recharge practices in countries such as Australia [47]. More recent ones pertain to Italy [48] or Oman [49]. In other regions of the world UAR has received less attention.

Previous work suggests that the Mezquital Valley might present hydraulic continuity with downgradient permeable formations, such as the volcanic deposits of the Alfajayucan aquifer [50]. Population nuclei located in these areas (San Agustín Village, Ixmiquilpan, Alfajayucan and others, see Figure 1c) are thus likely to be influenced by unregulated activities in terms of water quality and quantity.

Regional geology includes volcanic rocks and calcareous sediments (Figure 2). Lower Cretaceous rocks (El Doctor, Mexcala and Soyatal formations) are made up of limestone with interbedded clay horizons. A dolomite formation caps the sequence. Tectonic activity presents a NW-SE sequence of anticlines and synclines.

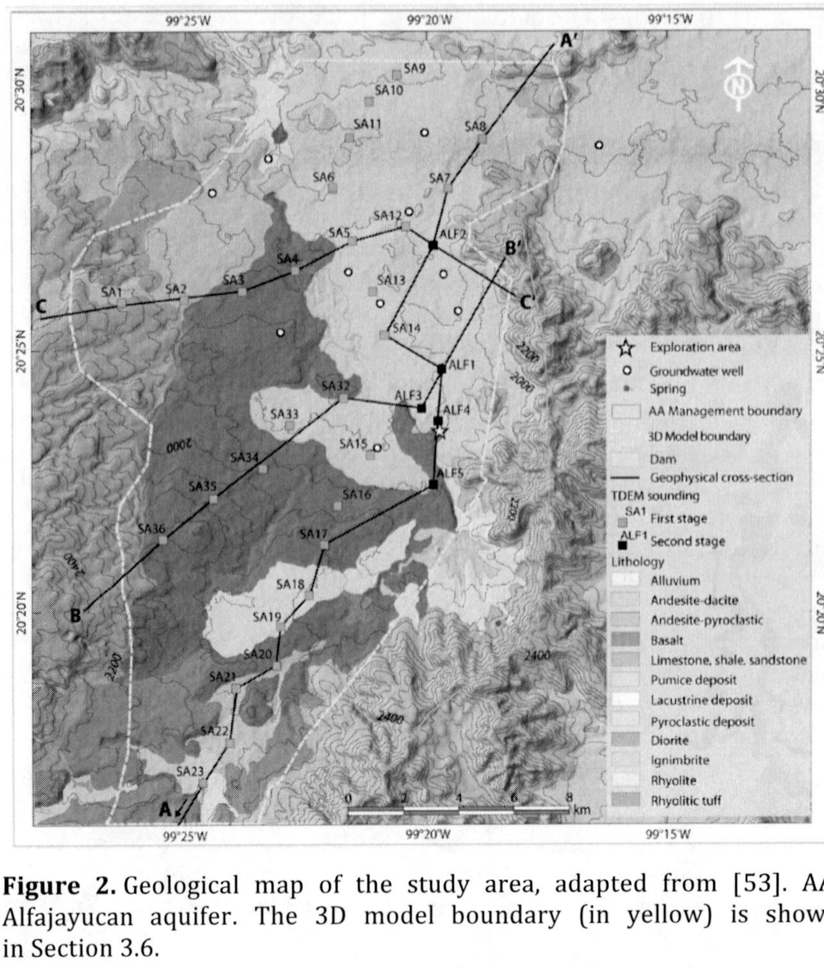

Figure 2. Geological map of the study area, adapted from [53]. AA: Alfajayucan aquifer. The 3D model boundary (in yellow) is shown in Section 3.6.

Middle and Upper Tertiary deposits such as the Tarango Formation (the main supply aquifer for Mexico City) are made up of lava flows and volcanic ash of basaltic-andesitic composition, with interbedded lacustrine deposits [37].

Tertiary volcanic rocks of the Donguinyó and Huichapan formations, related to the eruption of the Donguinyó-Huichapan caldera complex, predominate across the region. The eruptions released discrete pulses of intermediate flows and minor rhyolitic, ignimbrites, pumice fall and surge deposits [51].

From the regional standpoint (Mezquital Valley and surroundings areas), a multi-layer aquifer system has been proposed as a valid conceptual groundwater flow model. However, there are different conceptualizations. The British Geological Survey (BGS) and CONAGUA [37] delineated three overlapping aquifers: (a) the lava flows and coarse sediments of the Tarango Formation; (b) the Tertiary lava-pyroclastic deposits; and (c) the Cretaceous limestones. Other authors reported the existence of two aquifers based on water level measurements [52], while the National Water Commission [36] describes the Alfajayucan aquifer as a two-layer system.

The upper layer is made up of granular and fractured volcanic rocks up to ~400 m thick, while the lower (confined) layer is made up of limestone sediments and calcareous clays. Target units within the Alfajayucan aquifer correspond to the Donguinyó and Huichapan volcanic deposits.

The Mezquital Valley relies on groundwater for domestic supply. There are 456 wells, which extract ~98 Mm^3/year [52]. Approximately 17% of this volume corresponds to human consumption, 38% to agriculture, 33% to the industry and 12% to other uses [54].

On the other hand, 76 water sources were documented in 2012–2013 for the Alfajayucan regional aquifer. These include 45 boreholes, nine dug wells, 21 springs and one gallery. Groundwater abstraction is ~8 Mm^3/year, out of which 64% is for human consumption, 23% for agriculture and 13% for other uses [36]. Within the study area (Figure 2), 11 pumping wells were identified. Out of these, nine are currently inactive. There are significant uncertainties in relation to their depth, screening and hydraulic behavior.

2.2. Hydrogeophysical Investigation

Geophysical data was performed in two stages (Figure 2). The first one consisted of 26 soundings. These were acquired with a single loop configuration, 300 m by side. Five additional TDEM were acquired during the second stage with the same loop configuration, but with a size of 150 m side. All measurements were taken using a TerraTem equipment by Alpha Geoscience Co. (Clifton Park, NY, USA) maintaining an intensity of 7–8 amperes. After editing and averaging the decay curves, apparent resistivity curves were obtained for each TDEM station. One-dimensional (1D) models were computed using an Occam inversion scheme [55].

2.3. Borehole Drilling and Well Log Analysis

A 200-m, 12″-diamater borehole was drilled using a Gardner-Denver 14 W rig. The location was established based on the integration of geological and geophysical data (Figure 2). A static water table was found at a depth of 81 m. Geophysical logs were gathered so as to obtain the petrophysical properties of subsurface units. An 8144 Century Geophysical Co device was used for this purpose, with a voltage of 36 VDC and a logging speed of 9 m/min. This allowed to log variables such as temperature, spontaneous potential (SP), short and long normal resistivity, resistivity of the drilling fluid and gamma ray (GR).

Rock physics is based on sedimentary sequences [56], where oil and gas reservoirs usually take place. In volcanic environments, however, the radioactive nature of rocks makes the interpretation of these curves more challenging [57]. While in sedimentary materials the GR curve is controlled by clay content, the magma composition plays a major role in volcanic rocks. Based on the analysis of 254 samples, Stefansson et al. [58] concluded that there is a linear relationship between GR and silica content in volcanic materials. Though unimportant in the case at hand, it must be noted that this relationship needs to be handled with care in instances where silica content may be altered by the presence of clays. Hence, the borehole log could be completed and analyzed from a quantitative perspective. Completion was necessary because it was only possible to recover material from the upper 66 m of the borehole.

Lithology was inferred from silica content. Criteria established by [57] was used for this purpose. Basalt corresponds to a silica content of <52%, andesitic-basalt to 52%–57%, andesite to 57%–63% and dacite-rhyolite to >63%. Silica content was then checked against standard igneous petrology classifications, to ensure that the inferred lithologies were consistent with chemical composition.

Volcanic facies (lithofacies) were deduced based on silica content and GR data. These were in turn associated to their hydraulic properties (hydrofacies). On the other hand, groundwater salinity was estimated from the SP curve. Its response is controlled by electrochemical and electrokinetic variables, although the latter are generally negligible [56]. In-situ measurements were obtained using a Solinst 428 PVC bailer. Electric conductivity and total dissolved solids were measured by means of a Hannah HI 9828 multiparametric device. Field readings were then compared to theoretical salinity contents.

Total porosity is estimated by means of Archie's Law [59]. Archie's equation remains a method of choice for the estimation of porosity from petrophysical data [60]. Archie's equation is valid for saturated non-clayey media, and is expressed as follows [61]:

$$\rho = a\rho_W \phi^{-m} \tag{1}$$

where ρ corresponds to the resistivity of saturated rock [Ω·m], ρ_w is the resistivity of pore water [Ω·m], *m* the cementation coefficient [-] and *a* the tortuosity factor [-].

Equation (1) was used to develop a porosity curve (φ) for the saturated zone. Input data included the long normal resistivity curve and the estimated groundwater resistivity (as inferred from the SP log). The cementation and tortuosity coefficients were modified as per [62]. Four configurations were analyzed: (1) sandy media (m = 2.5, a = 0.62); (2) carbonate media (m = 2, a = 1); (3) vitreous media (m = 2, a = 1.85); and (4) m = a = 1. The lowest porosity values were obtained with configuration (4). This is the most consistent estimate with reported data for volcanic environments [63]. Finally, all information was integrated within a Schlumberger Petrel® platform [64] to perform petrophysical data analyses.

2.4. Well Yield and Aquifer Parameters

The borehole was equipped as a groundwater well and tested for aquifer parameters and expected yields. A 10″-diameter stainless steel screen casing (grid size 0.125″) and a gravel pack was fitted to the 100–200 m interval. The remainder of the outer annulus was sealed with a 4:1 cement-bentonite proportion.

The well was developed to remove all remaining drilling fluid and settle the gravel pack (procedures in [65]). Pressurized air was injected for 17 h, until a stable flow with negligible fine-particle content was obtained.

A 350-Horse Power submersible pump was installed at a depth of 100–150 m and used to carry out pumping tests. The pump was connected to a 150-m long 4″-diameter pipe. A 42-h 7-step pumping test was carried out to determine well yield, efficiency and aquifer parameters. Steps included: (1) flow rate = 534 m^3/day from 0 to 360 min; (2) 817 m^3/day from 360 to 720 min; (3) 1130 m^3/day from 720 to 1080 min; (4) 1332 m^3/day from 1080 to 1440 min; (5) 1532 m^3/day from 1440 to 1800 min; (6) 1714 m^3/day from 1800 to 2160 min; and (7) 1904 m^3/day from 2160 to 2520 min. Field measurements (elapsed time vs. drawdown vs. flow rate) are shown in Table 1.

Table 1. Field measurements derived from the step-drawdown test.

534 m^3/day [1]		817 m^3/day [1]		1130 m^3/day [1]		1332 m^3/day [1]		1532 m^3/day [1]		1714 m^3/day [1]		1904 m^3/day [1]	
Time (min)	s (m)	Time (min)	s (m)	Time (min)	s (m)	Time (min)	s (m)	Time (min)	s (m)	Time (min)	s (m)	Time (min)	s (m)
60	0.98	420	1.63	780	2.35	1140	2.89	1500	3.43	1860	3.75	2220	4.19
120	1.02	480	1.63	840	2.37	1200	2.87	1560	3.45	1920	3.76	2280	4.21
180	1.01	540	1.65	900	2.37	1260	2.95	1620	3.43	1980	3.76	2340	4.21
240	1	600	1.68	960	2.34	1320	2.96	1680	3.43	2040	3.76	2400	4.21
300	1.04	660	1.68	1020	2.32	1380	2.98	1740	3.44	2100	3.76	2460	4.22
360	0.99	720	1.68	1080	2.34	1440	2.98	1800	3.44	2160	3.76	2520	4.23

Notes: [1] Flow rate; s: Water table drawdown.

Results allowed for the evaluation of head losses associated to the aquifer properties and to the borehole itself. Classic interpretation models were used for this purpose [66,67]. These assume that theoretical drawdown and head losses can be estimated by means of Equation (2):

$$s_w = BQ + CQ^P \tag{2}$$

where s_w corresponds to the total drawdown inside the well [L], Q is well flow rate for each step [LT^{-3}], BQ is the theoretical drawdown component (the B coefficient is expressed in [TL^{-2}] units), and CQ^n is the nonlinear component of head losses (coefficient C is measured in [T^2L^{-5}] units). Hydraulic parameters were estimated by curve-matching procedures, using a nonlinear weighted least-squares parameter estimation and the Gauss-Newton linearization method.

Field data was fitted to the Theis model [68] for a pumping well in a confined aquifer, modified to include linear and nonlinear well losses in a step-drawdown test [69] (Equation (3)):

$$s_w = \frac{Q}{4\pi T}\left[w\left(\frac{r^2 S}{4tT}\right) + 2\sigma\right] + CQ^P \tag{3}$$

where w is the well function, i.e., the exponential integral function, r represents the well radius [L], S and T correspond to the storage coefficient [–] and aquifer transmissivity [L^2T^{-1}], respectively; t is the elapsed time [T] and σ the skin factor [–], estimated by means of Equation (3). The skin factor is an indicator of well damage and has been widely evaluated in the oil reservoir practice [70,71]. The skin factor causes an additional pressure drop, Δp_σ, across the altered zone in hydrocarbon wells [72]:

$$\Delta p_\sigma = \frac{141.2 q B_o \mu}{kh}\sigma \quad (4)$$

Equation (4) was adapted to cater for the conditions of groundwater systems. Considering that the oil-formation volumetric factor, B_o, and viscosity, μ, equals 1 in a single-flowing phase, and that matrix permeability (k)—reservoir thickness (h) product is equivalent to T, Equation (4) can be rearranged as follows:

$$\Delta h_\sigma = \frac{Q}{T}\sigma \quad (5)$$

where Δh_σ represents the additional drawdown caused by the skin factor [L]. Step test results were analyzed with AQTESOLV Pro, v4.5 (Hydrosolve Inc., Reston, VA, USA) [69].

2.5. Subsurface Models

2.5.1. Hydrogeoelectrical Model

TDEM resistivity models were interpolated in 10 m steps for the upper 100 m, 20 m between 100 and 200 m and 40 m for depths up to 400 m. A digital elevation model (DEM) was integrated in Petrel [64] as a reference for subsurface layers. Resistivity curves were upscaled and interpolated by means of a moving averages model with a fourth-power inverse distance algorithm. The hydrogeophysical resistivity model was used to correlate hydrofacies based on horizons and surfaces, as commonly done in the oil industry [73]. A square 800 × 800 m mesh was used to devise the hydrogeological units. Additionally, a vertical resistivity gradient was developed and used to identify subsurface electrofacies, associated to hydrostratigraphic changes.

2.5.2. Total Porosity Model

Borehole data was integrated with TDEM results using Archie's equation [59]. In the absence of predominantly clayey minerals, water resistivity is assumed homogeneous across the study area. Conductivity is assumed to be ionic at constant temperature and the effect of chemical thermalism is considered negligible.

2.5.3. 3D Model

Petrel [64], a 3D oil and gas reservoir modeling software, was used to integrate petrophysical data with TDEM results into a 3D visualization model. The model relies on 31 TDEM inverted resistivity curves, geophysical well logs, porosity information, groundwater depth in nine wells, surface geology, a digital elevation model (DEM), the inferred geometry of electrofacies and the geological information from the borehole drilling. Surface geology [51,53] was only useful to define the uppermost part of the top layer. All other layers were inferred from TDEM information. In other words, the model can only be used to infer subsurface geoelectrical properties.

The DEM was based on four DEM scenes from the Advanced Spaceborne Thermal Emission and Reflection Radiometer (ASTER GDEM v.2, NASA, California Institute of Technology, Pasadena, CA, USA) and then imported to Petrel as a shape file. DEM scenes present a native pixel resolution of 3 × 3 arcsec. Individual scenes were re-projected to a UTM coordinate system (zone 14 N) with a 90 m × 90 m pixel size. Bilinear resampling was used to account for realistic spatial deformation. This process was carried out using QGIS 2.12 Lyon (QGIS Development Team, Open Source Geospatial Foundation) [74].

The development of the 3D model consisted in the following steps:

1) Grid discretization and creation of geological horizons, based on the TDEM interpretation model. For that purpose, we built a three-layer model (layer-1 from 0 to 100 m; layer-2 from 100 to 200 m; layer-3 from 200 to 400 m), with a grid size of 2 km × 2 km, covering a surface area of ~300 km^2. Each layer was divided into 10 to 40 m-thick sublayers (e.g., geocellular grids) to gain vertical resolution;
2) Geoelectrical input data using resistivity population from each of the 31 TDEM inverted curves (see Section 2.2);
3) Creation of a 3D electrofacies model of the subsurface that enables the population all geocellular grids with discrete resistivity values of the electrofacies EF-1 (vadose zone), EF-2 (aquifer) and EF-3 (basement). For that purpose, a deterministic modeling approach was used by means of a moving averages algorithm, taking advantage of the Petrel Facies Modeling Module [64].

Since the model is built upon geoelectrical information, uncertainties can be expected to be higher in those areas which are located further away from the 32 control points (e.g., 31 TDEM soundings plus the borehole).

2.6. Groundwater Flow Evolution in the Mezquital Valley

Groundwater flow behavior in unmanaged recharge regions is a key factor when unregulated water disposal is a major recharge component for downgradient supply aquifers. In the case at hand, an investigation conducted by Geocalli [50] proposed that the Mezquital Valley might present hydraulic continuity with downgradient aquifers, including the study area (Figure 1c and Figure 2).

In order to assess potential hydraulic connectivity and to evaluate if a decrease in the unmanaged wastewater disposal may compromise downgradient aquifer replenishment, piezometric records were analyzed. These are considered a suitable proxy for temporal changes in the wastewater input from the Mezquital Valley.

Datasets from 1982 [50], 1998 [37] and 2007 [52] were taken into consideration. Water table depth contours were digitized from the original sources and the temporal changes of groundwater levels between 1982 and 1998, 1998 and 2007 and 1982 and 2007 were assessed by contouring the hydraulic evolution. Three new maps were created as a result, based on level measurements in 152, 59 and 30 production wells, respectively.

The groundwater evolution of each period of time was analyzed using conventional statistical tools. The spatial and statistical analysis was carried out in QGIS 2.12 Lyon [74].

3. RESULTS AND DISCUSSION

3.1. Hydrogeoelectrical Model

Figure 3 presents the hydrogeoelectrical 2D model (cross sections are shown in Figure 2). Electrofacies 1 (EF-1) corresponds to the uppermost unit. EF-1 presents a relatively homogeneous geometry. The average thickness is ~70 m, and resistivity ranges between 40 and 300 $\Omega \cdot m$, with a bimodal behavior (modes in 50 and 110 $\Omega \cdot m$). Gradients are negative and observed to change laterally. This unit is interpreted as the vadose zone of the system and correlated with a heterogeneous mix of lacustrine, alluvial and volcanic deposits.

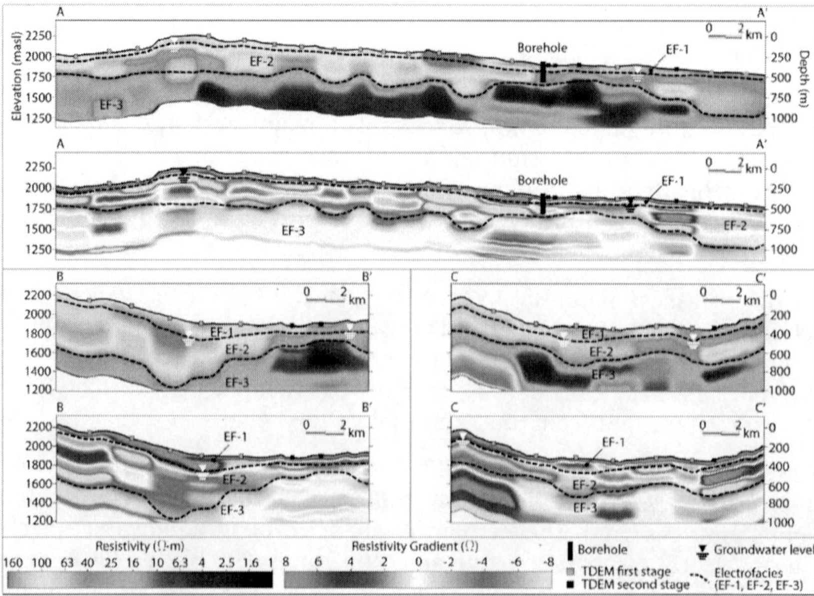

Figure 3. Subsurface hydrogeoelectrical model. EF: electrofacies. AA′, BB′ and CC′ refer to geophysical cross-sections shown in Figure 2.

Electrofacies 2 (EF-2) presents negative gradients and variable resistivities of 5–170, 70–90 and 20–80 Ω·m. Thickness ranges between 190 and 250 m (average = 220 m). Geoelectrical data suggests that geological heterogeneity is controlled by changes in volcanic facies. This unit is correlated to aquifer formations made up of Quaternary basaltic deposits, pyroclastic flows, volcanic breccia and sediments from Mio-Pliocene formations Donguinyó and Huichapan.

Electrofacies 3 (EF-3) provides a hydrogeophysical contrast with EF-1 and EF-2. Resistivity is comparatively low (4–20 Ω·m) at depths of 250–400 m. EF-3 consists of low-permeability clay sediments, marls and clayey limestones, probably related to the Soyatal Formation. Thus, unit EF-3 might represent the system basement. Our geoelectrical interpretation is consistent with those provided by other authors, which suggested that resistivity tends to diminish with increased clay content [47].

TDEM surveys allowed to identify EF-2 as the unit with a better aquifer potential in the region. Resistivity values (5–170; 70–90; 20–80 Ω·m) are consistent with the existing knowledge about the area's subsurface geology, as well as with standard geoelectrical values for similar geological formations [75,76]. Correlated materials (Huichapan and Donguinyó tuffs) include lithological contrasts associated to the eruption of lava and pyroclastic flows [51]. Unit EF-2 presents a marked resistivity contrast with the basement unit, EF-3.

3.2. Near-Borehole Petrophysical Response

3.2.1. Reconstructed Borehole Lithology

Figure 4 shows the results of geophysical logs carried out in the borehole. All readings except GR begin at the static groundwater level. The incomplete geological record (from 66 to 200 m) was reconstructed primarily from the GR curve. GR yielded values of ~60°–90° API (API = American Petroleum Institute) for the upper 40 m, growing to 90°–110° API between 40 and 60 m. The former is controlled by the presence of radioactive minerals in andesites and dry basalts. Increase with depth can be explained by the presence of a 20-m thick dacite horizon where an intensification in radioactivity was observed. From 60 to 200 m, values remain uniform at ~30°–60° API. Some radioactive peaks were also identified (e.g., 164–174 m). These were attributed to the presence of silica-rich rocks.

High GR readings usually hint at the presence of clay-rich materials. However, this does not appear to occur. A number studies in igneous terrains have demonstrated that a relationship exists between the intensity of GR response and the geochemical composition of volcanic rocks [57,77], instead of clayey minerals in sedimentary environments.

Criteria by [58] was used to obtain the SiO_2 content of volcanic rocks as a response of GR values. Lithology was then inferred from silica content. The reconstructed borehole lithology is shown in Figure 4.

A 48-m thick basalt unit was identified between 63 and 111 m. Basalt presents an estimated silica content of <52%. This unit is underlain by an andesitic breccia (~48%–62% SiO_2), andesitic flows (silica ~50%–65%, depths 111.5–121 m and 130–150 m) and acidic rocks (silica ~62%–95%, depth 163 to 174 m). This interbedded sequence can be correlated to EF-2. Thus, the borehole log reflects the presence of Huichapan and Donguinyó tuffs. Aguirre-Díaz et al. [24] report silica contents of 73.7%–74.5% and 55.6%–64.9% in outcrops of the Huichapan ignimbrite and Donguinyó tuffs, respectively, which are consistent with our results for the subsurface analysis.

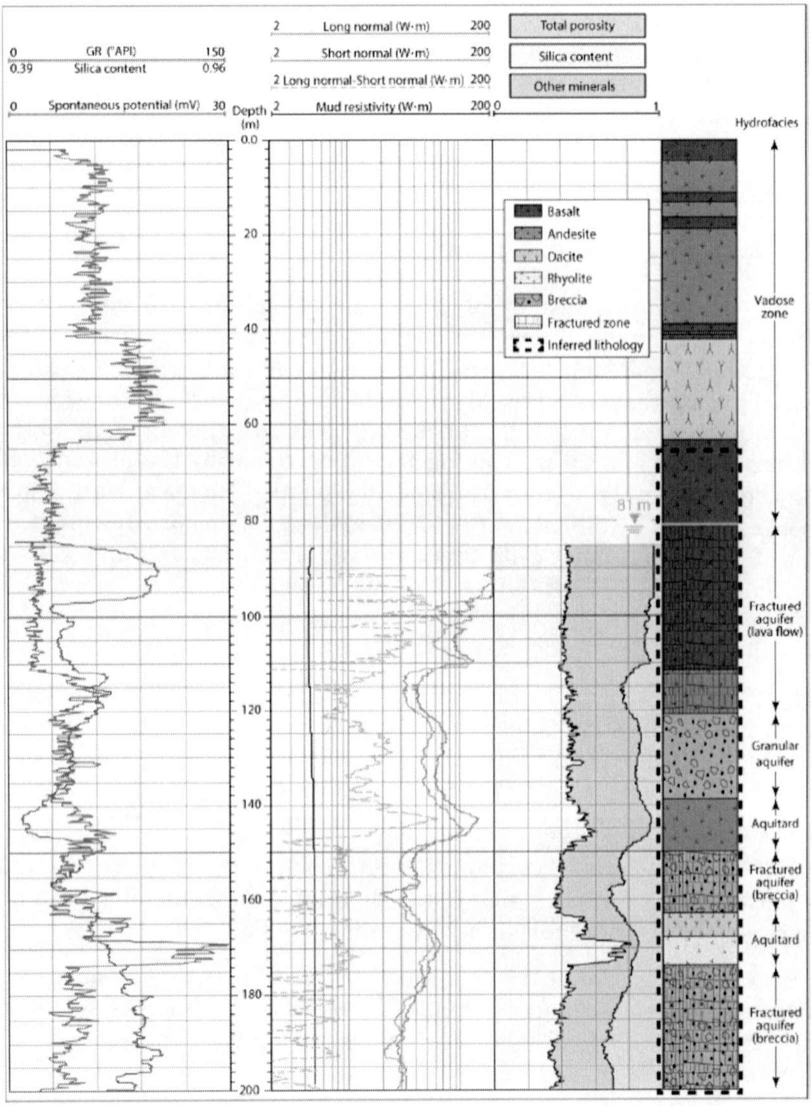

Figure 4. Geophysical logs acquired during drilling, reconstructed lithology and inferred vertical hydrofacies. GR: Gamma Ray; API: American Petroleum Institute.

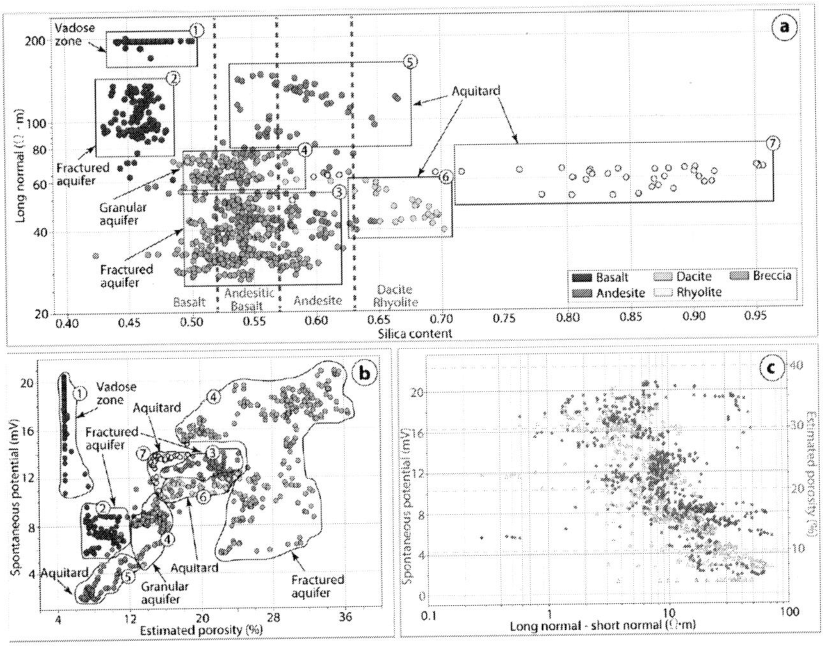

Figure 5. Cross-plots and petrophysical signatures. (**a**) Silica content vs. Long normal resistivity; (**b**) Porosity vs. Spontaneous potential; (**c**) Long normal—Short normal resistivity vs. Spontaneous potential vs. total porosity. For parts (**a**,**b**): (1) Massive basalt; (2) fractured basalt; (3) fractured andesite; (4) andesitic breccia; (5) massive andesite; (6) dacite; and (7) rhyolite.

3.2.2. Hydrofacies

Figure 5 shows the correlation between different petrophysical variables. These were used to understand the relationship between petrophysical response, groundwater occurrence and aquifer properties. Resistivity depends on compaction and fracturing, therefore, it is assumed that both can be obtained, at least qualitatively, from the resistive range of each lithological unit.

Seven lythofacies were identified attending to these criteria: (1) massive basalt; (2) fractured basalt; (3) fractured andesite; (4) andesitic breccia (both granular and fractured); (5) massive andesite; (6) dacite; and (7) rhyolite. Lythofacies provide the basis for the identification of the following hydrofacies:

a) Vadose zone (thickness ~81 m), made up of massive, scarcely fractured basalt. This is not only consistent with collected data, but also with direct observation of rock bits obtained from a depth of 64–66 m.
b) Fractured lava aquifer (thickness ~40 m). Fracturing is expected to be less intense towards the upper boundary. Total porosity is estimated at 7%–24%, resistivity is 40–150 Ω·m, radioactivity 10°–25°API and silica content <52%.
c) Granular aquifer in andesitic breccia and tuffs (thickness ~18 m). The granular aquifer presents some degree of fractured behavior. Total porosity is 12%–18%, resistivity is 25–55 Ω·m, radioactivity 45°–63°API and silica content 48%–62%.
d) Aquitard, made up of andesite and acidic rocks (thickness ~11 and 5–7 m, respectively). Total porosity is 7%–15%, resistivity is 40–143 Ω·m, radioactivity 45°–150°API and silica content 62%–70%.
e) Fractured breccia aquifer (thickness 13–26 m). Total porosity is 17%–26%, resistivity is 30–43 Ω·m, radioactivity 30°–50°API and silica content 50%–59%.

The vertical distribution of hydrofacies and lythofacies is shown in Figure 4. Petrophysical features are presented in Figure 5a,b.

3.3. Porosity and Qualitative Permeability Estimates

The permeability of hydrofacies was estimated on qualitative terms. Well logs establish that the resistivity of the drilling fluid (4–6 Ω·m) is much lower than that of the hydrofacies (25–200 Ω·m). Under these conditions, positive deflections in the SP curve could be interpreted as permeable sectors. In the absence of clay, the larger the deflection, the more permeable the rock would be. SP and resistivity curves should also present a positive correlation [78].

The SP curve presents a first deflection at ~84–96 m (Figure 4), within the sector identified as massive basalt. There is no clear correlation to establish whether this interval qualifies as permeable. Positive deflections were also found around 174 m. Here, there is a better correlation with resistivity curves. Hence, the SP hints at the presence of permeable rocks between 174 and 200 m. These correspond to the fractured breccia aquifer.

A porosity vs. SP plot was built to evaluate these results in a more systematic manner (Figure 5b). In general terms, porosity and SP correlate well. In volcanic rocks, this could be assimilated to a correlation between porosity and permeability [62], which in turn implies that SP is an indicator of permeability.

Permeability was also analyzed by means of resistivity curves. Previous studies establish that the gap between short and long resistivity curves can be used as an indicator of permeability [78]. However, in high permeability rocks, short and long resistivity curves will present very similar values as a result of deep-mud filtrate invasion. This may lead to spurious conclusions. Uncertainty was evaluated by studying the relationship between SP and the difference between long and short normal resistivity curves. The analysis aims at identifying a negative correlation between the SP curve and the other two. This would suggest that SP increases as the separation between the resistivity curves decreases (as a result of mud filtration). In turn, this could lead to identifying high permeability areas.

Figure 5c presents the results. A negative correlation comes through clearly. Maximum SP values (~15–20 mV) occur when the separation between the resistivity curves is minimal (0.1–2 Ω·m). This suggests that deep-mud filtrate invasion is actually taking place as a consequence of high permeability formations. Invasion areas can be identified visually in Figure 4. These are located at 150–163 m and 174–200 m. Deep-mud filtrate invasion has been previously explored in the oil industry [79], but little research has been carried out on the groundwater front.

Similarly, separation between the resistivity curves decreases with depth, while porosity increases. Therefore, a high permeability could be attributed to both sectors, corresponding to the fractured breccia aquifer (hydrofacies (e)). The porosity of this unit (17%–26%) is consistent with reports for similar geological formations, i.e., 10%–32% in [62].

Porosity is obtained from Archie's law, which in turn relies on resistivity logs. As a result, porosity may be overestimated. Although values from geophysical logs are consistent with the lithology, at the regional level there appears to be an overestimation of the subsurface porosity model. The model established a ~6%–42% interval for pore size distribution in the subsurface (Figure 6).

Previous works point out some of the limitations of estimating porosity from Archie's law without sonic logs or lab measurements. Alao et al. [80], for instance, explain that Archie's approach was originally developed for clay-free sedimentary rocks, and it fails to consider additional petrophysical features such as the presence of shale, fresh formation waters or thin-bed layers. Not considering these can lead to overestimate porosity and saturation. In turn, Markov et al. [81] reported that in carbonate rocks with a percolation threshold (connectivity and conductivity loss as the rock reaches critical void space) and chaotic microstructure (various pore systems of different types and scales), porosity cannot be estimated within the framework of Archie's Law.

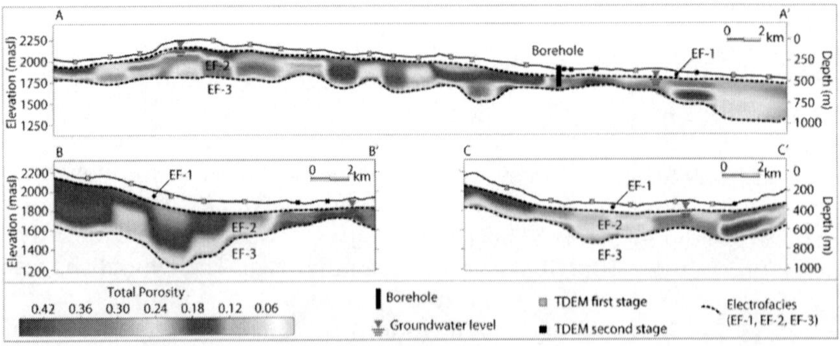

Figure 6. Subsurface porosity model. EF: electrofacies. AA′, BB′ and CC′ refer to geophysical cross-sections shown in Figure 2.

Finally, Wright et al. [62] argue that permeability, porosity and electrical conductivity may vary by several orders of magnitude within extrusive rocks. This is due to variable tortuosity, vesicle sizes, pore elongation and connected pathways during bubble expansion, eruption conditions and crystallization history. These authors conclude that Archie's two-constant scheme (m and a) could be inappropriate under these circumstances. All this means that the porosity model presented in this paper should be used with caution. Nevertheless, it is advocated as a useful tool to obtain first-approach estimates of porosity and further groundwater reserves.

3.4. Groundwater Salinity

Groundwater salinity for the upper permeable sector (174–200 m depth, within hydrofacies (e)) was estimated by means of petrophysical techniques [56]. Well logs present a positive SP of ~20.5 mV, while groundwater temperature is ~30 °C, the resistivity of the drilling fluid is ~5 Ω·m and the resistivity of the aquifer (R_w) is 9.6 Ω·m. Based on Shlumberger's log interpretation charts [82], a salinity value (NaCl) of ~400 ppm was calculated.

Direct measures of TDS during borehole construction yield 602–614 ppm. These values are representative of hydrofacies (c) and (d). The BGS and CONAGUA [37] reported TDS concentrations in springs and dug wells in Irrigation District 100, near Ixmiquilpan (9 km to the northeast of the borehole, Figure 1c). These range between 656–898 and 973–1280 ppm. More recently, Chávez et al. [83] reported TDS values within the Tula and Mezquital valleys of 865–1586 (wet season) to 861–1488 ppm (dry season) in abstraction wells, 1061–1488 (wet season) to 1111–1513 ppm (dry season) in springs, and 1434–1702 (wet season) to 1402–1597 (dry season) ppm in dug wells, respectively. Thus, the salinity estimates obtained during fieldwork suggest that wastewater infiltration has had no major impacts on aquifer salinity.

3.5. Expected Flow Rates, Well Efficiency and Aquifer Parameters

Step-drawdown test results are shown in Figure 7 and Table 2. These correspond to hydrofacies (b) and (c), that is, to the fractured lava flows and the granular breccia aquifer (100–150 m). The borehole yielded 6–22 L/s with a small drawdown (s_w = 1–4.29 m). Head losses attributable to the aquifer (*BQ*) account for most of the drawdown at each step (0.69 m ≤ *BQ* ≤ 2.22 m; *avg.* = 1.66 m). This means that 58%–70% of s_w is due to the aquifer component (*avg.* = 60.9%), while ~12%–19% is due to the well loss component (*avg.* = 15.9%). Additional nonlinear head losses at each step amounted to 0.12, 0.23, 0.37, 0.47, 0.58, 0.69 and 0.81 m, respectively.

As the borehole will supply downgradient towns in the short term (e.g., San Agustín Village, Figure 1c), test results appear encouraging. Well yield (534–1904 m^3/day, or 6–22 L/s; mean = ~1280 m^3/day, or ~15 L/s) exceeds by far San Agustin's water demand (~225 m^3/day, or 2.6 L/s). In larger urban cores, such as Alfajayucan (~19,000 inhabitants, according to INEGI (2013, unpublished), Figure 1c), the borehole will be able to roughly assist with ~25% of the current municipal demand.

Domestic demands are estimated by multiplying the number of inhabitants by a daily requirement of 250 L/person/day. This estimate is considered conservative and compliant with international standards. For instance, the WHO/UNICEF Joint Monitoring Programme, established a minimum dosage of only 20 L/person/day [84], to cover basic survival needs. This is consistent with estimates by several authors [85,86], although others suggest that the threshold should be raised to 50 [87]. In turn, World Health Organization (WHO) [88] cites minimal domestic requirements in the order of 50–100 L/person/day. From the volumetric standpoint, all this suggests that the newly-identified groundwater source (e.g., new groundwater well) provides a long term solution for a substantial share of the downgradient population of the Mezquital Valley. However, it is mandatory to carry out a detailed hydrogeochemical study to further assess water quality aspects.

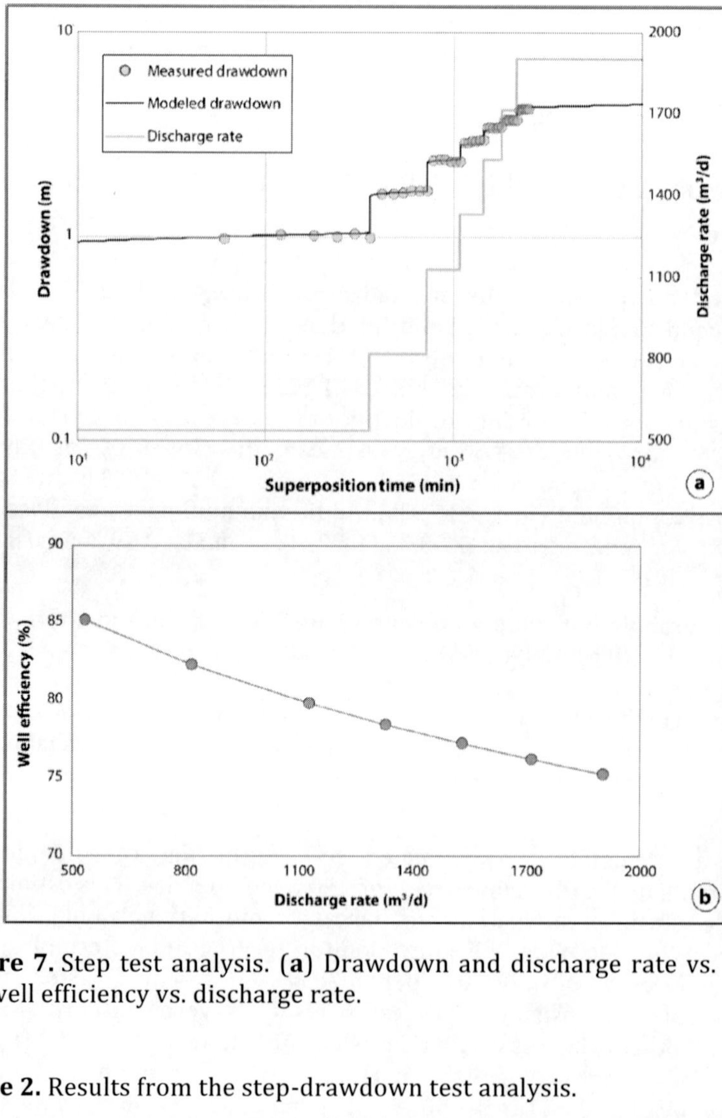

Figure 7. Step test analysis. (**a**) Drawdown and discharge rate vs. time; (**b**) well efficiency vs. discharge rate.

Table 2. Results from the step-drawdown test analysis.

Step	Q (m³/Day)	s_w (m)	q (L/s/m)	BQ (m)	CQ^P (m)	Δh_σ (m)	Aquifer Loss (%)	Well Loss (%)	w_{eff} (%)
1	534	1.00	6	0.69	0.12	0.46	69.23	12.05	85.18
2	817	1.68	6	1.06	0.23	0.71	62.98	13.55	82.29
3	1130	2.32	6	1.46	0.37	0.98	63.10	15.97	79.80
4	1332	3.00	5	1.72	0.47	1.16	57.49	15.79	78.45
5	1532	3.47	5	1.98	0.58	1.33	57.14	16.83	77.25
6	1714	3.77	5	2.22	0.69	1.49	58.86	18.34	76.24
7	1904	4.29	5	2.46	0.81	1.66	57.46	18.87	75.28
Min	534.00	1.00	5.11	0.69	0.12	0.46	57.14	12.05	75.28
Max	1904.00	4.29	6.18	2.46	0.81	1.66	69.23	18.87	85.18
Mean	1280.43	2.79	5.44	1.66	0.47	1.11	60.89	15.91	79.21
SD	489.44	1.18	0.40	0.63	0.25	0.43	4.48	2.45	3.52

Notes: Q: Flow rate; q: specific flow rate; BQ: theoretical drawdown; CQ^P: nonlinear well loss; Δh_σ: additional drawdown in the well related to the skin factor; w_{eff}: well efficiency; SD: standard deviation.

Regarding the hydraulic behavior of the well, the equation $1.86Q + 0.65Q^{1.5}$ ($B = 1.86$ [min/m^2]; $C = 0.65$ [min^2/m^5]; $P = 1.5$) allows for the estimation of the linear aquifer loss component, BC, and the nonlinear well loss, CQ^P, for any flow rate, Q. In this case, P is the order of nonlinear well losses. Jacob [89] suggested that well loss is approximately proportional to the square of the discharge. Thus, P can be assumed to be ~2 in most situations. Rorabaugh [67] showed that P may range from 1.5 to 3 in cases such as the one at hand. Walton [90] proposed an interpretation criteria for the nonlinear well coefficient, C. Values of 1900–3800 and >3800 [s^2/m^5] are indicators of mild and severe well deterioration, respectively.

A coefficient $C = 2365$ [s^2/m^5] was obtained for the borehole, suggesting suboptimal (but acceptable) behavior.

From the hydraulic point of view, the main implication is an additional drawdown within the well. This is related to turbulent flow at the aquifer–gravel pack interface, as well as to the skin effect. The latter, σ, refers to an alteration in near-borehole permeability, resulting from well completion effects, formation erosion, acidification, well development and mud filtrate invasion [91]. Renard [92] pointed out that σ is positive if the well is clogged and negative for the opposite case (stimulated well), while [93] reports that the skin factor may adopt any value, but that it rarely exceeds 20 in practice. In this case, a skin coefficient of +1.48 was obtained, leading to an additional drawdown of 0.4–1.6 m (*avg.* ≈ 1 m). This amounts to 16% of the total drawdown. Since mud filtrate invasion was earlier identified through the use of petrophysical signatures (Figure 4 and Figure 5c), these effects are best described as mild clogging due to incomplete well development.

Well efficiency (w_{eff}) was estimated by means of Equation (2) in the form $w_{eff} = BQ \;/\; BQ + CQ^P$ Table 2 shows the relative efficiencies of the seven discharge rates. As shown in Figure 7a, w_{eff} exceeds 85% for a rate of 534 m^3/day (6 L/s). It decreases as flow rate increases, up to ~75% for a rate of 1904 m^3/day (22 L/s). Although this efficiency is acceptable, hydraulic behavior and productivity can be improved by completing the well development using common techniques such as overpumping, surging, rawhiding or airlift pumping [65]. More novel methods, such as hydrojetting with coiled tubing and simultaneous pumping, have also been demonstrated to be a cost-effective cleaning procedure [94].

The Theis Model [68] was used to infer aquifer properties. A transmissivity (T) of 1700 m^2/day, a hydraulic conductivity (k) of 8–40 m/day and a storage coefficient (S) of 8.8×10^{-3} were obtained, with a mean residual sum squares (RSS) of 0.0023 m between measured and modeled drawdowns. Hydraulic conductivity was estimated by considering a saturated thickness range between 40 and 220 m. The former is referred to as the thickness of hydrofacies (b) while the latter is referred to the average thickness of electrofacies EF-2. In addition, the storage coefficient estimate must be handled with care because it was obtained from drawdown data within the pumping well. Specific capacity is 5–6 L/s/m drawdown. Provided that the well is further developed, these values suggest that flow rates may be increased considerably in case of need.

Aquifer properties are representative of hydrofacies (b) and (c). These correspond to fractured lava flows and volcanic breccia. Previous works report T = 864–4320 m^2/day, k = 0.035–103.7 m/day and $S = 2 \times 10^{-4}$–0.2 for basaltic rocks within the Alfajayucan aquifer [36,95], which are all consistent with the above results. These also show that hydrofacies are highly heterogeneous from hydraulic standpoint.

3.6. 3D Visualization and Integration Tool for Management Purposes

Field information was compiled into a Petrel 3D model [64]. Figure 8 shows several instances of the 3D visualization model. EF-2 represents a regional aquifer. As such, it may be developed at a larger scale. This aquifer is made up of three main hydrofacies, namely, a fractured lava aquifer, a granular breccia aquifer and a fractured breccia aquifer. These are partially isolated from one another due to the presence of intercalated aquitards. The spatial variation of hydrofacies cannot be inferred from a single borehole. Thus, these were only characterized around the drilling site.

3D methods are useful to appraise uncertainties on the physical attributes of the aquifer systems and greatly improve conceptual understanding of their behavior [29]. In this case, the 3D model provides a welcome visual aid to scientists and water managers. The model is useful for geostatistical analyses. It provides a reference for the study of the spatial variability of electrofacies or the geometry of the aquifer system, as well as to estimate the expected water table depth at any point within the model domain. It is also useful to inform further fieldwork, including new extraction wells.

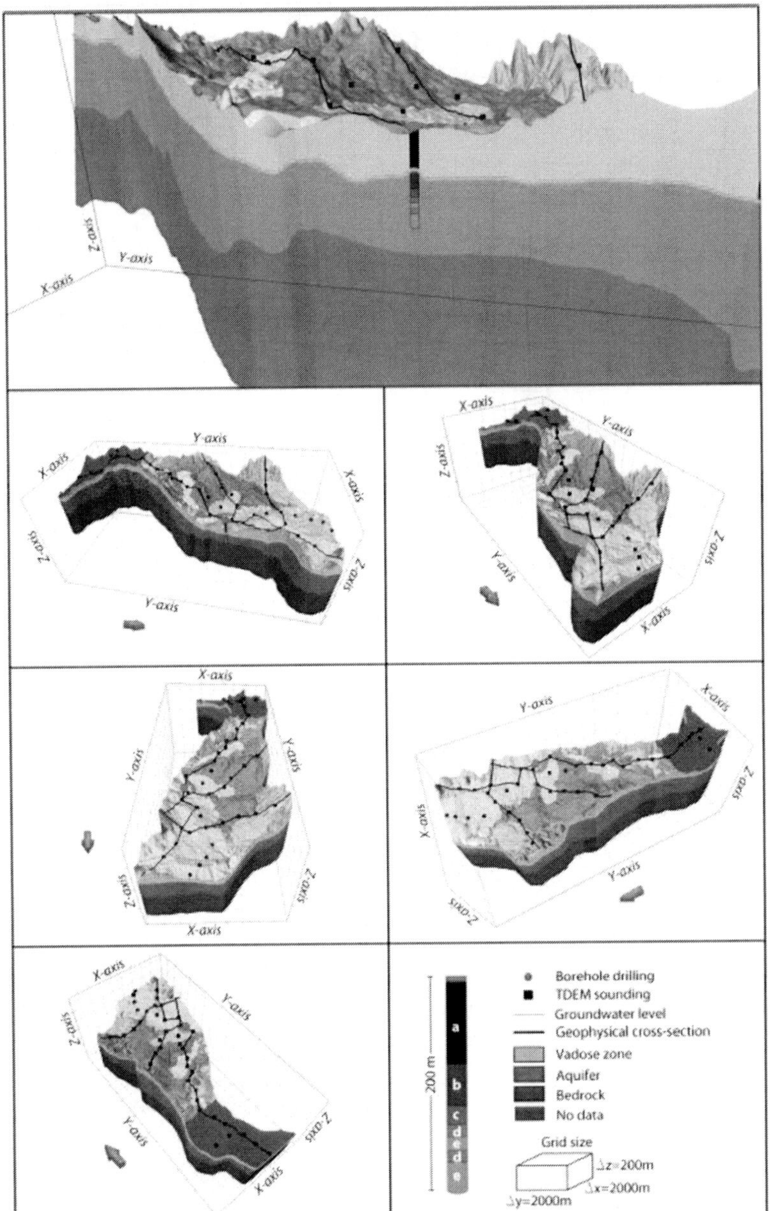

Figure 8. 3D visualization and integration model. (**a**) Vadose zone; (**b**) fractured lava aquifer; (**c**) granular aquifer in andesitic breccia; (**d**) aquitard; (**e**) fractured breccia aquifer. The colors in the uppermost layer refer to lithologies shown in Figure 2.

The model can be used to calculate the estimated reserves of the system. As it stands, the model spans an area of 301.62 km^2. EF-2 unit presents an average thickness of 220 m (Figure 3), with a porosity range between 15% and 21%. If the 8.8×10^{-3} storage coefficient is taken as a lower-bound estimate of drainage porosity, the aquifer could be expected to hold ~12,200 Mm^3 of groundwater, as well as to release a further ~580 Mm^3, as a result of physical drainage of interconnected pores. This means that the system's specific yield amounts to 4% of the total storage, an estimate which is roughly consistent with figures for the whole Mezquital Valley system (7%) [96]. While still rough, these figures may be taken as a baseline for detailed studies based on groundwater balance calculations [97] or numerical models [98].

Much like the oil industry, the water sector can benefit from the dynamic integration of data in 3D databases. This will contribute to a more robust understanding about aquifer systems related to unmanaged regions, thus providing a sound scientific base to underpin management decisions.

3.7. Groundwater Flow Evolution in the Unmanaged Region of the Mezquital Valley

Figure 9 provides an overview of the 1982–1998, 1998–2007 and 1982–2007 evolution intervals. Data shows large dispersions. For instance, changes of −60 to +60 m, −40 to +60 m and −65 to +55 m are observed for 1982–1998, 1998–2007 and 1982–2007, respectively.

There are three potential explanations for these large fluctuations. The first one has to do with the presence of spatially-isolated data points (Figure 9). Secondly, fluctuations may be attributed to the fact that, in some sectors of the Mezquital Valley, groundwater levels are controlled by abrupt topographic changes. This effect is common in shallow aquifers [99]. Finally, the Mezquital and Alfajayucan units have been described as multilayer systems with mutual interconnections [36].This suggests that the original datasets [37,50,52] mix shallow and deep levels.

The latter point is difficult to confirm. This is because there are significant uncertainties in terms of basic borehole attributes such as their depth or the location of the screen. However, field data combined with the experience of the authors in the study area suggest that most boreholes are screened from the water table to the bottom (some appear to be screened from throughout their whole length). This means that water table readings represent the average head of all aquifers. As a result, the piezometric analysis must be handled with some care.

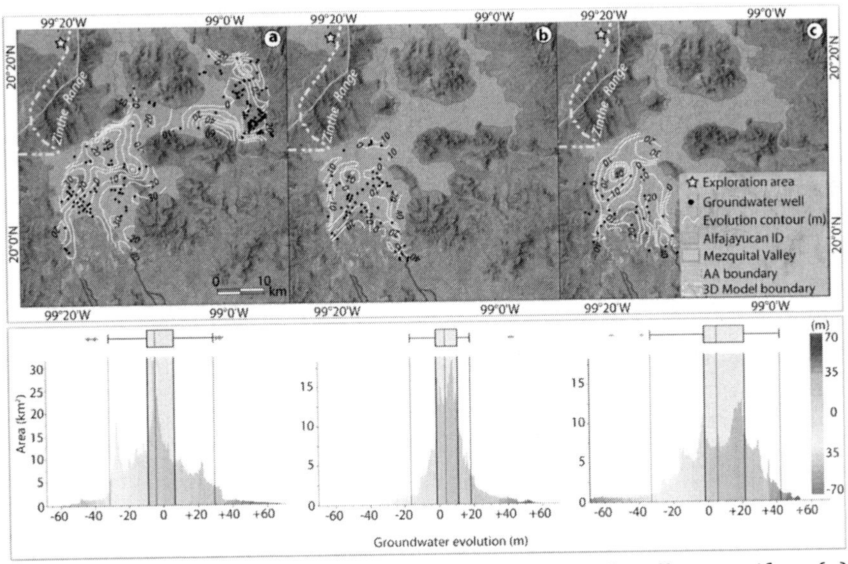

Figure 9. Groundwater evolution in the Mezquital valley aquifer. (**a**) 1982–1998; (**b**) 1998–2007; (**c**) 1982–2007. Negative contours (m) refer to a decline in potentiometric levels and opposite for positive values. ID: Irrigation district. Red crosses in the boxplots refer to outliers. AA: Alfajayucan aquifer.

Overall, the 1982–1998 interval histogram presents a bimodal behavior, modes being −5 and −30 m (Figure 9a). The boxplot interquartile range (Q_3–Q_1) which is a measure of statistical dispersion, provides a more representative range. This suggests that the water table fluctuated mostly between −8 m and +7 m over those 16 years.

Similar considerations apply to the other dispersive histograms (Figure 9b,c). However, the interquartile range oscillates within a considerable narrower interval, in the order of −1 to +22 for 1982–2007 and between −1 and +12 m for 1998–2007, respectively (consider that positive values refer to a rise in water table and opposite for negative values). Furthermore, the median of the temporal groundwater evolution is −4, +6 and +5 m, respectively, for the 1982–1998; 1998–2007 and 1982–2007 cases. This implies that the aquifer level has fluctuated by only ±20 cm/year, which is close enough to zero. In other words, the Mezquital Valley aquifer (MVA) is closed to steady-state conditions for the 25-year interval.

This argument is supported by independent evidence. Garfias-Quezada [100] analyzed monthly piezometric data from industrial wells located towards the SW of the MVA between 1984 and 2010. Hydrographs from five wells show that the water table depth has remained roughly constant. In some cases, these present a slightly upward trend.

For instance, one of the wells presented a water table depth of ~22 m in early 1984, and its fluctuation over 26 years was just 3% (21.9 ± 0.68 m, as referred to the median ± the standard deviation). Other wells registered similar behaviors (17.21 ± 2.29 m; 14.67 ± 0.44 m; 43.82 ± 1.12 m; 44.67 ± 1.02 m). Besides, linear trends for each hydrograph present near-zero gradients (10^{-4}–10^{-5}). This clearly implies that both the shallow (15–25 m) and deep piezometric heads (40–50 m) have remained stable.

This is remarkable because the MVA is a highly dynamic system in terms of the water balance. The National Water Commission [96] reported the following recharge and discharge volumes for the Mezquital Valley (2007–2012): abstractions from wells amount to ~137.7 Mm^3/year, springs yield ~85.1 Mm^3/year, baseflow through Tula River is estimated to be ~280 Mm^3/year, while direct evapotranspiration from shallow groundwater is in the order of 9.8 Mm^3/year. Conversely, induced recharge from irrigation, wastewater infiltration and network losses is about ~343.2 Mm^3/year, while rainfall infiltration amounts to ~49.5 Mm^3/year. These figures shows an aquifer storage variation of roughly +2.4 Mm^3/year. Given the size of the system and the magnitude of the different elements of the water budget, this figure is sufficiently close to zero. Thus, it is consistent with the outcomes of the piezometric analysis.

Whereas dropping water tables are relatively common across the world [101], long-term pseudo steady-state conditions such as these are relatively rare. Unmanaged recharge seems to regulate the system, thus preventing major changes in the aquifer balance. This provides a stark contrast with neighboring aquifer systems where there is no unmanaged recharge. For instance, the Cuautitlán-Pachuca aquifer presents a severe storage decrease over the same interval of −58 Mm^3/year [102]. This corresponds to a regional water table drop in the order of −0.5 to −2 m/year.

Based on 2007 records [52], groundwater flows from the south to the north of the Mezquital valley. Equipotential lines range from 2200 to 1980 m. a.s.l. (Figure 10b). These conditions are similar in the study area for 2012 [53], where flow gradients and dynamics are equivalent, albeit with a different range of equipotential values (1900–1720 m. a.s.l., Figure 10a). This suggests that there is hydraulic continuity between both regions. Thus, the Zhinte range (Figure 9 and Figure 10) does not constitute a no-flow boundary.

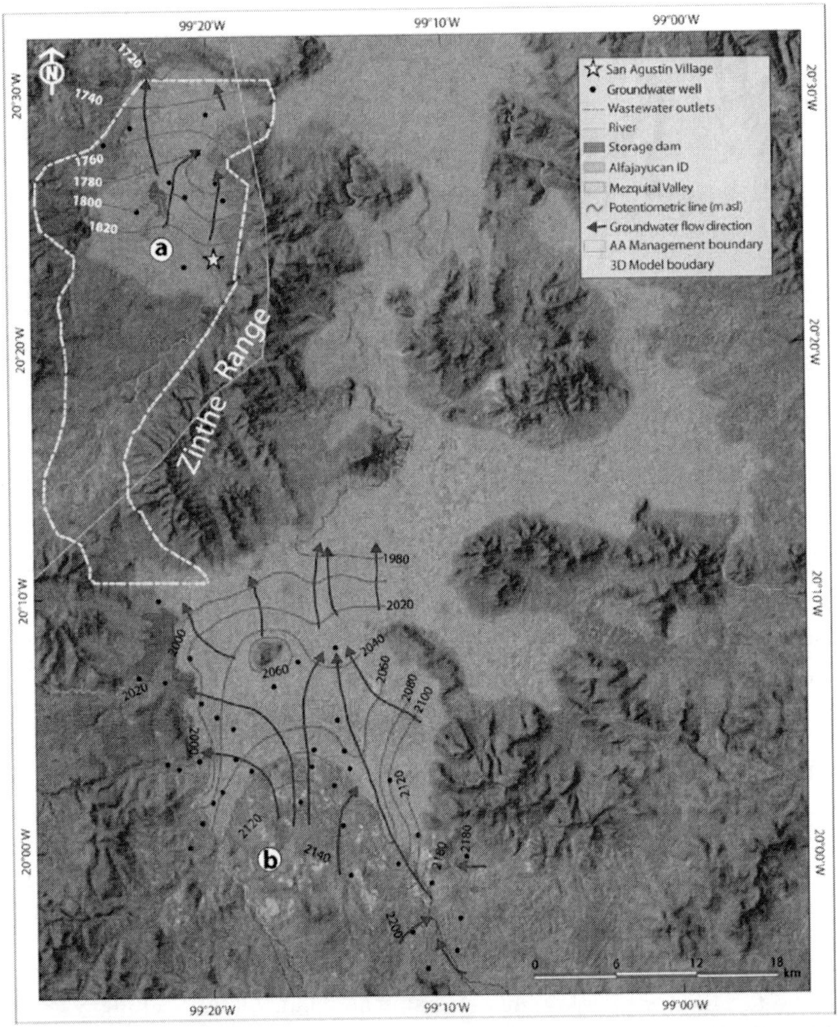

Figure 10. Groundwater flow net in (**a**) the Alfajayucan aquifer for 2012 (adapted from [53]); and (**b**) the Mezquital Valley aquifer for 2007 (adapted from [52]). ID: Irrigation district, AA: Alfajayucan aquifer.

In summary, pseudo steady-state conditions in the MVA, together with the spatial distribution of hydraulic gradients, suggests that the Alfajayucan aquifer is recharged with lateral flows from the wastewater irrigation of the Mezquital Valley. Thus, newly-developed water supplies within the study area are likely to be affected from the volumetric standpoint by upgradient treated wastewater disposal. In addition, the

MVA is likely to evolve into transient conditions, a state which it has not experienced for at least 30 years. The water treatment plant is expected to greatly improve the water quality for municipal use in the MVA and surrounding areas.

4. CONCLUSIONS

Hydrogeological exploration and characterization is still needed in many regions across the world, particularly in low-income countries and emerging economies. Within a context of unregulated wastewater-based recharge schemes, the search for new groundwater sources is essential to underpin the present and future life of local communities. Current water supply challenges call for integrated approaches. Interdisciplinary research combining geological, hydrogeophysical, petrophysical and hydraulic techniques, adapted from oil and gas reservoir modeling frameworks, need to be used jointly to evaluate the complexities of aquifer systems and optimize the development of new sources.

The proposed methodology contributes not only to inferring the hydrogeological parameters controlling aquifer dynamics, but also to increasing knowledge about the behavior of key physical properties in different lithologies. Thus, it can be exported to similar settings, so as to enhance our understanding of subsurface processes in unmanaged and managed recharge regions.

Regional and local-scale assessments are needed to adequately frame the outcomes and guarantee sustainable groundwater use in the long run. Practices such as unmanaged aquifer recharge constitute a mixed blessing. Cases like the Mezquital Valley show how it provides additional aquifer replenishment, and how this contributes to the equilibrium of groundwater systems. On the other hand, unmanaged aquifer recharge is likely to induce lasting hydrochemical changes in aquifers and ecosystems. The transition from unmanaged to managed recharge is expected to provide benefits to the Mezquital Valley inhabitants, and furthermore to generate new uncertainties in relation to aquifer dynamics and downgradient systems.

ACKNOWLEDGMENTS

The Earth Sciences Division and the Institute of Geophysics (UNAM) financially supported this research. The authors are grateful to Schlumberger for the donation of the Petrel® Platform (provisional academic license number PETREL_151139740_MAAAIACBH/aUA), during March-June 2015. We thank R. Castrejón for providing us technical guidelines during well logging analysis and to Fátima Espinosa, Alejandro Gómez, Francisco Miguel, Hermes Rochín, Alberto Garfias and Diego Ruiz, for helping out with field work and analysis. We thank Santiago Arellano Islas (Comision Estatal del Agua y Alcantarillado de Hidalgo) and the residents of San Agustín Village for their permission and support, respectively. The authors thank the reviewers and editor for helpful discussions which considerably improved the quality of the manuscript. This is the 9th contribution from the Hydrogeology Group (Earth Sciences Division), Faculty of Engineering, UNAM, Mexico.

REFERENCES

1. Oikonomidis, D.; Dimogianni, S.; Kazakis, N.; Voudouris, K. A GIS/Remote Sensing-based methodology for groundwater potentiality assessment in Tirnavos area, Greece. *J. Hydrol.* **2015**, *525*, 197–208.
2. Panigrahi, B.; Nayak, A.K.; Sharma, S.D. Application of remote sensing technology for groundwater potential evaluation. *Water Resour. Manag.* **1995**, *9*, 161–173.
3. Dawoud, M.A.; Raouf, A.R.A. Groundwater exploration and assessment in rural communities of Yobe State, Northern Nigeria. *Water Resour. Manag.* **2008**, *23*, 581–601.
4. Llamas, M.R.; Martínez-Santos, P. The silent revolution of intensive ground water use: Pros and cons. *Ground Water***2005**, *43*, 161.
5. Mallet, J.L. *Geomodeling*; Oxford University Press: New York, NY, USA, 2002.
6. Duca, C.; Abruzzi, D. Fully integrated hydrocarbon reservoir studies: Myth or reality? *Am. J. Appl. Sci.* **2010**, *7*, 1477–1486.
7. Warrlich, G.; Bosence, D.; Waltham, D.; Wood, C.; Boylan, A.; Badenas, B. 3D stratigraphic forward modelling for analysis and prediction of carbonate platform stratigraphies in exploration and production. *Mar. Pet. Geol.* **2008**, *25*, 35–58.
8. Sacchi, Q.; Salina Borello, E.; Weltje, G.J.; Dalman, R. Increasing the predictive power of geostatistical reservoir models by integration of geological constraints from stratigraphic forward modeling. *Mar. Pet. Geol.* **2016**, *69*, 112–126.
9. Dickey, P.A. *Petroleum Development Geology*; Pen Well Books: Tulsa, OK, USA, 1981.
10. Lake, L.; Carrol, B.H. *Reservoir Characterization*; Academic Press, Inc.: New York, NY, USA, 1988.
11. Musliu-Kehinde, O. Integrated Reservoir Characterization: A Case Study of an Onshore Reservoir in Niger Delta Basin. Master's Thesis, African University of Science and Technology, Abuja, Nigeria, 2011.
12. Guyoton, F.; Grasso, J.R.; Volant, P. Interrelation between induced seismic instabilities and complex geological structure. *Geophys. Res. Lett.* **1999**, *19*, 705–708.
13. Spikes, K.T. Model-based prediction of porosity and reservoir quality from P- and S-wave data. *Geophys. Res. Lett.***2003**, *30*, 1–4.
14. Falahat, R.; Shams, A.; MacBeth, C. Towards quantitative evaluation of gas injection using time-lapse seismic data. *Geophys. Prospect.* **2011**, *59*, 310–322.

15. Arnulf, A.F.; Singh, S.C.; Pye, J.W. Seismic evidence of a complex multi-lens melt reservoir beneath the 9° N Overlapping Spreading Center at the East Pacific Rise. *Geophys. Res. Lett.* **2014**, *41*, 6109–6115.
16. Arogunmati, A.; Harris, J.M. Quasi-continuous reservoir monitoring with surface seismic data. *Geophys. Prospect.***2014**, *62*, 117–132.
17. Grana, D.; Mukerji, T. Bayesian inversion of time-lapse seismic data for the estimation of static reservoir properties and dynamic property changes. *Geophys. Prospect.* **2015**, *63*, 637–655.
18. Wyllie, M.R.; Gregory, A.R.; Gardner, L.W. Elastic wave velocities in heterogeneous and porous media. *Geophysics***1956**, *27*, 569–589.
19. Pringle, J.K.; Westerman, A.R.; Clark, J.D.; Drinkwater, N.J.; Gardiner, A.R. 3D high-resolution digital models of outcrop analogue study sites to constrain reservoir model uncertainty: An example from Alport Castles, Derbyshire, UK. *Pet. Geosci.* **2004**, *10*, 343–352.
20. Bueno, J.F.; Drummond, R.D.; Vidal, A.C.; Sancevero, S.S. Constraining uncertainty in volumetric estimation: A case study from Namorado Field, Brazil. *J. Pet. Sci. Eng.* **2011**, *77*, 200–208.
21. Fokker, P.A.; Visser, K.; Peters, E.; Kunakbayeva, G.; Muntendam-Bos, A.G. Inversion of surface subsidence data to quantify reservoir compartmentalization: A field study. *J. Pet. Sci. Eng.* **2012**, *96–97*, 10–21.
22. Hosseini, S.A.; Lashgari, H.; Choi, J.W.; Nicot, J.P.; Lu, J.; Hovorka, S.D. Static and dynamic reservoir modeling for geological CO_2 sequestration at Cranfield, Mississippi, U.S.A. *Int. J. Greenh. Gas Control* **2013**, *18*, 449–462.
23. Bunch, M. A live test of automated facies prediction at wells for CO_2 storage projects. *Energy Procedia* **2014**, *63*, 3432–3446.
24. Pan, F.; McPherson, B.J.; Dai, Z.; Jia, W.; Lee, S.Y.; Ampomah, W.; Viswanathan, H.; Esser, R. Uncertainty analysis of carbon sequestration in an active CO_2-EOR field. *Int. J. Greenh. Gas Control* **2016**, *51*, 18–28.
25. Noad, J. The use of field analogues in the correlation and static reservoir methodology used in the Tern Field, Northern North Sea, UK. *Mar. Pet. Geol.* **2004**, *21*, 485–497.
26. Blöcher, M.G.; Zimmermann, G.; Moeck, I.; Brandt, W.; Hassanzadegan, A.; Magri, F. 3D numerical modeling of hydrothermal processes during the lifetime of a deep geothermal reservoir. *Geofluids* **2010**, *10*, 406–421.
27. Ross, M.; Parent, M.; Lefebvre, R. 3D geologic framework models for regional hydrogeology and land-use management: A case study from a Quaternary basin of southwestern Quebec, Canada. *Hydrogeol. J.* **2005**, *13*, 690–707.
28. Audouin, O.; Bodin, J.; Porel, G.; Bourbiaux, B. Flowpath structure in a limestone aquifer: Multi-borehole logging investigations at the hydrogeological experimental site of Poitiers, France. *Hydrogeol. J.* **2008**, *16*, 939–950.
29. Gill, B.; Cherry, D.; Adelana, M.; Chiang, X.; Mark, R. Using three-dimensional geological mapping methods to inform sustainable groundwater development in a volcanic landscape, Victoria, Australia. *Hydrogeol. J.* **2011**, *19*, 1349–1365.
30. Szabó, N.P.; Dobróka, M.; Turai, E.; Szűcs, P. Factor analysis of borehole logs for evaluating formation shaliness: A hydrogeophysical application for groundwater studies. *Hydrogeol. J.* **2013**, *22*, 511–526.
31. Turner, R.J.; Mansour, M.M.; Dearden, R. Improved understanding of groundwater flow in complex superficial deposits using three-dimensional geological-framework and groundwater models: An example from Glasgow, Scotland (UK). *Hydrogeol. J.* **2015**, *23*, 493–506.
32. Somaratne, N.; Mann, S. Integrated Use of Geological, Geophysical, Radiocarbon and Stable Isotopes Data for Tracing the Conduit Flow Paths in a Small Karstic Aquifer: Poocher Swamp Freshwater Lens, South Australia. *Environ. Nat. Resour. Res.* **2016**, *6*, 119.

33. El-Kaliouby, H.; Abdalla, O. Application of time-domain electromagnetic method in mapping saltwater intrusion of a coastal alluvial aquifer, North Oman. *J. Appl. Geophys.* **2015**, *115*, 59–64.
34. Porsani, J.L.; Almeida, E.R.; Bortolozo, C.A. TDEM survey in an area of seismicity induced by water wells in Paraná sedimentary basin, Northern São Paulo State, Brazil. *J. Appl. Geophys.* **2012**, *82*, 75–83.
35. Porsani, J.L.; Bortolozo, C.A.; Almeida, E.R.; Santos Sobrinho, N.S. TDEM survey in urban environmental for hydrogeological study at USP campus in São Paulo city, Brazil. *J. Appl. Geophys.* **2012**, *76*, 102–108.
36. Comisión Nacional del Agua (CONAGUA). *Actualización de la Disponibilidad de Agua en el Acuífero Chapantongo-Alfajayucan (1309), Estado de Hidalgo [Water Budget Update of the Chapantongo-Alfajayucan Aquifer (1309), Hidalgo State]*; CONAGUA: Mexico City, Mexico, 2013.
37. British Geological Survey (BGS); Comisión Nacional del Agua (CONAGUA). *Impact of Wastewater Reuse on Groundwater in the Mezquital Valley, Hidalgo State, Mexico*; BGS Final Report WC/98/42; CONAGUA: Mexico City, Mexico, 1998.
38. Lucho-Constantino, C.A.; Prieto-García, F.; Del Razo, L.M.; Rodríguez-Vázquez, R.; Poggi-Varaldo, H.M. Chemical fractionation of boron and heavy metals in soils irrigated with wastewater in central Mexico. *Agric. Ecosyst. Environ.* **2005**, *108*, 57–71.
39. Félix-Cañedo, T.E.; Durán-Álvarez, J.C.; Jiménez-Cisneros, B. The occurrence and distribution of a group of organic micropollutants in Mexico City's water sources. *Sci. Total Environ.* **2013**, *454–455*, 109–118.
40. Gibson, R.; Durán-Álvarez, J.C.; Estrada, K.L.; Chávez, A.; Jiménez Cisneros, B. Accumulation and leaching potential of some pharmaceuticals and potential endocrine disruptors in soils irrigated with wastewater in the Tula Valley, Mexico. *Chemosphere* **2010**, *81*, 1437–1445.
41. Downs, T.J.; Cifuentes-García, E.; Suffet, I.M. Risk screening for exposure to groundwater pollution in a wastewater irrigation district of the Mexico City region. *Environ. Health Perspect.* **1999**, *107*, 553–561.
42. Gallegos, E.; Warren, A.; Robles, E.; Campoy, E.; Calderon, A.; Sainz, M.G.; Bonilla, P.; Escolero, O. The effects of wastewater irrigation on groundwater quality in Mexico. *Water Sci. Technol.* **1999**, *40*, 45–52.
43. Comisión Nacional del Agua (CONAGUA). *Planta de Tratamiento de Aguas Residuales Atotonilco [Atotonilco Wastewater Treatment Plant]*; Federal Government, CONAGUA: Mexico City, Mexico, 2013.
44. Ward, J.; Dillon, P. *Robust Design of Managed Aquifer Recharge Policy in Australia*; The Commonwealth Scientific and Industrial Research Organisation (CSIRO): Canberra, Australia, 2009.
45. Lacher, L.J.; Turner, D.S.; Gungle, B.; Bushman, B.M.; Richter, H.E. Application of hydrologic tools and monitoring to support managed aquifer recharge decision making in the upper San Pedro River, Arizona, USA. *Water* **2011**, *6*, 3495–3527.
46. Megdal, S.B.; Dillon, P. Policy and economics of managed aquifer recharge and water banking. *Water* **2015**, *7*, 592–598.
47. Sinclair Knight Merz (SKM); The Commonwealth Scientific and Industrial Research Organisation (CSIRO). *Progress in Managed Aquifer Recharge in Australia*; Waterlines Report No. 73; National Water Commission: Canberra, Australia, 2012.
48. Rossetto, R.; Ansiati, A.; Barbagli, A.; Borsi, I.; Costabile, G.; Dietrich, P.; Mazzanti, G.; Picciaia, D.; Bonari, E. *Managing Induced Riverbank Filtration (IRF) at the Serchio Rivel Well Field, Tuscany, Italy (Italy)*; Geophysical Research Abstracts, 16, EGU2014-7395; EGU General Assembly: Viena, Austria, 2014.
49. Kacimov, A.R.; Zlotnik, V.; Al-Maktoumi, A.; Al-Abri, R. The green-Ampt one-dimensional infiltration from a ponded surface into a heterogeneous soil. *J. Irrig. Drain. Eng.* **2015**, *136*, 68–72.

50. Geocalli. *Actualización del Estudio Geohidrológico del Valle del Mezquital, Hidalgo [Update of the geohydrologic study of the Mezquital Valley, Hidalgo]*; Contract number GZA-81-68-GD; Secretaría de Agricultura and Recursos Hidráulicos (SARH): San Luis Potosi, Mexico, 1981.
51. Aguirre-Díaz, G.J.; López-Martínez, M. Geologic evolution of the Donguinyó-Huichapan caldera complex, central Mexican Volcanic Belt, Mexico. *J. Volcanol. Geotherm. Res.* **2009**, *179*, 133–148.
52. Lesser-carrillo, L.E.; Lesser-illades, J.M.; Arellano-Islas, S.; González-Posadas, D. Balance hídrico y calidad del agua subterránea en el acuífero del Valle del Mezquital, México central. *Rev. Mex. Cienc. Geol.* **2011**, *28*, 323–336.
53. Miguel-Cortés, F.; Rochín, G.H. Prospección Hidrogeológica y Factibilidad de Extracción de agua Subterránea en el Poblado de San Agustín Tlalixticapa, Hidalgo [Hydrogeological prospection and groundwater abstraction feasibility study, within the San Agustín Tlalixticapa Village, Hidalgo]. Bachelor's Thesis, Faculty of Engineering, Universidad Nacional Autónoma de México, México City, Mexico, 2010.
54. Jiménez, B.; Chávez, A. Quality assessment of an aquifer recharged with wastewater for its potential use as drinking source: "El Mezquital Valley" case. *Water Sci. Technol.* **2004**, *50*, 269–273.
55. Constable, S.C.; Parker, R.L.; Constable, C.G. Occam's inversion: A practical algorithm for generating smooth models from electromagnetic sounding data. *Geophysics* **1987**, *52*, 289–300.
56. Bassiouni, Z. *Theory, Measurement, and Interpretation of Well Logs*; Society of Petroleum Engineers (SPE): Woodlands, TX, USA, 1994.
57. Ran, Q.; Wang, Y.; Sun, Y.; Yan, L.; Tong, M. *Volcanic Gas Reservoir Characterization*; Elsevier: Amsterdam, The Netherlands, 2014.
58. Stefansson, V.; Gudlaugsson, S.T.; Gudmundsson, A. Silica content and Gamma Ray logs in volcanic rocks. In Proceedings of the World Geothermal Congress, Kyushu-Tohoku, Japan, 28 May–10 June 2000; pp. 2893–2897.
59. Archie, G.E. The electrical resistivity logs as an aid in determing some reservoir characteristics. *Trans. Am. Inst. Mech. Eng.* **1942**, *146*, 54–62.
60. Zhou, F.; Yao, G. Sensitivity analysis in permeability estimation using logging and injection-falloff test data for an anthracite coalbed methane reservoir in Southeast Qinshui Basin, China. *Int. J. Coal Geol.* **2014**, *131*, 41–51.
61. Rapti-Caputo, D.; Bratus, A.; Santarato, G. Strategic groundwater resources in the Tagliamento River basin (northern Italy): Hydrogeological investigation integrated with geophysical exploration. *Hydrogeol. J.* **2009**, *17*, 1393–1409.
62. Wright, H.M.N.; Cashman, K.V.; Gottesfeld, E.H.; Roberts, J.J. Pore structure of volcanic clasts: Measurements of permeability and electrical conductivity. *Earth Planet. Sci. Lett.* **2009**, *280*, 93–104.
63. Asfahani, J. Basalt Characterization by Means of Nuclear and Electrical Well Logging Techniques. Case Study from Southern Syria. *Appl. Radiat. Isot.* **2011**, *69*, 641–647.
64. Schlumberger. *Petrel E&P Software Platform 2014*; Release Notes; Schlumberger: Houston, TX, USA, 2014.
65. Driscoll, F.G. *Groundwater and Wells*, 2nd ed.; Johnson Screens: St. Paul, MN, USA, 1986.
66. Eden, R.N.; Hazel, C.P. *Computer and Graphical Analysis of Variable Discharge Pumping Tests of Wells*; Institution of Engineers Australia: Perth, Australia, 1973; pp. 5–10.
67. Rorabaugh, M.I. Graphical and theoretical analysis of step- drawdown tests of artesian wells. *Proc. Am. Soc. Civ. Eng.***1953**, *79*, 362.
68. Theis, C.V. The relation between the lowering of the Piezometric surface and the rate and duration of discharge of a well using ground-water storage. *Trans. Am. Geophys. Union* **1935**, *16*, 519.

69. Duffield, G.M. *AQTESOLV™ for Windows*, Version 4.5; HydroSOLV, Inc.: Reston, VA, USA, 2007.
70. Agarwal, R.G.; Al-Hussainy, R.; Ramey, H.J. An Investigation of Wellbore Storage and Skin Effect in Unsteady Liquid Flow. I: Analytical Treatment. *SPEJ* **1970**, *10*, 279–290.
71. Romero, D.J.; Valko, P.P.; Economides, M.J. Optimization of the Productivity Index and the Fracture Geometry of a Stimulated Well with Fracture Face and Choke Skins. In Proceedings of the International Symposium and Exhibition on Formation Damage Control, Society of Petroleum Engineers, San Antonio, TX, USA, 29 September–2 October 2002.
72. Bourdet, D. Well Test Analysis: The use of advanced interpretation models. In *Handbook of Petroleum Exploration and Production*; Elsevier: Amsterdam, The Netherlands, 2002.
73. Raef, A.E.; Mattern, F.; Philip, C.; Totten, M.W. 3D seismic attributes and well-log facies analysis for prospect identification and evaluation: Interpreted palaeoshoreline implications, Weirman Field, Kansas, USA. *J. Pet. Sci. Eng.* **2015**, *133*, 40–51.
74. QGIS Development Team. *QGIS Geographic Information System Developers Manual*; Open Source Geospatial Foundation Project, QGIS Version 2.12; QGIS Development Team: Lyon, France, 2015.
75. Han, T.; Best, A.I.; Sothcott, J.; North, L.J.; MacGregor, L.M. Relationships among low frequency (2 Hz) electrical resistivity, porosity, clay content and permeability in reservoir sandstones. *J. Appl. Geophys.* **2015**, *112*, 279–289.
76. Goldman, M.; Neubauer, F.M. Groundwater exploration using integrated geophysical techniques. *Surv. Geophys.* **1994**, *15*, 331–361.
77. Stefansson, V.; Gudmundsson, A.; Emmerman, R. *Gamma Ray Logging in Icelandic Rocks. The Log Analyst, November-December*; SPWLA: Houston, TX, USA, 1982; pp. 11–16.
78. Cannon, S. *Petrophysics: A Practical Guide*; John Wiley & Sons: Hoboken, NJ, USA, 2005.
79. Chi, S.; Torres-Verdin, C.; Wu, J.J.J.; Alpak, F.O. Assessment of Mud-Filtrate-Invasion Effects on Borehole Acoustic Logs and Radial Profiling of Formation Elastic Properties. *SPE Reserv. Eval. Eng.* **2006**, *9*, 553–564.
80. Alao, P.A.; Ata, A.I.; Nwoke, C.E. Subsurface and Petrophysical Studies of Shaly-Sand Reservoir Targets in Apete Field, Niger Delta. *ISRN Geophys.* **2013**, *2013*, 102450.
81. Markov, M.; Mousatov, A.; Kazatchenko, E.; Markova, I. Determination of electrical conductivity of double-porosity formations by using generalized differential effective medium approximation. *J. Appl. Geophys.* **2014**, *108*, 104–109.
82. Schulmberger. *Schulmberger Log Interpretation Charts*; Schulmberger: Houston, TX, USA, 2009.
83. Chávez, A.; Maya, C.; Gibson, R.; Jiménez, B. The removal of microorganisms and organic micropollutants from wastewater during infiltration to aquifers after irrigation of farmland in the Tula Valley, Mexico. *Environ. Pollut.* **2011**, *159*, 1354–1362.
84. World Health Organization (WHO); The United Nations Children's Fund (UNICEF). *Global Water Supply and Sanitation Assessment 2000 Report*; The WHO and UNICEF: Washington, DC, USA, 2000.
85. Carter, R.C.; Tyrrel, S.F.; Howsam, P. The impact and sustainability of water and sanitation programmes in developing countries. *Water Environ. Manag.* **1997**, *13*, 292–296.
86. Water and Environmental Health at London and Loughborough (WELL). *Guidance Manual on Water Supply and Sanitation Programmes*; Water Engineering DC: Loughborough, UK, 1998.

87. Gleick, P.H. Basic water requirements for human activities: Meeting basic needs. *Water Int.* **1996**, *21*, 83–92.
88. World Health Organization (WHO). *The Human Right to Water and Sanitation*; Media Brief; WHO: Geneva, Switzerland, 2011; pp. 1–8.
89. Jacob, C.E. Drawdown test to determine effective radius of artesian well. *Trans. Am. Soc. Civ. Eng.* **1947**, *112*, 1047–1064.
90. Walton, W.C. *Selected Analytical Methods for Well and Aquifer Evaluation*; Bulletin 49; Illinois State Water Survey: Urbana, IL, USA, 1962.
91. Ramey, H.J. Well-loss function and the skin effect: A review. In *Recent Trends in Hydrogeology*; Narasimhan, T.N., Ed.; Special Paper 189; Geological Society of America: Boulder, CO, USA, 1982; pp. 265–271.
92. Renard, P. Hydraulics of Well and Well Testing. In *Encyclopedia of Hydrological Sciences*; Anderson, M.G., McDonnell, J.J., Eds.; John Wiley & Sons, Ltd.: Chichester, UK, 2006.
93. Horne, R.N. *Modern Well Test Analysis: A Computer Aided Approach*; Petroway Inc.: Palo Alto, CA, USA, 1995.
94. Rosberg, J.E.; Bjelm, L. Well development by jetting using coiled tubing and simultaneous pumping. *Ground Water***2009**, *47*, 816–821.
95. SARH. *Servicios de Prospección y Levantamientos Geológicos y Geofísicos en la Zona de Alfajayucan-Chapantongo-Actopan, [Geological and Geophysical Prospection Studies in the Alfajayucan-Chapantongo-Actopan Area]*; Final Report; SARH: San Luis Potosi, Mexico, 1981.
96. Comisión Nacional del Agua (CONAGUA). *Actualización de la Disponibilidad de Agua en el Acuífero Valle del Mezquital (1310), Estado de Hidalgo [Water Budget Update of the Mezquital Valley Aquifer (1310), Hidalgo State]*; CONAGUA: Mexico City, Mexico, 2015.
97. Luijendijk, E.; Bruggeman, A. Groundwater resources in the Jabal Al Hass region, northwest Syria: An assessment of past use and future potential. *Hydrogeol. J.* **2008**, *16*, 511–530.
98. Jaworska-Szulc, B. Groundwater flow modelling of multi-aquifer systems for regional resources evaluation: The Gdansk hydrogeological system, Poland. *Hydrogeol. J.* **2009**, *1*, 1521–1542.
99. Haitjema, H.M.; Mitchell-Bruker, S. Are water tables a subdued replica of the topography? *Ground Water* **2005**, *43*, 781–786.
100. Garfias-Quezada, A. Modelación Numérica de Acuíferos en Diferencias Finitas Utilizando la Interfaz Libre PMWIN: Fundamentos Teóricos y caso de Aplicación en el Valle del Mezquital [Finite-Difference Groundwater Flow Modeling Using the Open Source Interface PWMIN: Theoretical Foundations and Application Case in the Mezquital Valley Aquifer]. Bachelor's Thesis, Faculty of Engineering, Universidad Nacional Autónoma de México, México City, Mexico, 2015.
101. Wada, Y.; van Beek, L.P.H.; van Kempen, C.M.; Reckman, J.W.T.M.; Vasak, S.; Bierkens, M.F.P. Global depletion of groundwater resources. *Geophys. Res. Lett.* **2010**, *37*, L20402.
102. Comisión Nacional del Agua (CONAGUA). *Actualización de la Disponibilidad de Agua en el Acuífero Cuautitlán-Pachuca (1508), Estado de México [Water Budget Update of the Cuautitlán-Pachuca Aquifer (1310), Hidalgo State]*; CONAGUA: Mexico City, Mexico, 2015.

CHAPTER 15

Water Supply Source Evaluation in Unmanaged Aquifer Recharge Zones: The Mezquital Valley (Mexico) Case Study

Antonio Hernández-Espriú [1,*], Claudia Arango-Galván [2], Alfonso Reyes-Pimentel [3], Pedro Martínez-Santos [4], Carlos Pita de la Paz [5], Sergio Macías-Medrano [1], Alberto Arias-Paz [1] and José Agustín Breña-Naranjo [6]

[1]*Hydrogeology Group, Earth Sciences Division, Faculty of Engineering, Universidad Nacional Autónoma de México (UNAM), C.P. 04510 Mexico City, Mexico*

[2]*Department of Geomagnetism and Exploration, Institute of Geophysics, Universidad Nacional Autónoma de México (UNAM), C.P. 04510 Mexico City, Mexico*

[3]*Comisión Nacional de Hidrocarburos, C.P. 03700 Ciudad de México, Mexico*

[4]*Department of Geodynamics, Faculty of Geological Sciences, Universidad Complutense de Madrid, 28040 Madrid, Spain*

[5]*GEOTEM Ingeniería, S.A. de C.V., C.P. 14640 Mexico D.F., Mexico*

[6]*Institute of Engineering, Universidad Nacional Autónoma de México (UNAM), C.P. 04510 Mexico City, Mexico*

ABSTRACT

The Mezquital Valley (MV) hosts the largest unmanaged aquifer recharge scheme in the world. The metropolitan area of Mexico City discharges ~60 m^3/s of raw wastewater into the valley, a substantial share of which infiltrates into the regional aquifer. In this work, we aim to develop a comprehensive approach, adapted from oil and gas reservoir modeling frameworks, to assess water supply sources located downgradient from unmanaged aquifer recharge zones. The methodology is demonstrated through its application to the Mezquital Valley region. Geological, geoelectrical, petrophysical and hydraulic information is combined into a 3D subsurface model and used to evaluate downgradient supply sources. Although hydrogeochemical variables are yet to be assessed, outcomes suggest that the newly-found groundwater sources may provide a long-term solution for water supply. Piezometric analyses based on 25-year records suggest that the MV is close to steady-state conditions. Thus, unmanaged recharge seems to have been regulating the groundwater balance for the last decades. The transition from unmanaged to managed recharge is expected to provide benefits to the MV inhabitants. It will also be likely to generate new uncertainties in relation to aquifer dynamics and downgradient systems.

Keywords: aquifer; groundwater exploration; water supply; well logs; geophysical prospection; TDEM; pumping test; unmanaged recharge; wastewater irrigation; Mezquital Valley

1. INTRODUCTION

Aquifer exploration and water supply evaluation is a key challenge in low- and mid-income countries [1,2,3], where the characterization of potential groundwater sources is still needed and where a spectacular increase in groundwater use has occurred in the past decades [4].

Groundwater exploration is usually most needed in arid regions. This is because scanty rainfall and high evapotranspiration rates tend to make surface sources unreliable. In these areas, aquifers play an essential role in ensuring sustainability, particularly whenever unregulated recharge schemes are involved. This paper focuses on developing a comprehensive and transferable approach to assessing groundwater supply sources. The methodology is illustrated through its application to a semiarid region subject to unmanaged aquifer recharge practices: the Mezquital Valley in central Mexico (Figure 1a). More specifically, we assess a representative sector of the so-called Alfajayucan aquifer (green boundary in Figure 1b,c), which is located downstream to the Mezquital Valley unmanaged region (Figure 1c).

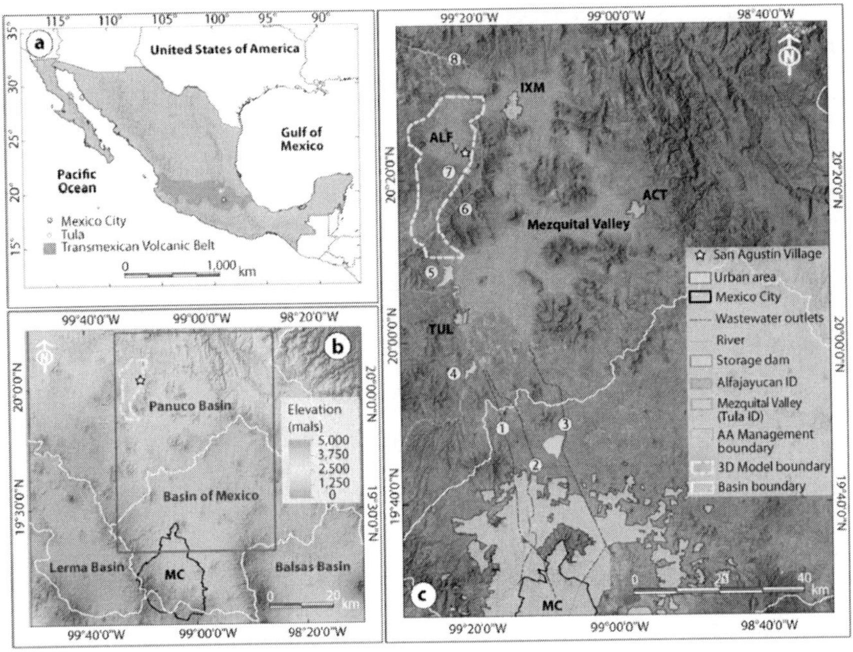

Figure 1. Geographical setting. (**a**) Mexico; (**b**) Mexico City (MC) and hydrologic watersheds; (**c**) study area. For part (**c**), MC: Mexico City, ACT: Actopan, TUL: Tula, IMX: Ixmiquilpan, ALF: Alfajayucan, ID: Irrigation District, AA: Alfajayucan aquifer (the management boundary is shown in green line, parts **b** and **c**), (1) Central Emitter; (2) Western wastewater outlet; (3) Grand Canal; (4) Requena Dam; (5) Endhó Dam; (6) Zinthe Mountain Range; (7) Rojo Gómez Dam; (8) Tula River. Note that the 3D Model boundary (yellow line in parts **b** and **c**) is shown in Section 3.6.

Evaluating groundwater resources in the vicinity of the world's largest unmanaged recharge region poses a major technical challenge. To address it, we develop an inclusive methodology based on geological reservoir modeling techniques. This encompasses all aspects related to stratigraphic, lithological and petrophysical properties of subsurface rocks, and serves the purpose of estimating the spatial distribution of target units and the volume of economic fluids [5]. Available information for integrative modeling combines static and dynamic data at different scales. Static records include geophysical surveys, well logs and geological data [6] whilst dynamic information provides evidence of the interactions between the static model and the fluid over time. This approach delivers an appropriate means to deal with incomplete datasets [7,8] and has been the standard in oil and gas reservoir modeling for at least three decades [9,10].

The aim of reservoir characterization is to understand and identify regional flow units, i.e., hydrofacies, as well as to assess the distribution of relevant properties in the vicinity of a borehole [11]. Static reservoir modeling is frequently coupled with complementary geophysical techniques. This is because drilling data is characterized by high vertical resolution–low horizontal resolution, whereas geophysical soundings present the opposite pattern [8]. In the case of the former, well logs provide a quantitative method to assess near-borehole petrophysical signatures. Within the latter, deep horizons (usually derived from seismic data) deliver the geometric, structural and stratigraphic properties of subsurface units [12,13,14,15,16,17,18]. Moreover, advances in three-dimensional (3D) mapping provide alternative means to developing visual representation of subsurface reservoirs and quantifying some of their key properties [7,19,20,21].

Static modeling frameworks have been successfully used to deal with a variety of practical problems. These include geological CO_2 sequestration and storage [22,23], uncertainty analysis in enhanced oil recovery processes [24], land subsidence related to subsurface gas development [21], correlation between oil field analogues [25] or the understanding of geothermal systems [26], among others.

Mainstream static reservoir modelling techniques, including 3D mapping or well logging, have gained recognition in the field of groundwater exploration in recent years [27,28,29,30,31]. However, integrated approaches (e.g., [32]) such as the one we present are comparatively scarce in the groundwater literature. More specifically, these have seldom been used for the purpose of characterizing groundwater bodies, and their application in the context of unmanaged aquifer recharge is practically non-existent.

A combination of geological, geoelectrical, petrophysical and hydraulic methods is used to gain an understanding of the interaction between lithological factors, hydrogeological properties and the spatial distribution of groundwater. Time domain electromagnetic (TDEM) techniques were used for subsurface modeling. TDEM has been widely used to characterize potential groundwater zones [33,34,35]. The TDEM-resistivity model was correlated with drilling data, and then interpreted jointly with well logs and pumping tests so as to deduce hydrogeological properties and borehole performance.

Within this context, the main contribution of this paper to the literature consists in demonstrating how these techniques may lead to locating and assessing reliable long-term sources of drinking water in unmanaged regions. This research presents additional value by dealing with a rather unique hydrogeological setting in the world.

2. MATERIALS AND METHODS

2.1. Geological and Hydrogeological Setting

The study area corresponds to the Alfajayucan aquifer system (Figure 1b,c), which in turn is located downstream from the Mezquital Valley, central Mexico (Figure 1c). Climate is predominantly semiarid. Rainfall amounts to 450 mm/year, while potential evapotranspiration is 2100 mm/year [36].

Mexico City discharges ~60 m^3/s of raw wastewater without formal treatment into the Mezquital Valley (Figure 1b,c). Wastewater is transported via a complex network that includes the Central Emitter, the Grand Canal and the Western wastewater outlets (Figure 1c). These are all part of a ~90,000 ha irrigation district that straddles across the Tula and Alfajayucan districts [36] (Figure 1c).

Artificial recharge within the Mezquital Valley is controlled by unmanaged wastewater discharge and estimated at 25 m^3/s, that is, over 13 times the natural recharge rate [36]. This has caused the water table to rise since the 1950s. New artesian wells and springs have appeared as a result. Some showcase substantial flowrates. Take for instance the Cerro Colorado spring (~600 L/s) [37].

Overall, the Mezquital Valley is the largest raw wastewater-based agricultural irrigation region in the world [37]. For nearly 110 years, raw wastewater has been used as a source of irrigation water and nutrients for crops. This has allowed local farmers to minimize production costs [38]. Numerous authors deal with the environmental impacts of wastewater irrigation [38], organic micropollutants [39], leaching potential of pharmaceutical components [40], health risk assessments [41] and groundwater pollution [42]. These and other environmental concerns led the Federal Government to build the Atotonilco water treatment plant in 2013 [43]. The plant was designed to treat ~35 m^3/s of wastewater by means of biological and physicochemical processes. Operations are expected to begin in 2016–2017.

The Mezquital Valley provides a unique example of unmanaged aquifer recharge (UAR). UAR is described as the disposal of water that results in unintentional aquifer replenishment [44], in contrast with managed recharge schemes [45,46]. There have been examples of longstanding unmanaged recharge practices in countries such as Australia [47]. More recent ones pertain to Italy [48] or Oman [49]. In other regions of the world UAR has received less attention.

Previous work suggests that the Mezquital Valley might present hydraulic continuity with downgradient permeable formations, such as the volcanic deposits of the Alfajayucan aquifer [50]. Population nuclei located in these areas (San Agustín Village, Ixmiquilpan, Alfajayucan and others, see Figure 1c) are thus likely to be influenced by unregulated activities in terms of water quality and quantity.

Regional geology includes volcanic rocks and calcareous sediments (Figure 2). Lower Cretaceous rocks (El Doctor, Mexcala and Soyatal formations) are made up of limestone with interbedded clay horizons. A dolomite formation caps the sequence. Tectonic activity presents a NW-SE sequence of anticlines and synclines.

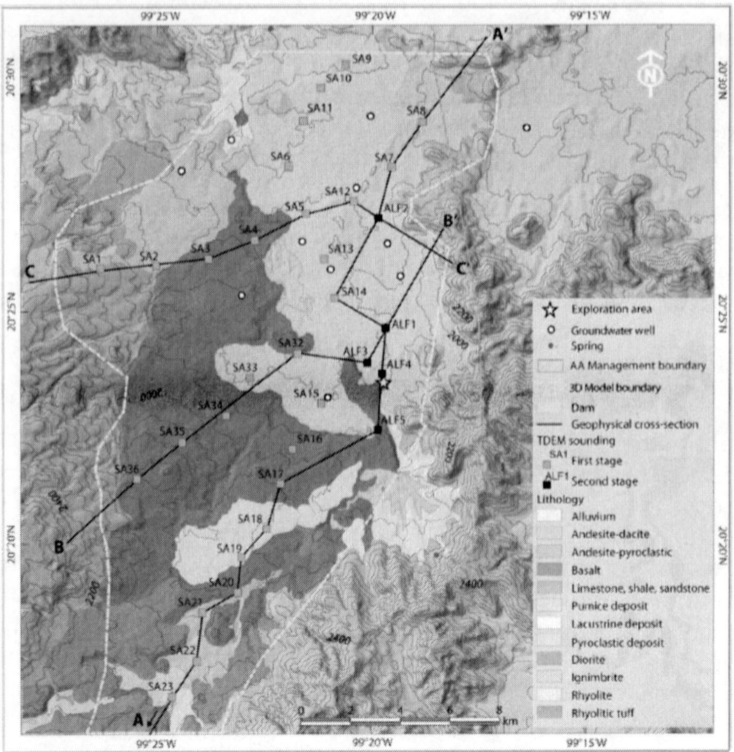

Figure 2. Geological map of the study area, adapted from [53]. AA: Alfajayucan aquifer. The 3D model boundary (in yellow) is shown in Section 3.6.

Middle and Upper Tertiary deposits such as the Tarango Formation (the main supply aquifer for Mexico City) are made up of lava flows and volcanic ash of basaltic-andesitic composition, with interbedded lacustrine deposits [37].

Tertiary volcanic rocks of the Donguinyó and Huichapan formations, related to the eruption of the Donguinyó-Huichapan caldera complex, predominate across the region. The eruptions released discrete pulses of intermediate flows and minor rhyolitic, ignimbrites, pumice fall and surge deposits [51].

From the regional standpoint (Mezquital Valley and surroundings areas), a multi-layer aquifer system has been proposed as a valid conceptual groundwater flow model. However, there are different conceptualizations. The British Geological Survey (BGS) and CONAGUA [37] delineated three overlapping aquifers: (a) the lava flows and coarse sediments of the Tarango Formation; (b) the Tertiary lava-pyroclastic deposits; and (c) the Cretaceous limestones. Other authors reported the existence of two aquifers based on water level measurements [52], while the National Water Commission [36] describes the Alfajayucan aquifer as a two-layer system.

The upper layer is made up of granular and fractured volcanic rocks up to ~400 m thick, while the lower (confined) layer is made up of limestone sediments and calcareous clays. Target units within the Alfajayucan aquifer correspond to the Donguinyó and Huichapan volcanic deposits.

The Mezquital Valley relies on groundwater for domestic supply. There are 456 wells, which extract ~98 Mm^3/year [52]. Approximately 17% of this volume corresponds to human consumption, 38% to agriculture, 33% to the industry and 12% to other uses [54].

On the other hand, 76 water sources were documented in 2012–2013 for the Alfajayucan regional aquifer. These include 45 boreholes, nine dug wells, 21 springs and one gallery. Groundwater abstraction is ~8 Mm^3/year, out of which 64% is for human consumption, 23% for agriculture and 13% for other uses [36]. Within the study area (Figure 2), 11 pumping wells were identified. Out of these, nine are currently inactive. There are significant uncertainties in relation to their depth, screening and hydraulic behavior.

2.2. Hydrogeophysical Investigation

Geophysical data was performed in two stages (Figure 2). The first one consisted of 26 soundings. These were acquired with a single loop configuration, 300 m by side. Five additional TDEM were acquired during the second stage with the same loop configuration, but with a size of 150 m side. All measurements were taken using a TerraTem equipment by Alpha Geoscience Co. (Clifton Park, NY, USA) maintaining an intensity of 7–8 amperes. After editing and averaging the decay curves, apparent resistivity curves were obtained for each TDEM station. One-dimensional (1D) models were computed using an Occam inversion scheme [55].

2.3. Borehole Drilling and Well Log Analysis

A 200-m, 12″-diamater borehole was drilled using a Gardner-Denver 14 W rig. The location was established based on the integration of geological and geophysical data (Figure 2). A static water table was found at a depth of 81 m. Geophysical logs were gathered so as to obtain the petrophysical properties of subsurface units. An 8144 Century Geophysical Co device was used for this purpose, with a voltage of 36 VDC and a logging speed of 9 m/min. This allowed to log variables such as temperature, spontaneous potential (SP), short and long normal resistivity, resistivity of the drilling fluid and gamma ray (GR).

Rock physics is based on sedimentary sequences [56], where oil and gas reservoirs usually take place. In volcanic environments, however, the radioactive nature of rocks makes the interpretation of these curves more challenging [57]. While in sedimentary materials the GR curve is controlled by clay content, the magma composition plays a major role in volcanic rocks. Based on the analysis of 254 samples, Stefansson et al. [58] concluded that there is a linear relationship between GR and silica content in volcanic materials. Though unimportant in the case at hand, it must be noted that this relationship needs to be handled with care in instances where silica content may be altered by the presence of clays. Hence, the borehole log could be completed and analyzed from a quantitative perspective. Completion was necessary because it was only possible to recover material from the upper 66 m of the borehole.

Lithology was inferred from silica content. Criteria established by [57] was used for this purpose. Basalt corresponds to a silica content of <52%, andesitic-basalt to 52%–57%, andesite to 57%–63% and dacite-rhyolite to >63%. Silica content was then checked against standard igneous petrology classifications, to ensure that the inferred lithologies were consistent with chemical composition.

Volcanic facies (lithofacies) were deduced based on silica content and GR data. These were in turn associated to their hydraulic properties (hydrofacies). On the other hand, groundwater salinity was estimated from the SP curve. Its response is controlled by electrochemical and electrokinetic variables, although the latter are generally negligible [56].

In-situ measurements were obtained using a Solinst 428 PVC bailer. Electric conductivity and total dissolved solids were measured by means of a Hannah HI 9828 multiparametric device. Field readings were then compared to theoretical salinity contents.

Total porosity is estimated by means of Archie's Law [59]. Archie's equation remains a method of choice for the estimation of porosity from petrophysical data [60]. Archie's equation is valid for saturated non-clayey media, and is expressed as follows [61]:

$$\rho = a\rho_w\phi^{-m} \tag{1}$$

where ρ corresponds to the resistivity of saturated rock [Ω·m], ρ_w is the resistivity of pore water [Ω·m], *m* the cementation coefficient [–] and *a* the tortuosity factor [–].

Equation (1) was used to develop a porosity curve (φ) for the saturated zone. Input data included the long normal resistivity curve and the estimated groundwater resistivity (as inferred from the SP log). The cementation and tortuosity coefficients were modified as per [62]. Four configurations were analyzed: (1) sandy media ($m = 2.5$, $a = 0.62$); (2) carbonate media ($m = 2$, $a = 1$); (3) vitreous media ($m = 2$, $a = 1.85$); and (4) $m = a = 1$. The lowest porosity values were obtained with configuration (4). This is the most consistent estimate with reported data for volcanic environments [63]. Finally, all information was integrated within a Schlumberger Petrel® platform [64] to perform petrophysical data analyses.

2.4. Well Yield and Aquifer Parameters

The borehole was equipped as a groundwater well and tested for aquifer parameters and expected yields. A 10″-diameter stainless steel screen casing (grid size 0.125″) and a gravel pack was fitted to the 100–200 m interval. The remainder of the outer annulus was sealed with a 4:1 cement-bentonite proportion.

The well was developed to remove all remaining drilling fluid and settle the gravel pack (procedures in [65]). Pressurized air was injected for 17 h, until a stable flow with negligible fine-particle content was obtained.

A 350-Horse Power submersible pump was installed at a depth of 100–150 m and used to carry out pumping tests. The pump was connected to a 150-m long 4″-diameter pipe. A 42-h 7-step pumping test was carried out to determine well yield, efficiency and aquifer parameters. Steps included: (1) flow rate = 534 m^3/day from 0 to 360 min; (2) 817 m^3/day from 360 to 720 min; (3) 1130 m^3/day from 720 to 1080 min; (4) 1332 m^3/day from 1080 to 1440 min; (5) 1532 m^3/day from 1440 to 1800 min; (6)

1714 m^3/day from 1800 to 2160 min; and (7) 1904 m^3/day from 2160 to 2520 min. Field measurements (elapsed time vs. drawdown vs. flow rate) are shown in Table 1.

Table 1. Field measurements derived from the step-drawdown test.

534 m^3/day [1]		817 m^3/day [1]		1130 m^3/day [1]		1332 m^3/day [1]		1532 m^3/day [1]		1714 m^3/day [1]		1904 m^3/day [1]	
Time (min)	s (m)	Time (min)	s (m)	Time (min)	s (m)	Time (min)	s (m)	Time (min)	s (m)	Time (min)	s (m)	Time (min)	s (m)
60	0.98	420	1.63	780	2.35	1140	2.89	1500	3.43	1860	3.75	2220	4.19
120	1.02	480	1.63	840	2.37	1200	2.87	1560	3.45	1920	3.76	2280	4.21
180	1.01	540	1.65	900	2.37	1260	2.95	1620	3.43	1980	3.76	2340	4.21
240	1	600	1.68	960	2.34	1320	2.96	1680	3.43	2040	3.76	2400	4.21
300	1.04	660	1.68	1020	2.32	1380	2.98	1740	3.44	2100	3.76	2460	4.22
360	0.99	720	1.68	1080	2.34	1440	2.98	1800	3.44	2160	3.76	2520	4.23

Notes: [1] Flow rate; s: Water table drawdown.

Results allowed for the evaluation of head losses associated to the aquifer properties and to the borehole itself. Classic interpretation models were used for this purpose [66,67]. These assume that theoretical drawdown and head losses can be estimated by means of Equation (2):

$$s_w = BQ + CQ^P \quad (2)$$

where s_w corresponds to the total drawdown inside the well [L], Q is well flow rate for each step [LT^{-3}], BQ is the theoretical drawdown component (the B coefficient is expressed in [TL^{-2}] units), and CQ^n is the nonlinear component of head losses (coefficient C is measured in [T^2L^{-5}] units). Hydraulic parameters were estimated by curve-matching procedures, using a nonlinear weighted least-squares parameter estimation and the Gauss-Newton linearization method.

Field data was fitted to the Theis model [68] for a pumping well in a confined aquifer, modified to include linear and nonlinear well losses in a step-drawdown test [69] (Equation (3)):

$$s_w = \frac{Q}{4\pi T}\left[w\left(\frac{r^2 S}{4tT}\right) + 2\sigma\right] + CQ^P \quad (3)$$

where w is the well function, i.e., the exponential integral function, r represents the well radius [L], S and T correspond to the storage coefficient [–] and aquifer transmissivity [L^2T^{-1}], respectively; t is the elapsed time [T] and σ the skin factor [–], estimated by means of Equation (3). The skin factor is an indicator of well damage and has been widely evaluated in the oil reservoir practice [70,71]. The skin factor causes an additional pressure drop, Δp_σ, across the altered zone in hydrocarbon wells [72]:

$$\Delta p_{\sigma} = \frac{141.2 q B_{o} \mu}{kh} \sigma \tag{4}$$

Equation (4) was adapted to cater for the conditions of groundwater systems. Considering that the oil-formation volumetric factor, B_o, and viscosity, μ, equals 1 in a single-flowing phase, and that matrix permeability (k)—reservoir thickness (h) product is equivalent to T, Equation (4) can be rearranged as follows:

$$\Delta h_{\sigma} = \frac{Q}{T} \sigma \tag{5}$$

where Δh_σ represents the additional drawdown caused by the skin factor [L]. Step test results were analyzed with AQTESOLV Pro, v4.5 (Hydrosolve Inc., Reston, VA, USA) [69].

2.5. Subsurface Models

2.5.1. Hydrogeoelectrical Model

TDEM resistivity models were interpolated in 10 m steps for the upper 100 m, 20 m between 100 and 200 m and 40 m for depths up to 400 m. A digital elevation model (DEM) was integrated in Petrel [64] as a reference for subsurface layers. Resistivity curves were upscaled and interpolated by means of a moving averages model with a fourth-power inverse distance algorithm. The hydrogeophysical resistivity model was used to correlate hydrofacies based on horizons and surfaces, as commonly done in the oil industry [73]. A square 800 × 800 m mesh was used to devise the hydrogeological units. Additionally, a vertical resistivity gradient was developed and used to identify subsurface electrofacies, associated to hydrostratigraphic changes.

2.5.2. Total Porosity Model

Borehole data was integrated with TDEM results using Archie's equation [59]. In the absence of predominantly clayey minerals, water resistivity is assumed homogeneous across the study area. Conductivity is assumed to be ionic at constant temperature and the effect of chemical thermalism is considered negligible.

2.5.3. 3D Model

Petrel [64], a 3D oil and gas reservoir modeling software, was used to integrate petrophysical data with TDEM results into a 3D visualization model. The model relies on 31 TDEM inverted resistivity curves, geophysical well logs, porosity information, groundwater depth in nine wells, surface geology, a digital elevation model (DEM), the inferred geometry of electrofacies and the geological information from the borehole drilling. Surface geology [51,53] was only useful to define the uppermost part of the top layer. All other layers were inferred from TDEM information. In other words, the model can only be used to infer subsurface geoelectrical properties.

The DEM was based on four DEM scenes from the Advanced Spaceborne Thermal Emission and Reflection Radiometer (ASTER GDEM v.2, NASA, California Institute of Technology, Pasadena, CA, USA) and then imported to Petrel as a shape file. DEM scenes present a native pixel resolution of 3 × 3 arcsec. Individual scenes were re-projected to a UTM coordinate system (zone 14 N) with a 90 m × 90 m pixel size. Bilinear resampling was used to account for realistic spatial deformation. This process was carried out using QGIS 2.12 Lyon (QGIS Development Team, Open Source Geospatial Foundation) [74].

The development of the 3D model consisted in the following steps:

(1) Grid discretization and creation of geological horizons, based on the TDEM interpretation model. For that purpose, we built a three-layer model (layer-1 from 0 to 100 m; layer-2 from 100 to 200 m; layer-3 from 200 to 400 m), with a grid size of 2 km × 2 km, covering a surface area of ~300 km^2. Each layer was divided into 10 to 40 m-thick sublayers (e.g., geocellular grids) to gain vertical resolution;

(2) Geoelectrical input data using resistivity population from each of the 31 TDEM inverted curves (see Section 2.2);

(3) Creation of a 3D electrofacies model of the subsurface that enables the population all geocellular grids with discrete resistivity values of the electrofacies EF-1 (vadose zone), EF-2 (aquifer) and EF-3 (basement). For that purpose, a deterministic modeling approach was used by means of a moving averages algorithm, taking advantage of the Petrel Facies Modeling Module [64].

Since the model is built upon geoelectrical information, uncertainties can be expected to be higher in those areas which are located further away from the 32 control points (e.g., 31 TDEM soundings plus the borehole).

2.6. Groundwater Flow Evolution in the Mezquital Valley

Groundwater flow behavior in unmanaged recharge regions is a key factor when unregulated water disposal is a major recharge component for downgradient supply aquifers. In the case at hand, an investigation conducted by Geocalli [50] proposed that the Mezquital Valley might present hydraulic continuity with downgradient aquifers, including the study area (Figure 1c and Figure 2).

In order to assess potential hydraulic connectivity and to evaluate if a decrease in the unmanaged wastewater disposal may compromise downgradient aquifer replenishment, piezometric records were analyzed. These are considered a suitable proxy for temporal changes in the wastewater input from the Mezquital Valley.

Datasets from 1982 [50], 1998 [37] and 2007 [52] were taken into consideration. Water table depth contours were digitized from the original sources and the temporal changes of groundwater levels between 1982 and 1998, 1998 and 2007 and 1982 and 2007 were assessed by contouring the hydraulic evolution. Three new maps were created as a result, based on level measurements in 152, 59 and 30 production wells, respectively.

The groundwater evolution of each period of time was analyzed using conventional statistical tools. The spatial and statistical analysis was carried out in QGIS 2.12 Lyon [74].

3. RESULTS AND DISCUSSION

3.1. Hydrogeoelectrical Model

Figure 3 presents the hydrogeoelectrical 2D model (cross sections are shown in Figure 2). Electrofacies 1 (EF-1) corresponds to the uppermost unit. EF-1 presents a relatively homogeneous geometry. The average thickness is ~70 m, and resistivity ranges between 40 and 300 Ω·m, with a bimodal behavior (modes in 50 and 110 Ω·m). Gradients are negative and observed to change laterally. This unit is interpreted as the vadose zone of the system and correlated with a heterogeneous mix of lacustrine, alluvial and volcanic deposits.

Electrofacies 2 (EF-2) presents negative gradients and variable resistivities of 5–170, 70–90 and 20–80 Ω·m. Thickness ranges between 190 and 250 m (average = 220 m). Geoelectrical data suggests that geological heterogeneity is controlled by changes in volcanic facies. This unit is correlated to aquifer formations made up of Quaternary basaltic deposits, pyroclastic flows, volcanic breccia and sediments from Mio-Pliocene formations Donguinyó and Huichapan.

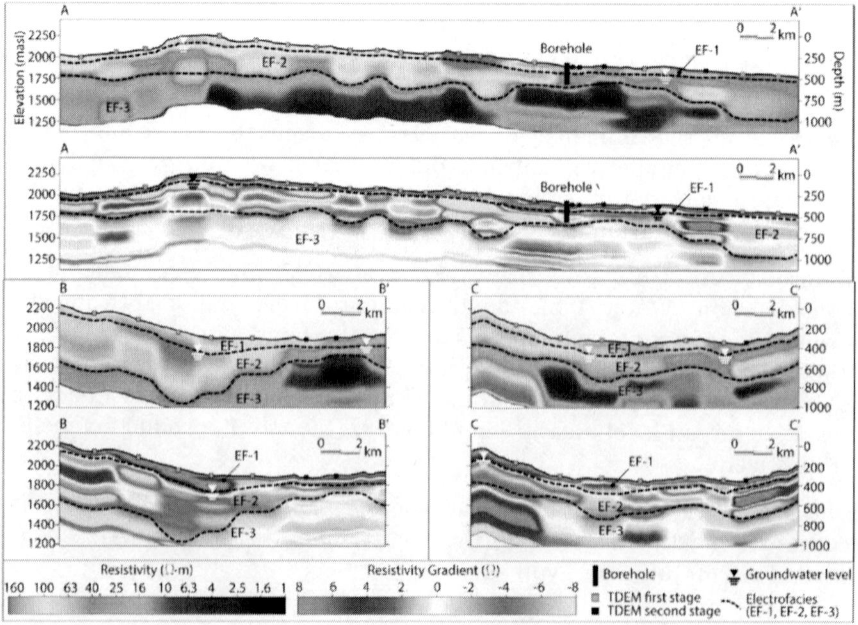

Figure 3. Subsurface hydrogeoelectrical model. EF: electrofacies. AA′, BB′ and CC′ refer to geophysical cross-sections shown in Figure 2.

Electrofacies 3 (EF-3) provides a hydrogeophysical contrast with EF-1 and EF-2. Resistivity is comparatively low (4–20 Ω·m) at depths of 250–400 m. EF-3 consists of low-permeability clay sediments, marls and clayey limestones, probably related to the Soyatal Formation. Thus, unit EF-3 might represent the system basement. Our geoelectrical interpretation is consistent with those provided by other authors, which suggested that resistivity tends to diminish with increased clay content [47].

TDEM surveys allowed to identify EF-2 as the unit with a better aquifer potential in the region. Resistivity values (5–170; 70–90; 20–80 Ω·m) are consistent with the existing knowledge about the area's subsurface geology, as well as with standard geoelectrical values for similar geological formations [75,76]. Correlated materials (Huichapan and Donguinyó tuffs) include lithological contrasts associated to the eruption of lava and pyroclastic flows [51]. Unit EF-2 presents a marked resistivity contrast with the basement unit, EF-3.

3.2. Near-Borehole Petrophysical Response

3.2.1. Reconstructed Borehole Lithology

Figure 4 shows the results of geophysical logs carried out in the borehole. All readings except GR begin at the static groundwater level. The incomplete geological record (from 66 to 200 m) was reconstructed primarily from the GR curve. GR yielded values of ~60°–90° API (API = American Petroleum Institute) for the upper 40 m, growing to 90°–110° API between 40 and 60 m. The former is controlled by the presence of radioactive minerals in andesites and dry basalts. Increase with depth can be explained by the presence of a 20-m thick dacite horizon where an intensification in radioactivity was observed. From 60 to 200 m, values remain uniform at ~30°–60° API. Some radioactive peaks were also identified (e.g., 164–174 m). These were attributed to the presence of silica-rich rocks.

High GR readings usually hint at the presence of clay-rich materials. However, this does not appear to occur. A number studies in igneous terrains have demonstrated that a relationship exists between the intensity of GR response and the geochemical composition of volcanic rocks [57,77], instead of clayey minerals in sedimentary environments.

Criteria by [58] was used to obtain the SiO_2 content of volcanic rocks as a response of GR values. Lithology was then inferred from silica content. The reconstructed borehole lithology is shown in Figure 4.

A 48-m thick basalt unit was identified between 63 and 111 m. Basalt presents an estimated silica content of <52%. This unit is underlain by an andesitic breccia (~48%–62% SiO_2), andesitic flows (silica ~50%–65%, depths 111.5–121 m and 130–150 m) and acidic rocks (silica ~62%–95%, depth 163 to 174 m). This interbedded sequence can be correlated to EF-2. Thus, the borehole log reflects the presence of Huichapan and Donguinyó tuffs. Aguirre-Díaz et al. [24] report silica contents of 73.7%–74.5% and 55.6%–64.9% in outcrops of the Huichapan ignimbrite and Donguinyó tuffs, respectively, which are consistent with our results for the subsurface analysis.

3.2.2. Hydrofacies

Figure 5 shows the correlation between different petrophysical variables. These were used to understand the relationship between petrophysical response, groundwater occurrence and aquifer properties. Resistivity depends on compaction and fracturing, therefore, it is assumed that both can be obtained, at least qualitatively, from the resistive range of each lithological unit.

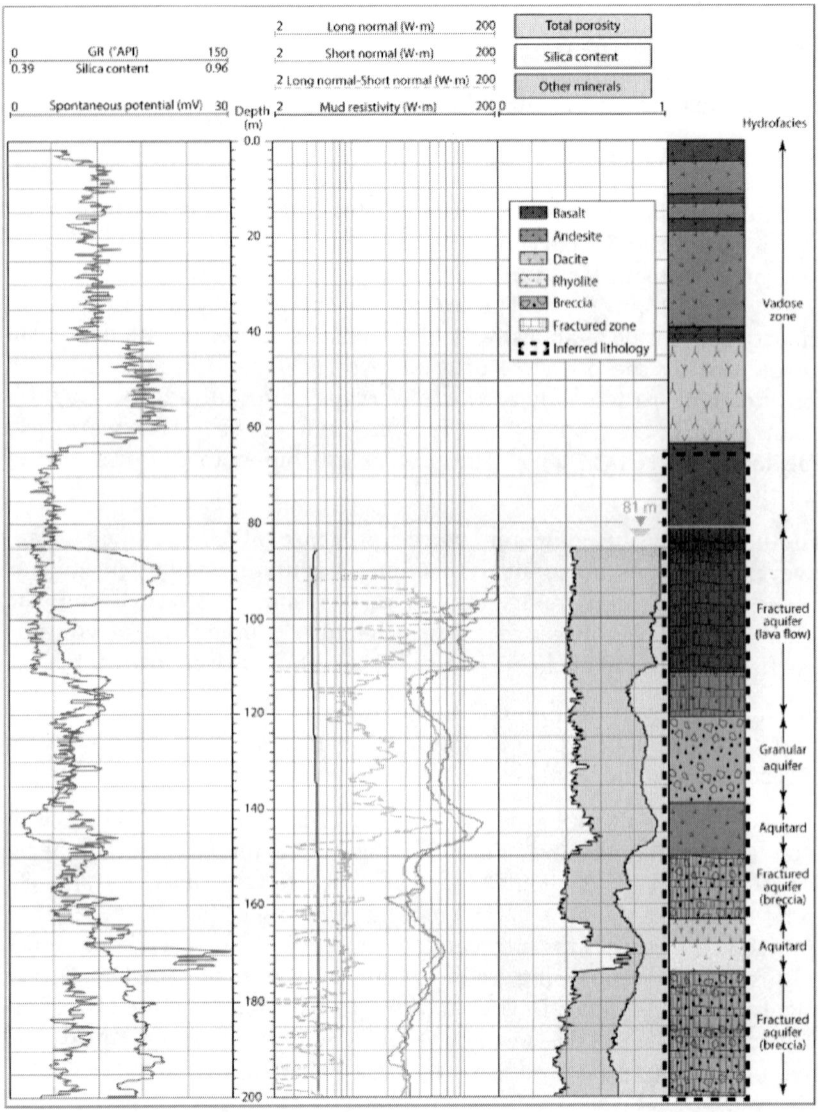

Figure 4. Geophysical logs acquired during drilling, reconstructed lithology and inferred vertical hydrofacies. GR: Gamma Ray; API: American Petroleum Institute.

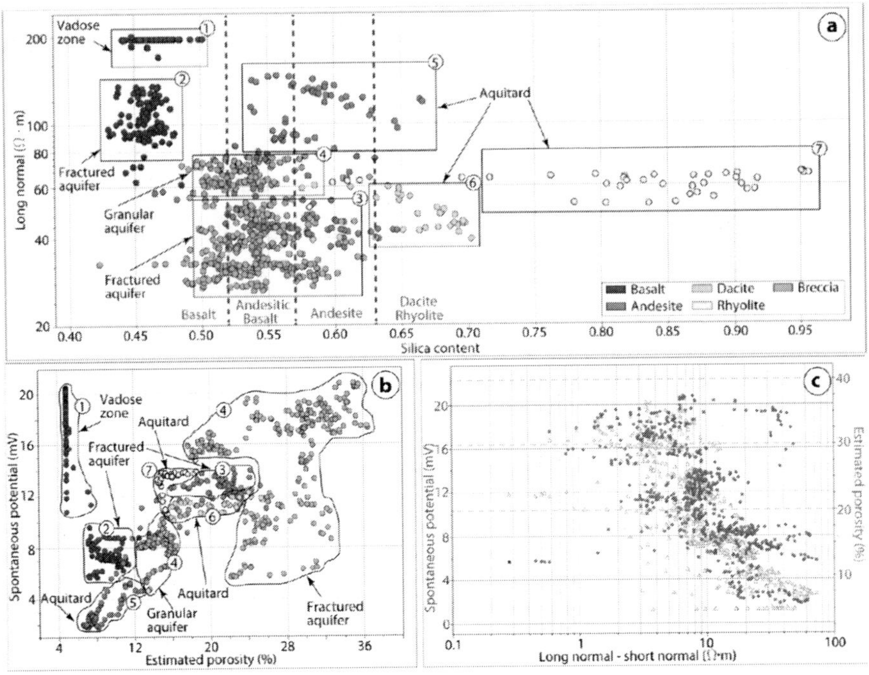

Figure 5. Cross-plots and petrophysical signatures. (**a**) Silica content vs. Long normal resistivity; (**b**) Porosity vs. Spontaneous potential; (**c**) Long normal—Short normal resistivity vs. Spontaneous potential vs. total porosity. For parts (**a**,**b**): (1) Massive basalt; (2) fractured basalt; (3) fractured andesite; (4) andesitic breccia; (5) massive andesite; (6) dacite; and (7) rhyolite.

Seven lythofacies were identified attending to these criteria: (1) massive basalt; (2) fractured basalt; (3) fractured andesite; (4) andesitic breccia (both granular and fractured); (5) massive andesite; (6) dacite; and (7) rhyolite. Lythofacies provide the basis for the identification of the following hydrofacies:

(a) Vadose zone (thickness ~81 m), made up of massive, scarcely fractured basalt. This is not only consistent with collected data, but also with direct observation of rock bits obtained from a depth of 64–66 m.

(b) Fractured lava aquifer (thickness ~40 m). Fracturing is expected to be less intense towards the upper boundary. Total porosity is estimated at 7%–24%, resistivity is 40–150 Ω·m, radioactivity 10°–25°API and silica content <52%.

(c) Granular aquifer in andesitic breccia and tuffs (thickness ~18 m). The granular aquifer presents some degree of fractured behavior. Total porosity is 12%–18%, resistivity is 25–55 Ω·m, radioactivity 45°–63°API and silica content 48%–62%.

(d) Aquitard, made up of andesite and acidic rocks (thickness ~11 and 5–7 m, respectively). Total porosity is 7%–15%, resistivity is 40–143 Ω·m, radioactivity 45°–150°API and silica content 62%–70%.

(e) Fractured breccia aquifer (thickness 13–26 m). Total porosity is 17%–26%, resistivity is 30–43 Ω·m, radioactivity 30°–50°API and silica content 50%–59%.

The vertical distribution of hydrofacies and lythofacies is shown in Figure 4. Petrophysical features are presented in Figure 5a,b.

3.3. Porosity and Qualitative Permeability Estimates

The permeability of hydrofacies was estimated on qualitative terms. Well logs establish that the resistivity of the drilling fluid (4–6 Ω·m) is much lower than that of the hydrofacies (25–200 Ω·m). Under these conditions, positive deflections in the SP curve could be interpreted as permeable sectors. In the absence of clay, the larger the deflection, the more permeable the rock would be. SP and resistivity curves should also present a positive correlation [78].

The SP curve presents a first deflection at ~84–96 m (Figure 4), within the sector identified as massive basalt. There is no clear correlation to establish whether this interval qualifies as permeable. Positive deflections were also found around 174 m. Here, there is a better correlation with resistivity curves. Hence, the SP hints at the presence of permeable rocks between 174 and 200 m. These correspond to the fractured breccia aquifer.

A porosity vs. SP plot was built to evaluate these results in a more systematic manner (Figure 5b). In general terms, porosity and SP correlate well. In volcanic rocks, this could be assimilated to a correlation between porosity and permeability [62], which in turn implies that SP is an indicator of permeability.

Permeability was also analyzed by means of resistivity curves. Previous studies establish that the gap between short and long resistivity curves can be used as an indicator of permeability [78]. However, in high permeability rocks, short and long resistivity curves will present very similar values as a result of deep-mud filtrate invasion. This may lead to spurious conclusions. Uncertainty was evaluated by studying the relationship between SP and the difference between long and short normal resistivity curves. The analysis aims at identifying a negative correlation between the SP curve and the other two. This would suggest that SP increases as the separation between the resistivity curves decreases (as a result of mud filtration). In turn, this could lead to identifying high permeability areas.

Figure 5c presents the results. A negative correlation comes through clearly. Maximum SP values (~15–20 mV) occur when the separation between the resistivity curves is minimal (0.1–2 Ω·m). This suggests that deep-mud filtrate invasion is actually taking place as a consequence of high permeability formations. Invasion areas can be identified visually in Figure 4. These are located at 150–163 m and 174–200 m. Deep-mud filtrate invasion has been previously explored in the oil industry [79], but little research has been carried out on the groundwater front.

Similarly, separation between the resistivity curves decreases with depth, while porosity increases. Therefore, a high permeability could be attributed to both sectors, corresponding to the fractured breccia aquifer (hydrofacies (e)). The porosity of this unit (17%–26%) is consistent with reports for similar geological formations, i.e., 10%–32% in [62].

Porosity is obtained from Archie's law, which in turn relies on resistivity logs. As a result, porosity may be overestimated. Although values from geophysical logs are consistent with the lithology, at the regional level there appears to be an overestimation of the subsurface porosity model. The model established a ~6%–42% interval for pore size distribution in the subsurface (Figure 6).

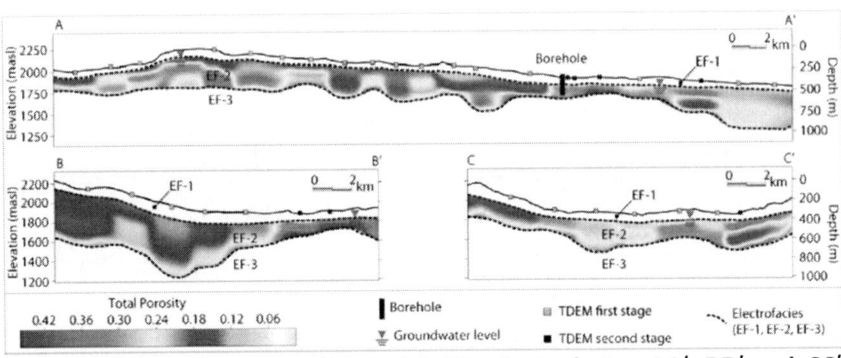

Figure 6. Subsurface porosity model. EF: electrofacies. AA′, BB′ and CC′ refer to geophysical cross-sections shown in Figure 2.

Previous works point out some of the limitations of estimating porosity from Archie's law without sonic logs or lab measurements. Alao et al. [80], for instance, explain that Archie's approach was originally developed for clay-free sedimentary rocks, and it fails to consider additional petrophysical features such as the presence of shale, fresh formation waters or thin-bed layers. Not considering these can lead to overestimate porosity and saturation. In turn, Markov et al. [81] reported that in carbonate rocks with a percolation threshold (connectivity and conductivity loss as the rock reaches critical void space) and chaotic microstructure (various pore systems of different types and scales), porosity cannot be estimated within the framework of Archie's Law.

Finally, Wright et al. [62] argue that permeability, porosity and electrical conductivity may vary by several orders of magnitude within extrusive rocks. This is due to variable tortuosity, vesicle sizes, pore elongation and connected pathways during bubble expansion, eruption conditions and crystallization history. These authors conclude that Archie's two-constant scheme (m and a) could be inappropriate under these circumstances. All this means that the porosity model presented in this paper should be used with caution. Nevertheless, it is advocated as a useful tool to obtain first-approach estimates of porosity and further groundwater reserves.

3.4. Groundwater Salinity

Groundwater salinity for the upper permeable sector (174–200 m depth, within hydrofacies (e)) was estimated by means of petrophysical techniques [56]. Well logs present a positive SP of ~20.5 mV, while groundwater temperature is ~30 °C, the resistivity of the drilling fluid is ~5 Ω·m and the resistivity of the aquifer (R_w) is 9.6 Ω·m. Based on Shlumberger's log interpretation charts [82], a salinity value (NaCl) of ~400 ppm was calculated.

Direct measures of TDS during borehole construction yield 602–614 ppm. These values are representative of hydrofacies (c) and (d). The BGS and CONAGUA [37] reported TDS concentrations in springs and dug wells in Irrigation District 100, near Ixmiquilpan (9 km to the northeast of the borehole, Figure 1c). These range between 656–898 and 973–1280 ppm. More recently, Chávez et al. [83] reported TDS values within the Tula and Mezquital valleys of 865–1586 (wet season) to 861–1488 ppm (dry season) in abstraction wells, 1061–1488 (wet season) to 1111–1513 ppm (dry season) in springs, and 1434–1702 (wet season) to 1402–1597 (dry season) ppm in dug wells, respectively. Thus, the salinity estimates obtained during fieldwork suggest that wastewater infiltration has had no major impacts on aquifer salinity.

3.5. Expected Flow Rates, Well Efficiency and Aquifer Parameters

Step-drawdown test results are shown in Figure 7 and Table 2. These correspond to hydrofacies (b) and (c), that is, to the fractured lava flows and the granular breccia aquifer (100–150 m). The borehole yielded 6–22 L/s with a small drawdown (s_w = 1–4.29 m). Head losses attributable to the aquifer (*BQ*) account for most of the drawdown at each step (0.69 m ≤ *BQ* ≤ 2.22 m; *avg.* = 1.66 m). This means that 58%–70% of s_w is due to the aquifer component (*avg.* = 60.9%), while ~12%–19% is due to the well loss component (*avg.* = 15.9%). Additional nonlinear head losses at each step amounted to 0.12, 0.23, 0.37, 0.47, 0.58, 0.69 and 0.81 m, respectively.

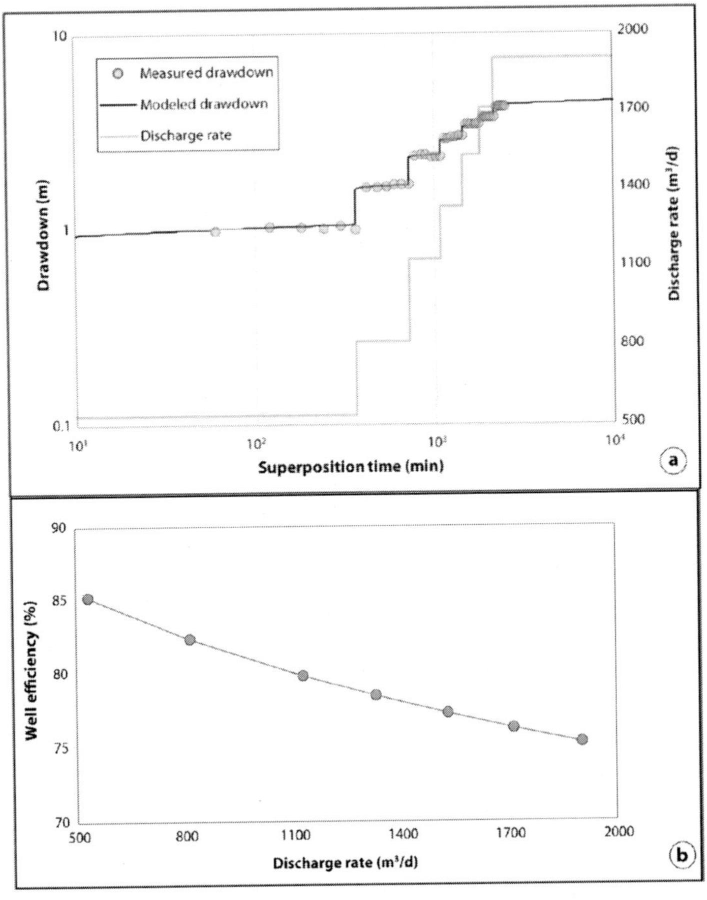

Figure 7. Step test analysis. (**a**) Drawdown and discharge rate vs. time; (**b**) well efficiency vs. discharge rate.

Table 2. Results from the step-drawdown test analysis.

Step	Q (m^3/Day)	s_w (m)	q (L/s/m)	BQ (m)	CQ^P (m)	Δh_σ (m)	Aquifer Loss (%)	Well Loss (%)	w_{eff} (%)
1	534	1.00	6	0.69	0.12	0.46	69.23	12.05	85.18
2	817	1.68	6	1.06	0.23	0.71	62.98	13.55	82.29
3	1130	2.32	6	1.46	0.37	0.98	63.10	15.97	79.80
4	1332	3.00	5	1.72	0.47	1.16	57.49	15.79	78.45
5	1532	3.47	5	1.98	0.58	1.33	57.14	16.83	77.25
6	1714	3.77	5	2.22	0.69	1.49	58.86	18.34	76.24
7	1904	4.29	5	2.46	0.81	1.66	57.46	18.87	75.28
Min	534.00	1.00	5.11	0.69	0.12	0.46	57.14	12.05	75.28
Max	1904.00	4.29	6.18	2.46	0.81	1.66	69.23	18.87	85.18
Mean	1280.43	2.79	5.44	1.66	0.47	1.11	60.89	15.91	79.21
SD	489.44	1.18	0.40	0.63	0.25	0.43	4.48	2.45	3.52

Notes: Q: Flow rate; q: specific flow rate; BQ: theoretical drawdown; CQ^P: nonlinear well loss; Δh_σ: additional drawdown in the well related to the skin factor; w_{eff}: well efficiency; SD: standard deviation.

As the borehole will supply downgradient towns in the short term (e.g., San Agustín Village, Figure 1c), test results appear encouraging. Well yield (534–1904 m^3/day, or 6–22 L/s; mean = ~1280 m^3/day, or ~15 L/s) exceeds by far San Agustin's water demand (~225 m^3/day, or 2.6 L/s). In larger urban cores, such as Alfajayucan (~19,000 inhabitants, according to INEGI (2013, unpublished), Figure 1c), the borehole will be able to roughly assist with ~25% of the current municipal demand.

Domestic demands are estimated by multiplying the number of inhabitants by a daily requirement of 250 L/person/day. This estimate is considered conservative and compliant with international standards. For instance, the WHO/UNICEF Joint Monitoring Programme, established a minimum dosage of only 20 L/person/day [84], to cover basic survival needs. This is consistent with estimates by several authors [85,86], although others suggest that the threshold should be raised to 50 [87]. In turn, World Health Organization (WHO) [88] cites minimal domestic requirements in the order of 50–100 L/person/day. From the volumetric standpoint, all this suggests that the newly-identified groundwater source (e.g., new groundwater well) provides a long term solution for a substantial share of the downgradient population of the Mezquital Valley. However, it is mandatory to carry out a detailed hydrogeochemical study to further assess water quality aspects.

Regarding the hydraulic behavior of the well, the equation $1.86Q + 0.65Q^{1.5}$ (B = 1.86 [min/m^2]; C = 0.65 [min^2/m^5]; P = 1.5) allows for the estimation of the linear aquifer loss component, BC, and the nonlinear well loss, CQ^P, for any flow rate, Q. In this case, P is the order of nonlinear well losses. Jacob [89] suggested that well loss is approximately proportional to the square of the discharge. Thus, P can be assumed to be ~2 in most situations. Rorabaugh [67] showed that P may range from 1.5 to 3 in cases such as the one at hand. Walton [90] proposed an interpretation criteria for the nonlinear well coefficient, C. Values of 1900–3800 and >3800 [s^2/m^5] are indicators of mild and severe well deterioration, respectively.

A coefficient C = 2365 [s^2/m^5] was obtained for the borehole, suggesting suboptimal (but acceptable) behavior.

From the hydraulic point of view, the main implication is an additional drawdown within the well. This is related to turbulent flow at the aquifer–gravel pack interface, as well as to the skin effect. The latter, σ, refers to an alteration in near-borehole permeability, resulting from well completion effects, formation erosion, acidification, well development and mud filtrate invasion [91]. Renard [92] pointed out that σ is positive if the well is clogged and negative for the opposite case (stimulated well), while [93] reports that the skin factor may adopt any value, but that it rarely exceeds 20 in practice. In this case, a skin coefficient of +1.48 was obtained, leading to an additional drawdown of 0.4–1.6 m (*avg.* ≈ 1 m). This amounts to 16% of the total drawdown. Since mud filtrate invasion was earlier identified through the use of petrophysical signatures (Figure 4 and Figure 5c), these effects are best described as mild clogging due to incomplete well development.

Well efficiency (w_{eff}) was estimated by means of Equation (2) in the form weff=BQ/BQ+CQP. Table 2 shows the relative efficiencies of the seven discharge rates. As shown in Figure 7a, w_{eff} exceeds 85% for a rate of 534 m^3/day (6 L/s). It decreases as flow rate increases, up to ~75% for a rate of 1904 m^3/day (22 L/s). Although this efficiency is acceptable, hydraulic behavior and productivity can be improved by completing the well development using common techniques such as overpumping, surging, rawhiding or airlift pumping [65]. More novel methods, such as hydrojetting with coiled tubing and simultaneous pumping, have also been demonstrated to be a cost-effective cleaning procedure [94].

The Theis Model [68] was used to infer aquifer properties. A transmissivity (T) of 1700 m^2/day, a hydraulic conductivity (k) of 8–40 m/day and a storage coefficient (S) of 8.8 × 10^{-3} were obtained, with a mean residual sum squares (RSS) of 0.0023 m between measured and modeled drawdowns. Hydraulic conductivity was estimated by considering a saturated thickness range between 40 and 220 m. The former is referred to as the thickness of hydrofacies (b) while the latter is referred to the average thickness of electrofacies EF-2. In addition, the storage coefficient estimate must be handled with care because it was obtained from drawdown data within the pumping well. Specific capacity is 5–6 L/s/m drawdown. Provided that the well is further developed, these values suggest that flow rates may be increased considerably in case of need.

Aquifer properties are representative of hydrofacies (b) and (c). These correspond to fractured lava flows and volcanic breccia. Previous works report T = 864–4320 m^2/day, k = 0.035–103.7 m/day and S = 2 × 10^{-4}–0.2

for basaltic rocks within the Alfajayucan aquifer [36,95], which are all consistent with the above results. These also show that hydrofacies are highly heterogeneous from hydraulic standpoint.

3.6. 3D Visualization and Integration Tool for Management Purposes

Field information was compiled into a Petrel 3D model [64]. Figure 8 shows several instances of the 3D visualization model. EF-2 represents a regional aquifer. As such, it may be developed at a larger scale. This aquifer is made up of three main hydrofacies, namely, a fractured lava aquifer, a granular breccia aquifer and a fractured breccia aquifer. These are partially isolated from one another due to the presence of intercalated aquitards. The spatial variation of hydrofacies cannot be inferred from a single borehole. Thus, these were only characterized around the drilling site.

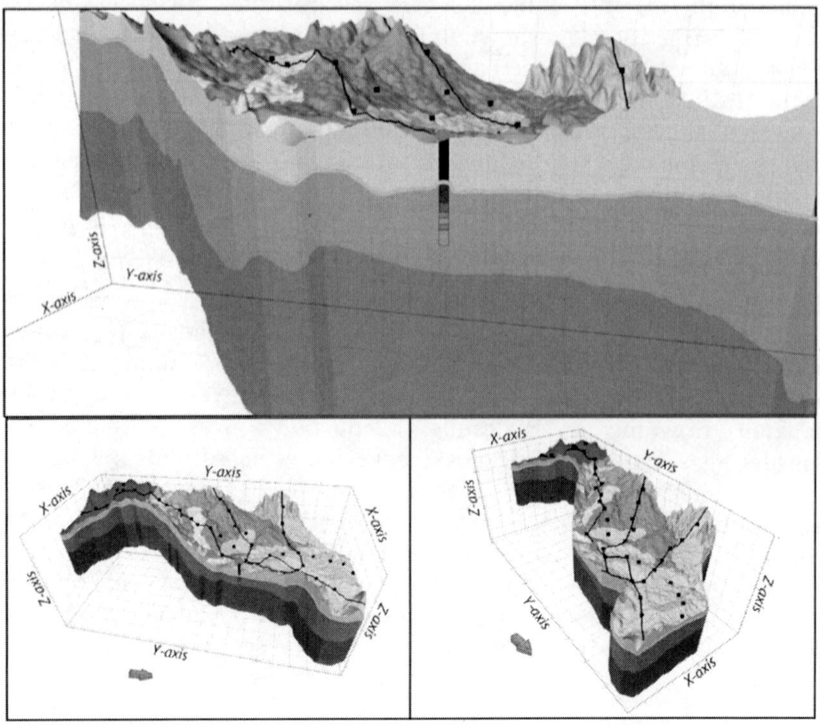

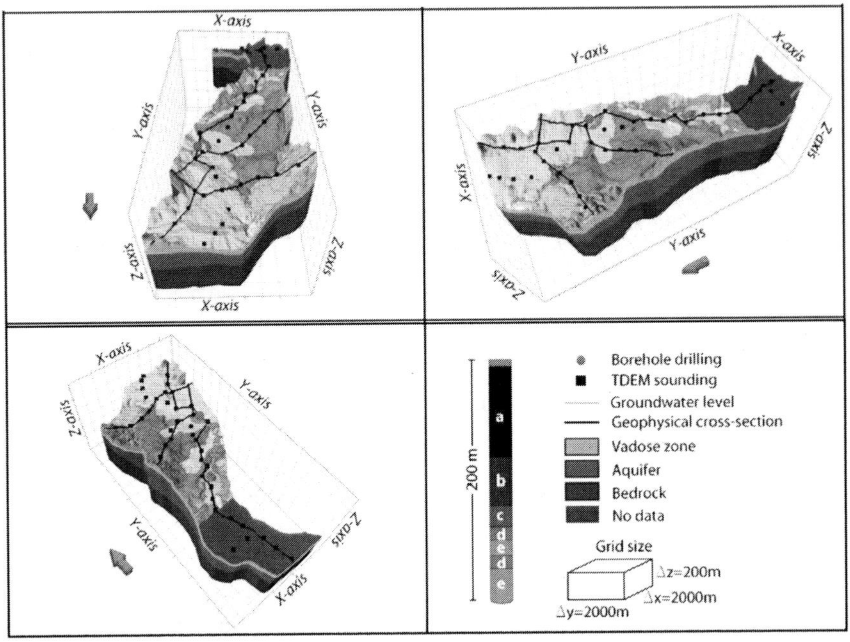

Figure 8. 3D visualization and integration model. (**a**) Vadose zone; (**b**) fractured lava aquifer; (**c**) granular aquifer in andesitic breccia; (**d**) aquitard; (**e**) fractured breccia aquifer. The colors in the uppermost layer refer to lithologies shown in Figure 2.

3D methods are useful to appraise uncertainties on the physical attributes of the aquifer systems and greatly improve conceptual understanding of their behavior [29]. In this case, the 3D model provides a welcome visual aid to scientists and water managers. The model is useful for geostatistical analyses. It provides a reference for the study of the spatial variability of electrofacies or the geometry of the aquifer system, as well as to estimate the expected water table depth at any point within the model domain. It is also useful to inform further fieldwork, including new extraction wells.

The model can be used to calculate the estimated reserves of the system. As it stands, the model spans an area of 301.62 km^2. EF-2 unit presents an average thickness of 220 m (Figure 3), with a porosity range between 15% and 21%. If the 8.8×10^{-3} storage coefficient is taken as a lower-bound estimate of drainage porosity, the aquifer could be expected to hold ~12,200 Mm^3 of groundwater, as well as to release a further ~580 Mm^3, as a result of physical drainage of interconnected pores. This means that the system's specific yield amounts to 4% of the total storage, an estimate which is roughly consistent with figures for the whole Mezquital

Valley system (7%) [96]. While still rough, these figures may be taken as a baseline for detailed studies based on groundwater balance calculations [97] or numerical models [98].

Much like the oil industry, the water sector can benefit from the dynamic integration of data in 3D databases. This will contribute to a more robust understanding about aquifer systems related to unmanaged regions, thus providing a sound scientific base to underpin management decisions.

3.7. Groundwater Flow Evolution in the Unmanaged Region of the Mezquital Valley

Figure 9 provides an overview of the 1982–1998, 1998–2007 and 1982–2007 evolution intervals. Data shows large dispersions. For instance, changes of −60 to +60 m, −40 to +60 m and −65 to +55 m are observed for 1982–1998, 1998–2007 and 1982–2007, respectively.

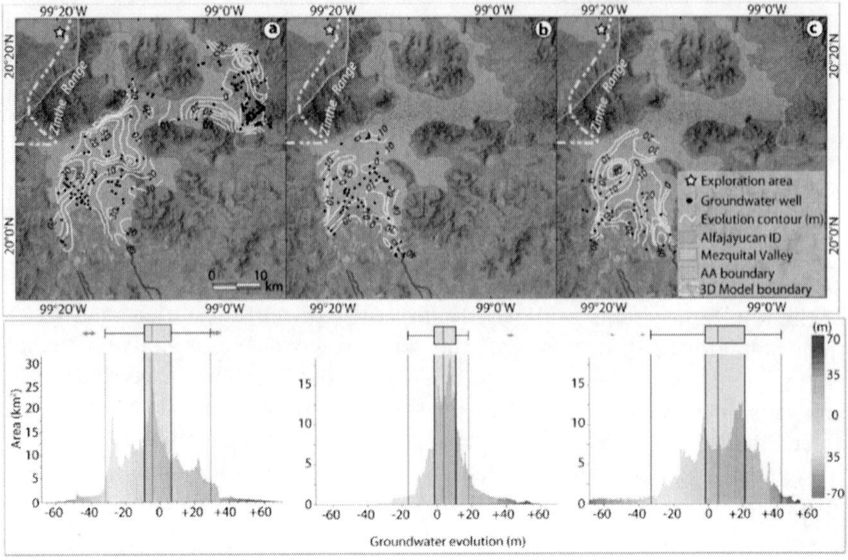

Figure 9. Groundwater evolution in the Mezquital valley aquifer. (**a**) 1982–1998; (**b**) 1998–2007; (**c**) 1982–2007. Negative contours (m) refer to a decline in potentiometric levels and opposite for positive values. ID: Irrigation district. Red crosses in the boxplots refer to outliers. AA: Alfajayucan aquifer.

There are three potential explanations for these large fluctuations. The first one has to do with the presence of spatially-isolated data points (Figure 9). Secondly, fluctuations may be attributed to the fact that, in some sectors of the Mezquital Valley, groundwater levels are controlled by abrupt topographic changes. This effect is common in shallow aquifers [99]. Finally, the Mezquital and Alfajayucan units have been described as multilayer systems with mutual interconnections [36].This suggests that the original datasets [37,50,52] mix shallow and deep levels.

The latter point is difficult to confirm. This is because there are significant uncertainties in terms of basic borehole attributes such as their depth or the location of the screen. However, field data combined with the experience of the authors in the study area suggest that most boreholes are screened from the water table to the bottom (some appear to be screened from throughout their whole length). This means that water table readings represent the average head of all aquifers. As a result, the piezometric analysis must be handled with some care.

Overall, the 1982–1998 interval histogram presents a bimodal behavior, modes being −5 and −30 m (Figure 9a). The boxplot interquartile range (Q_3–Q_1) which is a measure of statistical dispersion, provides a more representative range. This suggests that the water table fluctuated mostly between −8 m and +7 m over those 16 years.

Similar considerations apply to the other dispersive histograms (Figure 9b,c). However, the interquartile range oscillates within a considerable narrower interval, in the order of −1 to +22 for 1982–2007 and between −1 and +12 m for 1998–2007, respectively (consider that positive values refer to a rise in water table and opposite for negative values). Furthermore, the median of the temporal groundwater evolution is −4, +6 and +5 m, respectively, for the 1982–1998; 1998–2007 and 1982–2007 cases. This implies that the aquifer level has fluctuated by only ±20 cm/year, which is close enough to zero. In other words, the Mezquital Valley aquifer (MVA) is closed to steady-state conditions for the 25-year interval.

This argument is supported by independent evidence. Garfias-Quezada [100] analyzed monthly piezometric data from industrial wells located towards the SW of the MVA between 1984 and 2010. Hydrographs from five wells show that the water table depth has remained roughly constant. In some cases, these present a slightly upward trend.

For instance, one of the wells presented a water table depth of ~22 m in early 1984, and its fluctuation over 26 years was just 3% (21.9 ± 0.68 m, as referred to the median ± the standard deviation). Other wells registered similar behaviors (17.21 ± 2.29 m; 14.67 ± 0.44 m; 43.82 ± 1.12 m; 44.67 ± 1.02 m). Besides, linear trends for each hydrograph present near-zero gradients (10^{-4}–10^{-5}). This clearly implies that both the shallow (15–25 m) and deep piezometric heads (40–50 m) have remained stable.

This is remarkable because the MVA is a highly dynamic system in terms of the water balance. The National Water Commission [96] reported the following recharge and discharge volumes for the Mezquital Valley (2007–2012): abstractions from wells amount to ~137.7 Mm^3/year, springs yield ~85.1 Mm^3/year, baseflow through Tula River is estimated to be ~280 Mm^3/year, while direct evapotranspiration from shallow groundwater is in the order of 9.8 Mm^3/year. Conversely, induced recharge from irrigation, wastewater infiltration and network losses is about ~343.2 Mm^3/year, while rainfall infiltration amounts to ~49.5 Mm^3/year. These figures shows an aquifer storage variation of roughly +2.4 Mm^3/year. Given the size of the system and the magnitude of the different elements of the water budget, this figure is sufficiently close to zero. Thus, it is consistent with the outcomes of the piezometric analysis.

Whereas dropping water tables are relatively common across the world [101], long-term pseudo steady-state conditions such as these are relatively rare. Unmanaged recharge seems to regulate the system, thus preventing major changes in the aquifer balance. This provides a stark contrast with neighboring aquifer systems where there is no unmanaged recharge. For instance, the Cuautitlán-Pachuca aquifer presents a severe storage decrease over the same interval of −58 Mm^3/year [102]. This corresponds to a regional water table drop in the order of −0.5 to −2 m/year.

Based on 2007 records [52], groundwater flows from the south to the north of the Mezquital valley. Equipotential lines range from 2200 to 1980 m. a.s.l. (Figure 10b). These conditions are similar in the study area for 2012 [53], where flow gradients and dynamics are equivalent, albeit with a different range of equipotential values (1900–1720 m. a.s.l., Figure 10a). This suggests that there is hydraulic continuity between both regions. Thus, the Zhinte range (Figure 9 and Figure 10) does not constitute a no-flow boundary.

In summary, pseudo steady-state conditions in the MVA, together with the spatial distribution of hydraulic gradients, suggests that the Alfajayucan aquifer is recharged with lateral flows from the wastewater irrigation of the Mezquital Valley. Thus, newly-developed water supplies within the study area are likely to be affected from the volumetric standpoint by upgradient treated wastewater disposal. In addition, the MVA is likely to evolve into transient conditions, a state which it has not experienced for at least 30 years. The water treatment plant is expected to greatly improve the water quality for municipal use in the MVA and surrounding areas.

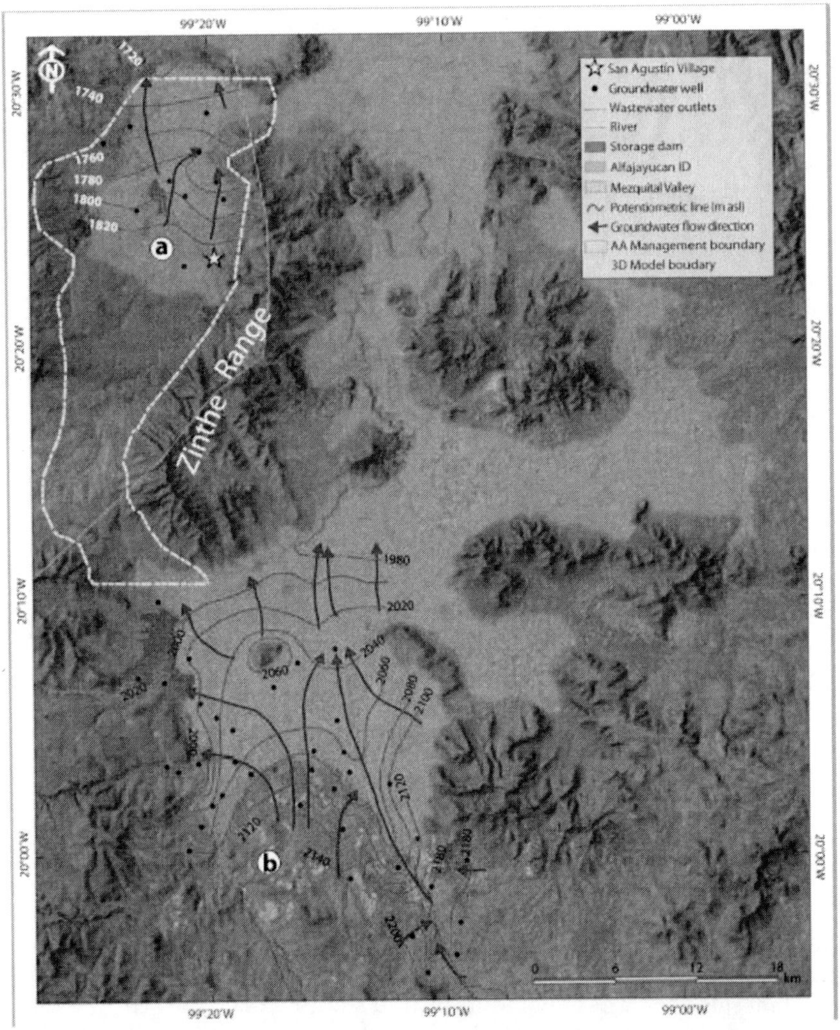

Figure 10. Groundwater flow net in (**a**) the Alfajayucan aquifer for 2012 (adapted from [53]); and (**b**) the Mezquital Valley aquifer for 2007 (adapted from [52]). ID: Irrigation district, AA: Alfajayucan aquifer.

4. CONCLUSIONS

Hydrogeological exploration and characterization is still needed in many regions across the world, particularly in low-income countries and emerging economies. Within a context of unregulated wastewater-based recharge schemes, the search for new groundwater sources is essential to underpin the present and future life of local communities. Current water supply challenges call for integrated approaches. Interdisciplinary research combining geological, hydrogeophysical, petrophysical and hydraulic techniques, adapted from oil and gas reservoir modeling frameworks, need to be used jointly to evaluate the complexities of aquifer systems and optimize the development of new sources.

The proposed methodology contributes not only to inferring the hydrogeological parameters controlling aquifer dynamics, but also to increasing knowledge about the behavior of key physical properties in different lithologies. Thus, it can be exported to similar settings, so as to enhance our understanding of subsurface processes in unmanaged and managed recharge regions.

Regional and local-scale assessments are needed to adequately frame the outcomes and guarantee sustainable groundwater use in the long run. Practices such as unmanaged aquifer recharge constitute a mixed blessing. Cases like the Mezquital Valley show how it provides additional aquifer replenishment, and how this contributes to the equilibrium of groundwater systems. On the other hand, unmanaged aquifer recharge is likely to induce lasting hydrochemical changes in aquifers and ecosystems. The transition from unmanaged to managed recharge is expected to provide benefits to the Mezquital Valley inhabitants, and furthermore to generate new uncertainties in relation to aquifer dynamics and downgradient systems.

ACKNOWLEDGMENTS

The Earth Sciences Division and the Institute of Geophysics (UNAM) financially supported this research. The authors are grateful to Schlumberger for the donation of the Petrel® Platform (provisional academic license number PETREL_151139740_MAAAIACBH/aUA), during March-June 2015. We thank R. Castrejón for providing us technical guidelines during well logging analysis and to Fátima Espinosa, Alejandro Gómez, Francisco Miguel, Hermes Rochín, Alberto Garfias and Diego Ruiz, for helping out with field work and analysis. We thank Santiago Arellano Islas (Comision Estatal del Agua y Alcantarillado de Hidalgo) and the residents of San Agustín Village for their permission and support, respectively. The authors thank the reviewers and editor for helpful discussions which considerably improved the quality of the manuscript. This is the 9th contribution from the Hydrogeology Group (Earth Sciences Division), Faculty of Engineering, UNAM, Mexico.

REFERENCES

1. Oikonomidis, D.; Dimogianni, S.; Kazakis, N.; Voudouris, K. A GIS/Remote Sensing-based methodology for groundwater potentiality assessment in Tirnavos area, Greece. *J. Hydrol.* **2015**, *525*, 197–208.
2. Panigrahi, B.; Nayak, A.K.; Sharma, S.D. Application of remote sensing technology for groundwater potential evaluation. *Water Resour. Manag.* **1995**, *9*, 161–173.
3. Dawoud, M.A.; Raouf, A.R.A. Groundwater exploration and assessment in rural communities of Yobe State, Northern Nigeria. *Water Resour. Manag.* **2008**, *23*, 581–601.
4. Llamas, M.R.; Martínez-Santos, P. The silent revolution of intensive ground water use: Pros and cons. *Ground Water***2005**, *43*, 161.
5. Mallet, J.L. *Geomodeling*; Oxford University Press: New York, NY, USA, 2002.
6. Duca, C.; Abruzzi, D. Fully integrated hydrocarbon reservoir studies: Myth or reality? *Am. J. Appl. Sci.* **2010**, *7*, 1477–1486.
7. Warrlich, G.; Bosence, D.; Waltham, D.; Wood, C.; Boylan, A.; Badenas, B. 3D stratigraphic forward modelling for analysis and prediction of carbonate platform stratigraphies in exploration and production. *Mar. Pet. Geol.* **2008**, *25*, 35–58.
8. Sacchi, Q.; Salina Borello, E.; Weltje, G.J.; Dalman, R. Increasing the predictive power of geostatistical reservoir models by integration of geological constraints from stratigraphic forward modeling. *Mar. Pet. Geol.* **2016**, *69*, 112–126.
9. Dickey, P.A. *Petroleum Development Geology*; Pen Well Books: Tulsa, OK, USA, 1981.
10. Lake, L.; Carrol, B.H. *Reservoir Characterization*; Academic Press, Inc.: New York, NY, USA, 1988.
11. Musliu-Kehinde, O. Integrated Reservoir Characterization: A Case Study of an Onshore Reservoir in Niger Delta Basin. Master's Thesis, African University of Science and Technology, Abuja, Nigeria, 2011.
12. Guyoton, F.; Grasso, J.R.; Volant, P. Interrelation between induced seismic instabilities and complex geological structure. *Geophys. Res. Lett.* **1999**, *19*, 705–708.
13. Spikes, K.T. Model-based prediction of porosity and reservoir quality from P- and S-wave data. *Geophys. Res. Lett.***2003**, *30*, 1–4.
14. Falahat, R.; Shams, A.; MacBeth, C. Towards quantitative evaluation of gas injection using time-lapse seismic data. *Geophys. Prospect.* **2011**, *59*, 310–322.
15. Arnulf, A.F.; Singh, S.C.; Pye, J.W. Seismic evidence of a complex multi-lens melt reservoir beneath the 9° N Overlapping Spreading Center at the East Pacific Rise. *Geophys. Res. Lett.* **2014**, *41*, 6109–6115.
16. Arogunmati, A.; Harris, J.M. Quasi-continuous reservoir monitoring with surface seismic data. *Geophys. Prospect.***2014**, *62*, 117–132.
17. Grana, D.; Mukerji, T. Bayesian inversion of time-lapse seismic data for the estimation of static reservoir properties and dynamic property changes. *Geophys. Prospect.* **2015**, *63*, 637–655.
18. Wyllie, M.R.; Gregory, A.R.; Gardner, L.W. Elastic wave velocities in heterogeneous and porous media. *Geophysics***1956**, *27*, 569–589.
19. Pringle, J.K.; Westerman, A.R.; Clark, J.D.; Drinkwater, N.J.; Gardiner, A.R. 3D high-resolution digital models of outcrop analogue study sites to constrain reservoir model uncertainty: An example from Alport Castles, Derbyshire, UK. *Pet. Geosci.* **2004**, *10*, 343–352.
20. Bueno, J.F.; Drummond, R.D.; Vidal, A.C.; Sancevero, S.S. Constraining uncertainty in volumetric estimation: A case study from Namorado Field, Brazil. *J. Pet. Sci. Eng.* **2011**, *77*, 200–208.

21. Fokker, P.A.; Visser, K.; Peters, E.; Kunakbayeva, G.; Muntendam-Bos, A.G. Inversion of surface subsidence data to quantify reservoir compartmentalization: A field study. *J. Pet. Sci. Eng.* **2012**, *96–97*, 10–21.
22. Hosseini, S.A.; Lashgari, H.; Choi, J.W.; Nicot, J.P.; Lu, J.; Hovorka, S.D. Static and dynamic reservoir modeling for geological CO_2 sequestration at Cranfield, Mississippi, U.S.A. *Int. J. Greenh. Gas Control* **2013**, *18*, 449–462.
23. Bunch, M. A live test of automated facies prediction at wells for CO_2 storage projects. *Energy Procedia* **2014**, *63*, 3432–3446.
24.]Pan, F.; McPherson, B.J.; Dai, Z.; Jia, W.; Lee, S.Y.; Ampomah, W.; Viswanathan, H.; Esser, R. Uncertainty analysis of carbon sequestration in an active CO_2-EOR field. *Int. J. Greenh. Gas Control* **2016**, *51*, 18–28.
25. Noad, J. The use of field analogues in the correlation and static reservoir methodology used in the Tern Field, Northern North Sea, UK. *Mar. Pet. Geol.* **2004**, *21*, 485–497.
26. Blöcher, M.G.; Zimmermann, G.; Moeck, I.; Brandt, W.; Hassanzadegan, A.; Magri, F. 3D numerical modeling of hydrothermal processes during the lifetime of a deep geothermal reservoir. *Geofluids* **2010**, *10*, 406–421.
27. Ross, M.; Parent, M.; Lefebvre, R. 3D geologic framework models for regional hydrogeology and land-use management: A case study from a Quaternary basin of southwestern Quebec, Canada. *Hydrogeol. J.* **2005**, *13*, 690–707.
28. Audouin, O.; Bodin, J.; Porel, G.; Bourbiaux, B. Flowpath structure in a limestone aquifer: Multi-borehole logging investigations at the hydrogeological experimental site of Poitiers, France. *Hydrogeol. J.* **2008**, *16*, 939–950.
29. Gill, B.; Cherry, D.; Adelana, M.; Chiang, X.; Mark, R. Using three-dimensional geological mapping methods to inform sustainable groundwater development in a volcanic landscape, Victoria, Australia. *Hydrogeol. J.* **2011**, *19*, 1349–1365.
30. Szabó, N.P.; Dobróka, M.; Turai, E.; Szűcs, P. Factor analysis of borehole logs for evaluating formation shaliness: A hydrogeophysical application for groundwater studies. *Hydrogeol. J.* **2013**, *22*, 511–526.
31. Turner, R.J.; Mansour, M.M.; Dearden, R. Improved understanding of groundwater flow in complex superficial deposits using three-dimensional geological-framework and groundwater models: An example from Glasgow, Scotland (UK). *Hydrogeol. J.* **2015**, *23*, 493–506.
32. Somaratne, N.; Mann, S. Integrated Use of Geological, Geophysical, Radiocarbon and Stable Isotopes Data for Tracing the Conduit Flow Paths in a Small Karstic Aquifer: Poocher Swamp Freshwater Lens, South Australia. *Environ. Nat. Resour. Res.* **2016**, *6*, 119.
33. El-Kaliouby, H.; Abdalla, O. Application of time-domain electromagnetic method in mapping saltwater intrusion of a coastal alluvial aquifer, North Oman. *J. Appl. Geophys.* **2015**, *115*, 59–64.
34. Porsani, J.L.; Almeida, E.R.; Bortolozo, C.A. TDEM survey in an area of seismicity induced by water wells in Paraná sedimentary basin, Northern São Paulo State, Brazil. *J. Appl. Geophys.* **2012**, *82*, 75–83.
35. Porsani, J.L.; Bortolozo, C.A.; Almeida, E.R.; Santos Sobrinho, N.S. TDEM survey in urban environmental for hydrogeological study at USP campus in São Paulo city, Brazil. *J. Appl. Geophys.* **2012**, *76*, 102–108.
36. Comisión Nacional del Agua (CONAGUA). *Actualización de la Disponibilidad de Agua en el Acuífero Chapantongo-Alfajayucan (1309), Estado de Hidalgo [Water Budget Update of the Chapantongo-Alfajayucan Aquifer (1309), Hidalgo State]*; CONAGUA: Mexico City, Mexico, 2013.
37. British Geological Survey (BGS); Comisión Nacional del Agua (CONAGUA). *Impact of Wastewater Reuse on Groundwater in the Mezquital Valley, Hidalgo State, Mexico*; BGS Final Report WC/98/42; CONAGUA: Mexico City, Mexico, 1998.
38. Lucho-Constantino, C.A.; Prieto-García, F.; Del Razo, L.M.; Rodríguez-Vázquez, R.; Poggi-Varaldo, H.M. Chemical fractionation of boron and heavy metals in soils

irrigated with wastewater in central Mexico. *Agric. Ecosyst. Environ.* **2005**, *108*, 57–71.
39. Félix-Cañedo, T.E.; Durán-Álvarez, J.C.; Jiménez-Cisneros, B. The occurrence and distribution of a group of organic micropollutants in Mexico City's water sources. *Sci. Total Environ.* **2013**, *454–455*, 109–118.
40. Gibson, R.; Durán-Álvarez, J.C.; Estrada, K.L.; Chávez, A.; Jiménez Cisneros, B. Accumulation and leaching potential of some pharmaceuticals and potential endocrine disruptors in soils irrigated with wastewater in the Tula Valley, Mexico. *Chemosphere* **2010**, *81*, 1437–1445.
41. Downs, T.J.; Cifuentes-García, E.; Suffet, I.M. Risk screening for exposure to groundwater pollution in a wastewater irrigation district of the Mexico City region. *Environ. Health Perspect.* **1999**, *107*, 553–561.
42. Gallegos, E.; Warren, A.; Robles, E.; Campoy, E.; Calderon, A.; Sainz, M.G.; Bonilla, P.; Escolero, O. The effects of wastewater irrigation on groundwater quality in Mexico. *Water Sci. Technol.* **1999**, *40*, 45–52.
43. Comisión Nacional del Agua (CONAGUA). *Planta de Tratamiento de Aguas Residuales Atotonilco [Atotonilco Wastewater Treatment Plant]*; Federal Government, CONAGUA: Mexico City, Mexico, 2013.
44. Ward, J.; Dillon, P. *Robust Design of Managed Aquifer Recharge Policy in Australia*; The Commonwealth Scientific and Industrial Research Organisation (CSIRO): Canberra, Australia, 2009.
45. Lacher, L.J.; Turner, D.S.; Gungle, B.; Bushman, B.M.; Richter, H.E. Application of hydrologic tools and monitoring to support managed aquifer recharge decision making in the upper San Pedro River, Arizona, USA. *Water* **2011**, *6*, 3495–3527.
46. Megdal, S.B.; Dillon, P. Policy and economics of managed aquifer recharge and water banking. *Water* **2015**, *7*, 592–598.
47. Sinclair Knight Merz (SKM); The Commonwealth Scientific and Industrial Research Organisation (CSIRO). *Progress in Managed Aquifer Recharge in Australia*; Waterlines Report No. 73; National Water Commission: Canberra, Australia, 2012.
48. Rossetto, R.; Ansiati, A.; Barbagli, A.; Borsi, I.; Costabile, G.; Dietrich, P.; Mazzanti, G.; Picciaia, D.; Bonari, E. *Managing Induced Riverbank Filtration (IRF) at the Serchio Rivel Well Field, Tuscany, Italy (Italy)*; Geophysical Research Abstracts, 16, EGU2014-7395; EGU General Assembly: Viena, Austria, 2014.
49. Kacimov, A.R.; Zlotnik, V.; Al-Maktoumi, A.; Al-Abri, R. The green-Ampt one-dimensional infiltration from a ponded surface into a heterogeneous soil. *J. Irrig. Drain. Eng.* **2015**, *136*, 68–72.
50. Geocalli. *Actualización del Estudio Geohidrológico del Valle del Mezquital, Hidalgo [Update of the geohydrologic study of the Mezquital Valley, Hidalgo]*; Contract number GZA-81-68-GD; Secretaría de Agricultura and Recursos Hidráulicos (SARH): San Luis Potosi, Mexico, 1981.
51. Aguirre-Díaz, G.J.; López-Martínez, M. Geologic evolution of the Donguinyó-Huichapan caldera complex, central Mexican Volcanic Belt, Mexico. *J. Volcanol. Geotherm. Res.* **2009**, *179*, 133–148.
52. Lesser-carrillo, L.E.; Lesser-illades, J.M.; Arellano-Islas, S.; González-Posadas, D. Balance hídrico y calidad del agua subterránea en el acuífero del Valle del Mezquital, México central. *Rev. Mex. Cienc. Geol.* **2011**, *28*, 323–336.
53. Miguel-Cortés, F.; Rochín, G.H. Prospección Hidrogeológica y Factibilidad de Extracción de agua Subterránea en el Poblado de San Agustín Tlalixticapa, Hidalgo [Hydrogeological prospection and groundwater abstraction feasibility study, within the San Agustín Tlalixticapa Village, Hidalgo]. Bachelor's Thesis, Faculty of Engineering, Universidad Nacional Autónoma de México, México City, Mexico, 2010.
54. Jiménez, B.; Chávez, A. Quality assessment of an aquifer recharged with wastewater for its potential use as drinking source: "El Mezquital Valley" case. *Water Sci. Technol.* **2004**, *50*, 269–273.

55. Constable, S.C.; Parker, R.L.; Constable, C.G. Occam's inversion: A practical algorithm for generating smooth models from electromagnetic sounding data. *Geophysics* **1987**, *52*, 289–300.
56. Bassiouni, Z. *Theory, Measurement, and Interpretation of Well Logs*; Society of Petroleum Engineers (SPE): Woodlands, TX, USA, 1994.
57. Ran, Q.; Wang, Y.; Sun, Y.; Yan, L.; Tong, M. *Volcanic Gas Reservoir Characterization*; Elsevier: Amsterdam, The Netherlands, 2014.
58. Stefansson, V.; Gudlaugsson, S.T.; Gudmundsson, A. Silica content and Gamma Ray logs in volcanic rocks. In Proceedings of the World Geothermal Congress, Kyushu-Tohoku, Japan, 28 May–10 June 2000; pp. 2893–2897.
59. Archie, G.E. The electrical resistivity logs as an aid in determing some reservoir characteristics. *Trans. Am. Inst. Mech. Eng.* **1942**, *146*, 54–62.
60. Zhou, F.; Yao, G. Sensitivity analysis in permeability estimation using logging and injection-falloff test data for an anthracite coalbed methane reservoir in Southeast Qinshui Basin, China. *Int. J. Coal Geol.* **2014**, *131*, 41–51.
61. Rapti-Caputo, D.; Bratus, A.; Santarato, G. Strategic groundwater resources in the Tagliamento River basin (northern Italy): Hydrogeological investigation integrated with geophysical exploration. *Hydrogeol. J.* **2009**, *17*, 1393–1409.
62. Wright, H.M.N.; Cashman, K.V.; Gottesfeld, E.H.; Roberts, J.J. Pore structure of volcanic clasts: Measurements of permeability and electrical conductivity. *Earth Planet. Sci. Lett.* **2009**, *280*, 93–104.
63. Asfahani, J. Basalt Characterization by Means of Nuclear and Electrical Well Logging Techniques. Case Study from Southern Syria. *Appl. Radiat. Isot.* **2011**, *69*, 641–647.
64. Schlumberger. *Petrel E&P Software Platform 2014*; Release Notes; Schlumberger: Houston, TX, USA, 2014.
65. Driscoll, F.G. *Groundwater and Wells*, 2nd ed.; Johnson Screens: St. Paul, MN, USA, 1986.
66. Eden, R.N.; Hazel, C.P. *Computer and Graphical Analysis of Variable Discharge Pumping Tests of Wells*; Institution of Engineers Australia: Perth, Australia, 1973; pp. 5–10.
67. Rorabaugh, M.I. Graphical and theoretical analysis of step- drawdown tests of artesian wells. *Proc. Am. Soc. Civ. Eng.* **1953**, *79*, 362.
68. Theis, C.V. The relation between the lowering of the Piezometric surface and the rate and duration of discharge of a well using ground-water storage. *Trans. Am. Geophys. Union* **1935**, *16*, 519.
69. Duffield, G.M. *AQTESOLV™ for Windows*, Version 4.5; HydroSOLV, Inc.: Reston, VA, USA, 2007.
70. Agarwal, R.G.; Al-Hussainy, R.; Ramey, H.J. An Investigation of Wellbore Storage and Skin Effect in Unsteady Liquid Flow. I: Analytical Treatment. *SPEJ* **1970**, *10*, 279–290.
71. Romero, D.J.; Valko, P.P.; Economides, M.J. Optimization of the Productivity Index and the Fracture Geometry of a Stimulated Well with Fracture Face and Choke Skins. In Proceedings of the International Symposium and Exhibition on Formation Damage Control, Society of Petroleum Engineers, San Antonio, TX, USA, 29 September–2 October 2002.
72. Bourdet, D. Well Test Analysis: The use of advanced interpretation models. In *Handbook of Petroleum Exploration and Production*; Elsevier: Amsterdam, The Netherlands, 2002.
73. Raef, A.E.; Mattern, F.; Philip, C.; Totten, M.W. 3D seismic attributes and well-log facies analysis for prospect identification and evaluation: Interpreted palaeoshoreline implications, Weirman Field, Kansas, USA. *J. Pet. Sci. Eng.* **2015**, *133*, 40–51.
74. QGIS Development Team. *QGIS Geographic Information System Developers Manual*; Open Source Geospatial Foundation Project, QGIS Version 2.12; QGIS Development Team: Lyon, France, 2015.

75. Han, T.; Best, A.I.; Sothcott, J.; North, L.J.; MacGregor, L.M. Relationships among low frequency (2 Hz) electrical resistivity, porosity, clay content and permeability in reservoir sandstones. *J. Appl. Geophys.* **2015**, *112*, 279–289.
76. Goldman, M.; Neubauer, F.M. Groundwater exploration using integrated geophysical techniques. *Surv. Geophys.***1994**, *15*, 331–361.
77. Stefansson, V.; Gudmundsson, A.; Emmerman, R. *Gamma Ray Logging in Icelandic Rocks. The Log Analyst, November-December*; SPWLA: Houston, TX, USA, 1982; pp. 11–16.
78. Cannon, S. *Petrophysics: A Practical Guide*; John Wiley & Sons: Hoboken, NJ, USA, 2005.
79. Chi, S.; Torres-Verdin, C.; Wu, J.J.J.; Alpak, F.O. Assessment of Mud-Filtrate-Invasion Effects on Borehole Acoustic Logs and Radial Profiling of Formation Elastic Properties. *SPE Reserv. Eval. Eng.* **2006**, *9*, 553–564.
80. Alao, P.A.; Ata, A.I.; Nwoke, C.E. Subsurface and Petrophysical Studies of Shaly-Sand Reservoir Targets in Apete Field, Niger Delta. *ISRN Geophys.* **2013**, *2013*, 102450.
81. Markov, M.; Mousatov, A.; Kazatchenko, E.; Markova, I. Determination of electrical conductivity of double-porosity formations by using generalized differential effective medium approximation. *J. Appl. Geophys.* **2014**, *108*, 104–109.
82. Schulmberger. *Schulmberger Log Interpretation Charts*; Schulmberger: Houston, TX, USA, 2009.
83. Chávez, A.; Maya, C.; Gibson, R.; Jiménez, B. The removal of microorganisms and organic micropollutants from wastewater during infiltration to aquifers after irrigation of farmland in the Tula Valley, Mexico. *Environ. Pollut.***2011**, *159*, 1354–1362.
84. World Health Organization (WHO); The United Nations Children's Fund (UNICEF). *Global Water Supply and Sanitation Assessment 2000 Report*; The WHO and UNICEF: Washington, DC, USA, 2000.
85. Carter, R.C.; Tyrrel, S.F.; Howsam, P. The impact and sustainability of water and sanitation programmes in developing countries. *Water Environ. Manag.* **1997**, *13*, 292–296.
86. Water and Environmental Health at London and Loughborough (WELL). *Guidance Manual on Water Supply and Sanitation Programmes*; Water Engineering DC: Loughborough, UK, 1998.
87. Gleick, P.H. Basic water requirements for human activities: Meeting basic needs. *Water Int.* **1996**, *21*, 83–92.
88. World Health Organization (WHO). *The Human Right to Water and Sanitation*; Media Brief; WHO: Geneva, Switzerland, 2011; pp. 1–8.
89. Jacob, C.E. Drawdown test to determine effective radius of artesian well. *Trans. Am. Soc. Civ. Eng.* **1947**, *112*, 1047–1064.
90. Walton, W.C. *Selected Analytical Methods for Well and Aquifer Evaluation*; Bulletin 49; Illinois State Water Survey: Urbana, IL, USA, 1962.
91. Ramey, H.J. Well-loss function and the skin effect: A review. In *Recent Trends in Hydrogeology*; Narasimhan, T.N., Ed.; Special Paper 189; Geological Society of America: Boulder, CO, USA, 1982; pp. 265–271.
92. Renard, P. Hydraulics of Well and Well Testing. In *Encyclopedia of Hydrological Sciences*; Anderson, M.G., McDonnell, J.J., Eds.; John Wiley & Sons, Ltd.: Chichester, UK, 2006.
93. Horne, R.N. *Modern Well Test Analysis: A Computer Aided Approach*; Petroway Inc.: Palo Alto, CA, USA, 1995.
94. Rosberg, J.E.; Bjelm, L. Well development by jetting using coiled tubing and simultaneous pumping. *Ground Water***2009**, *47*, 816–821.
95. SARH. *Servicios de Prospección y Levantamientos Geológicos y Geofísicos en la Zona de Alfajayucan-Chapantongo-Actopan, [Geological and Geophysical Prospection Studies in the Alfajayucan-Chapantongo-Actopan Area]*; Final Report; SARH: San Luis Potosi, Mexico, 1981.

96. Comisión Nacional del Agua (CONAGUA). *Actualización de la Disponibilidad de Agua en el Acuífero Valle del Mezquital (1310), Estado de Hidalgo [Water Budget Update of the Mezquital Valley Aquifer (1310), Hidalgo State]*; CONAGUA: Mexico City, Mexico, 2015.
97. Luijendijk, E.; Bruggeman, A. Groundwater resources in the Jabal Al Hass region, northwest Syria: An assessment of past use and future potential. *Hydrogeol. J.* **2008**, *16*, 511–530.
98. Jaworska-Szulc, B. Groundwater flow modelling of multi-aquifer systems for regional resources evaluation: The Gdansk hydrogeological system, Poland. *Hydrogeol. J.* **2009**, *1*, 1521–1542.
99. Haitjema, H.M.; Mitchell-Bruker, S. Are water tables a subdued replica of the topography? *Ground Water* **2005**, *43*, 781–786.
100. Garfias-Quezada, A. Modelación Numérica de Acuíferos en Diferencias Finitas Utilizando la Interfaz Libre PMWIN: Fundamentos Teóricos y caso de Aplicación en el Valle del Mezquital [Finite-Difference Groundwater Flow Modeling Using the Open Source Interface PWMIN: Theoretical Foundations and Application Case in the Mezquital Valley Aquifer]. Bachelor's Thesis, Faculty of Engineering, Universidad Nacional Autónoma de México, México City, Mexico, 2015.
101. Wada, Y.; van Beek, L.P.H.; van Kempen, C.M.; Reckman, J.W.T.M.; Vasak, S.; Bierkens, M.F.P. Global depletion of groundwater resources. *Geophys. Res. Lett.* **2010**, *37*, L20402.
102. Comisión Nacional del Agua (CONAGUA). *Actualización de la Disponibilidad de Agua en el Acuífero Cuautitlán-Pachuca (1508), Estado de México [Water Budget Update of the Cuautitlán-Pachuca Aquifer (1310), Hidalgo State]*; CONAGUA: Mexico City, Mexico, 2015.

Index